Fundamentals of Seismic Loading on Structures

Fundamentals of Seismic Loading on Structures

Tapan K. Sen

A John Wiley and Sons, Ltd, Publication

This edition first published 2009

Registered office
John Wiley & Sons Ltd, The Atrium, Southern Gate, Chichester, West Sussex,
PO19 8SQ, United Kingdom

For details of our global editorial offices, for customer services and for information about how to apply for permission to reuse the copyright material in this book please see our website at www.wiley.com.

Library of Congress Cataloging-in-Publication Data

Sen, Tapan K.
Fundamentals of seismic loading on structures / by Tapan K. Sen.
p. cm.
Includes bibliographical references and index.
ISBN 978-0-470-01755-5 (cloth)
1. Earthquake resistant design. 2. Earthquake hazard analysis.
3. Strains and stresses. I. Title.
TA658.44.S44 2009
624.1′762—dc22
2008044377

A catalogue record for this book is available from the British Library.

ISBN: 978-0-470-01755-5

Set in 10/12pt Times by Integra Software Services Pvt. Ltd, Pondicherry, India
Printed in Singapore by Markono

To My Parents

Contents

Preface

Since ancient times, earthquakes have taken a heavy toll on human life and property, made worse on a psychological level by the unexpected nature of the event. Mitigation of the damage caused by earthquakes is a prime requirement in many parts of the world today.

As part of the design process, assessment of seismic loading is of fundamental importance and is the key step before any building design activity can be undertaken. Generally this also requires the assessment of an appropriate structural loading.

Much effort has been devoted in the last fifty years towards understanding the causes of earthquakes and defining the loading on buildings associated with such events. Great advances have been made in the field of plate tectonics followed by developments in geotechnical engineering and soil mechanics. Though much remains to be followed up and quantified, understanding in these areas has reached a stage where the findings and procedures may now be integrated into a rational determination of 'earthquake loadings' for the design of structures.

There are many books on earthquake engineering and structural dynamics on the market, so why the need for another book?

I have been a practising structural engineer for many years and now have the good fortune to be teaching the subject at postgraduate level at City University in London. Interaction with students has helped shape my perspective on this issue, and it has become apparent that a book which blends the relevant aspects of structural dynamics with ground motion assessment and hazard analysis, incorporating examples from practical design projects, should have its own merit and fill a niche in the market. Definition of the loading was the difficult part for most students. Their expectations shaped the contents of the book and its emphasis in a very large measure. The final contents of the text reflect my own interpretation of the vast amount of information that is available today.

This book is written from the perspective of a practising engineer at a level suitable for graduate students and young engineers and is intended to be as brief and concise as possible rather than a lengthy text, while translating research findings into a project environment. A significant portion of my effort has been devoted to making the design process and associated mathematics accessible to a wider community of students, in an easily understandable format.

With this aim, I have taken worked examples from design projects with actual numerical values. I hope that it will give the reader a perspective on the engineering challenges and help develop a working knowledge of the complex subject that is earthquake engineering.

Earthquake engineering is multi-faceted with inputs from many disciplines. The designer must grasp the interrelationship of structural dynamics, geotechnical aspects of soils, material properties and detailing of joints, to highlight just a few matters. The contents have been laid out with the aim of maintaining the focus on seismic loading in the broader context of overall design, highlighting the essentials and basics, to help the student find their way through the maze created by the explosion of information in this field.

To this end, it is organized into four major sections as shown in the flow diagram on page xix. After an introductory survey of the science of earthquakes and the origin and nature of seismic waves, the mathematical techniques needed for dynamic analysis of building structures are explained under the section 'Analysis'.

The initial two chapters (5 and 6) under 'Seismic Loading' deal with the characteristics of ground motion, and proceed to attenuation relationships, application of the response spectrum method to building structures, and development of site-dependent response spectra, with a worked example from a design project.

In Chapter 7, special emphasis is given to explaining probabilistic methods, which are now widely used in the assessment of seismic hazard. The use of Monte Carlo Simulation as a numerical procedure for the construction of uniform hazard spectra is illustrated with many worked examples. Following on from uniform hazard spectra, a simple numerical procedure for deaggregation is outlined.

The philosophy behind development of the design codes and incorporation of loading provisions is explored next (Chapter 8). Worked examples using Eurocode 8 and IBC 2000 are included.

Inelastic design concepts, soil-structure interaction issues and the phenomenon of liquefaction are each discussed with reference to behaviour observed at earthquake damaged sites. This will help illustrate a few important concepts in our approach to seismic loading and design. Assessment of liquefaction potential and more recent developments which are currently in practical use are discussed with worked examples.

Finally Chapter 12 gives a very short introduction to performance based earthquake engineering which, it is hoped, will help the reader with a way forward.

Simplicity of presentation has been my aim throughout the book, and if this succeeds in appealing to the student then I will be amply rewarded.

Tapan K. Sen,
London,
August 2008
e-mail: tapan.sen@btinternet.com

Acknowledgements

This book has been in preparation for more than five years. The enormity of the task dawned on me only after Wiley accepted the proposal. I knew the task of putting together information that would be relevant from the graduate students' perspective would not be easy. Almost all weekends and evenings were devoted, sifting through the enormous volume of research material that exists on the subject. The spontaneous outpouring of encouragement from colleagues, old friends, fellow professionals and family members kept me going.

Through this long journey of mine, I have received help from many people. I am indebted to Professor T.A. Wyatt, for his valuable suggestions and guidance and also for reviewing several chapters.

For detailed critiques of individual chapters I am grateful to Dr. P. Sircar, Dr. Navil Shetty, Dr. S. Bhattacharya, Dr. M.M. Mitra, Dr. P. Chakraborty, Professor J. Humar, Ms. Juliane Buchheister, Mr. J.K. Mitra and Mr. K. Ghosh.

Mrs. Shikha Sen Gupta meticulously checked through the chapters, ensuring accuracy of technical content, numerical inputs and calculations. I am most grateful for her dedicated help.

The book would not have been completed had Caitriona MacNeill not stepped in, to organise and seek copyright permissions also making available valuable research material throughout; a significant contribution. Most of all, I wish to acknowledge the serious and painstaking effort of Caitriona, who proofread the manuscript, not once, but twice. I am privileged and greatly indebted to her.

Professor S. Prakash provided valuable suggestions; Murray Russell read through large chunks of the text suggesting improvements in presentation, style and conciseness. I am extremely thankful for this.

I have also had the benefit of discussions with and help from many friends and colleagues and would like to thank particularly: A. Bedrossian, M. Hesar, Peter Carr, Prof. R.N. Dutta, P. Handidjaja, Prof. Leung Chun Fai, Ms. Gopa Roy, Debbie Warman, Miranda O' Brien, R. Sen Gupta, A.P. Sen, Dr. S.F. Yasseri. S. Mawer, Dr. R. Sabatino, Dr. S. Sengupta, Dr. S. Bannerjee, Peter Loveridge, Rob Sears, T. Leggo, Dr. S.K. Ghosh, Miss Karabi and Malini Basu, Prof. P.J. Dowling and Prof. L.F. Boswell.

Debra Francis and Dr. S. Bhattacharya also helped in a very large measure with the many technical papers and reports for which I am thankful.

Grateful thanks are due to Ramesh Shanbagh, Sitangshu M. Basu and U.B. Dasgupta, who all helped with preparing the illustrations. At a later stage, Stuart Holmes proved invaluable in making the photographs come alive with his skill and knowledge of the subject. I am grateful for all his help.

Despite all this help, there are bound to be some mistakes; errors of understanding or errors of printing. They are entirely my fault and I offer my sincere apologies.

The Wiley team of Wendy, Debbie, Nicky and Liz were highly commendable in their support and special thanks to Wendy who assured me of help, if needed, with the illustrations.

And finally, undertaking a book of this nature cannot be sustained without the support of the family and I would like to thank them for their forbearance over the past few years. I would not have been able to devote the time at home if my wife, Mitra, had not been rock solid in holding the family together. To my sons Sudipto and Saugata, who helped in various ways, my heartfelt thanks. Young Saugata was always obliging and declared this to be very much a family effort.

Tapan K. Sen,
London,
December 2008

Organisation of the Book

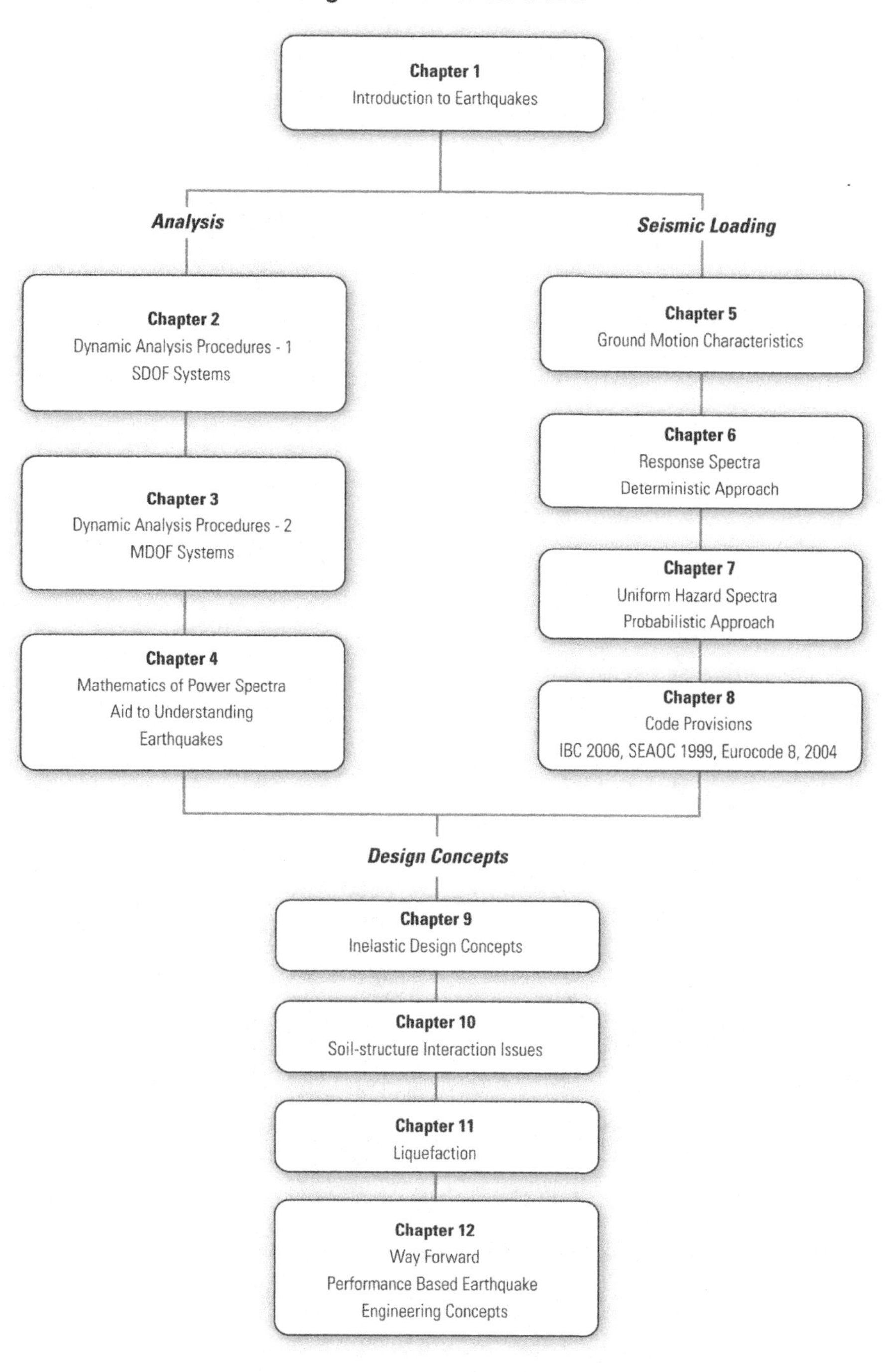

1

Introduction to Earthquakes

1.1 A Historical Perspective

The earthquake is among the most dreaded of all natural disasters, exacting a devastating toll on human life. In the last 100 years alone there have been many major earthquakes, including San Francisco (1906), Tokyo (1923), Alaska (1964), Iran (1968), Mexico (1985) (Figure 1.3), Kobe (1995) (Figure 1.4) and Turkey (1998), to name but a few. The devastating tsunami of 26 December 2004, that struck the coastlines of the Indian Ocean, was caused by an underwater earthquake. Most recently the destructive force of the earthquake was felt in Sichuan Province, China where in May 2008 an estimated 87,000 were killed and 5 million lost their homes.

During these earthquakes hundreds of thousands of lives were lost and billions of dollars of damage sustained to property, and the physical suffering and mental anguish of earthquake survivors are beyond contemplation.

Several earthquakes, of the many that have occurred in the past century, have featured more prominently in terms of the development of the seismic theories and hazard mitigation procedures that we now have at this time. One of these earthquakes, which triggered a great interest in the scientific community, was the 'Great' 1906 San Francisco earthquake. The San Francisco earthquake (18 April 1906) measured between VII and IX on the Modified Mercalli Intensity scale and confounded geologists with its large horizontal displacements and great rupture length, along the northernmost 296 miles (477 km) of the San Andreas fault. The San Francisco earthquake was a momentous event of its era. More than 28,000 buildings were destroyed, damaged or affected by the earthquake (Figure 1.1) and casualties totalled approximately 3000. The earthquake is still noted for the raging fire it caused, which burned for almost three days (Figure 1.2). In its immediate aftermath a State Earthquake Investigation Commission, consisting of some of the most distinguished US geologists of the time, was appointed to bring together the work of scientific investigations and observations following the San Francisco earthquake and its final report is seen as a landmark document in terms of geological and seismological research. Henry Fielding Reid, Professor of Geology at John Hopkins University, Baltimore developed the 'Elastic Rebound Theory' (discussed later) from his studies of displacements and strain in the surrounding crust following the

---

Fundamentals of Seismic Loading on Structures Tapan Sen

Figure 1.1 Aftermath of the San Francisco earthquake (1906). Wreckage of the Emporium and James Flood Building on Market Street. Reproduced from http://www.sfmuseum.org 18/07/08, the Virtual Museum of the City of San Francisco.

Figure 1.2 San Francisco (1906) Huge crowds watch Market Street fire fight. Reproduced from http://www.sfmuseum.org 18/07/08, the Virtual Museum of the City of San Francisco.

earthquake. For the first time people began to realize the knock-on effects earthquakes have on our planet.

On average 17,000 persons per year were killed as a result of earthquake activity in the twentieth century. Among the deadliest earthquakes of the past are those occurring in China, including the Gansu (1920) and Xining (1927) whose death tolls were 200,000 each and the

Figure 1.3 Image of damage during the Mexico City (1985) earthquake. Reproduced from http://www.drgeorgepc.com/Tsunami1985Mexico.html 21/07/08 by permission of Dr George Pararas-Carayaniss.

Figure 1.4 A scene of destruction caused by Kobe earthquake (1995). Reproduced from http://.bristol.ac.uk/civilengineering/research/structures/eerc/realearthquakes/kobe-fire.gif 13/12/07 by permission of Dr A.J. Crewe, University of Bristol.

Tangshan (1976) which resulted in 255,000 deaths. The European earthquake with the highest loss of life was the Messina earthquake in Italy (1908), which resulted in 70,000 deaths.

The enormity of the scale of these earthquakes is justification enough to understand them and therefore to design structures that can most effectively resist them. First we must understand where and why earthquakes happen. Most severe earthquakes occur where the Earth's tectonic plates meet along plate boundaries. For example as two plates move towards each other, one plate can be pushed down under the other into the Earth's mantle. This is a destructive plate boundary and if the plates become locked together immense pressure builds up in the surrounding rocks. When this pressure is released shock waves are produced. These are called seismic waves and radiate outward from the source of the earthquake known as the epicentre, causing the ground to vibrate. Earthquakes are also very common on conservative plate boundaries, where the two plates slide past each other. So-called *intraplate* earthquakes occur away from plate boundaries, along fault zones in the interior of the plate.

The effects of earthquakes vary greatly due to a range of different factors, including the magnitude of the earthquake, and the level of population and economic development in the affected area. Earthquakes in seismic prone areas have an impact on our lives, on our buildings and the very landscape of our planet.

The severity of an earthquake is described both in terms of its *magnitude* and its *intensity*. These two frequently confused concepts refer to different, but related, observations. Magnitude characterizes the size of an earthquake by measuring indirectly the energy released. By contrast, intensity indicates the local effects and potential for damage produced by an earthquake on the Earth's surface as it affects population, structures, and natural features.

Basic consequences of earthquakes include collapsed buildings, fires (San Francisco, 1906), tsunamis (Indonesia, 2004) and landslides. Again these impacts depend on magnitude and intensity. Smaller tremors, that can occur either before or after the main shock, accompany most large earthquakes; these are termed foreshocks and aftershocks. Aftershocks can be felt from halfway around the world. While almost all earthquakes have aftershocks, foreshocks occur in only about 10 % of events. The force of an earthquake is usually distributed over a small area, but in large earthquakes it can spread out over the entire planet.

Since seismologists cannot directly predict when the next earthquake will happen, they rely on numerical experiments to analyse seismic waves and to accurately assess the magnitude and intensity of earthquakes. Such analyses allow scientists to estimate the locations and likelihoods of future earthquakes, helping to identify areas of greatest risk and to ensure the safety of people and buildings located in such hazardous areas.

1.1.1 Seismic Areas of the World

Figure 1.5 shows the epicentres of the 358,214 earthquakes that occurred in the years 1963–1998. Many places at which the earthquakes occurred, match with the boundaries of the tectonic plates. There are 15 major tectonic plates on Earth:

1. African Plate
2. Antarctic Plate
3. Arabian Plate
4. Australian Plate
5. Caribbean Plate
6. Cocos Plate
7. Eurasian Plate
8. Indian Plate
9. Juan de Fuca Plate

10. Nazca Plate
11. North American Plate
12. Pacific Plate
13. Philippine Plate
14. Scotia Plate
15. South American Plate

Figure 1.5 illustrates clearly that the Oceanic regions of the world have well-revealed seismicity following the lines of the plate boundaries. The major tectonic plates of the earth are shown in Figure 1.14.

As well as the major tectonic plates, there are many smaller sub plates known as *platelets*. These smaller plates often move which could be due to movement of the larger plates. The presence of platelets means that small but nonetheless damaging earthquakes can potentially occur almost anywhere in the world. These unexpected earthquakes, though less likely to occur, can be extremely destructive as in the 1960 Agadir earthquake in Africa, where none of the buildings in the affected area had been designed to be earthquake-resistant.

1.1.2 Types of Failure

Normally infrastructure damage during earthquakes is a result of structural inadequacy (Northridge, 1994), foundation failure (Mexico, 1985; Kobe, 1995), or a combination of both. Where foundation failure occurs the soil supporting the foundation plays an important role. The behaviour of foundations often depends on the soil's response to the shaking of the ground. (Figures 1.6 to 1.8)

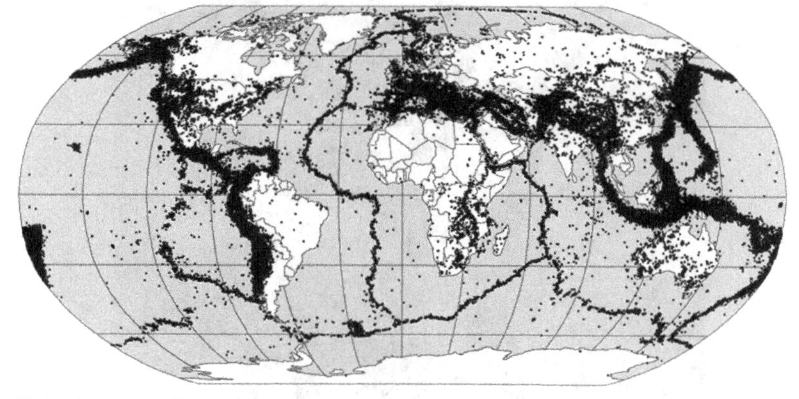

Figure 1.5 Epicentral Locations around the World (1963–1998). Courtesy of NASA. The NASA home page is http://www.nasa.gov 01/12/07.

The following examples of geotechnical damage to the built environment can be cited:

1. Failures of earth structures such as dams, embankments, landfill and waste sites. Failure of dams often causes flooding. During the Bhuj (India) earthquake of 2001 four dams (Fatehgarh, Kaswati, Suvi and Tapar) suffered severe damage. Fortunately their reservoirs were practically empty at the time of the earthquake and as a result there were no flood disasters.
2. Soil liquefaction resulting in widespread destruction to road networks and foundations. It is often observed that raft and piled foundations collapse without any damage to the superstructure.
3. Damage to underground or buried structures such as tunnels, box-culverts, underground storage facilities, buckling of pipelines, lifting of manholes etc. During the Turkish earthquake of 1999, the Bolu Tunnel suffered severe damage.
4. Damage to foundations due to large ground displacements owing to liquefaction-induced lateral spreading. Such foundations include underground retaining walls, quay walls and pile-supported wharfs.
5. Damage to foundations due to fault movement.

Some examples of damage to our environment due to geotechnical effects are shown in Figures 1.6 to 1.8.

Figure 1.6 Damage to Fatehgarh dam. Reproduced from http://gees.usc.edu/GEER/Bhuj/image15.gif 18/07/08 by permission of NSF-sponsored GEER (Geo-Engineering Earthquake Reconnaissance) Association.

Figure 1.7 Liquefaction-induced damage. Image of leaning apartment houses following Niigata Earthquake, Japan 1964. Reproduced from http://www.ngdc.noaa.gov/seg/hazard/slideset/1/1_25_slide.shtml 01/06/08 by permission of the National Geophysical Data Center.

Figure 1.8 Subsidence of a running track by more than three metres. Reproduced from http://www.whfreeman.com/bolt/content/bt00/figure2.jpg 08/04/08 by kind permission of Beverley Bolt.

1.1.3 Fault Movement and its Destructive Action

Fault movements, which may occur, especially at plate boundaries are shown in Figure 1.9. Fault movements can be very destructive if there are structures passing through them, as

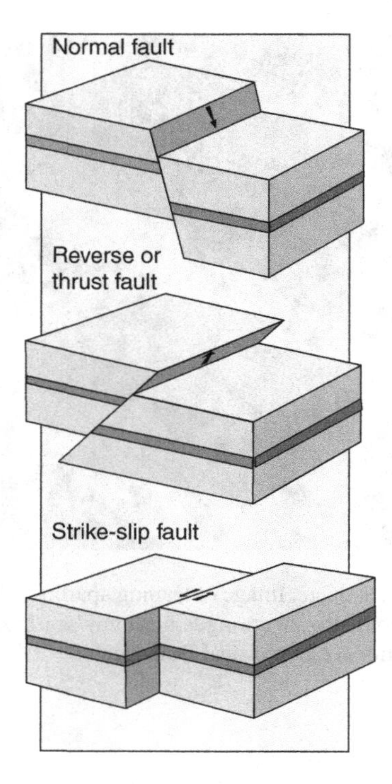

Figure 1.9 Fault categories. Courtesy of the U.S. Geological Survey. The USGS home page is http://www.usgs.gov 18/07/2007.

occurred in the 1999 Taiwanese and Turkish earthquakes. Faults are usually defined as a form of discontinuity in the bedrock and are associated with relative displacement of two large blocks of rock masses. Faults are broadly subdivided into three categories depending on their relative movement (Figure 1.9).

1. *Normal fault*: in normal faulting, one block (often termed as the hanging wall block) moves down relative to the other block (often termed as footwall block). The fault plane usually makes a high angle with the surface.
2. *Reverse fault* (also known as *thrust fault*): in reverse faulting, the hanging wall block moves up relative to the footwall block. The fault plane usually makes a low angle with the surface.
3. *Strike-slip fault*: in this fault, the two blocks move either to the left or to the right relative to one another.

1.2 The Nature of Earthquakes

As with many natural phenomena the origins of an earthquake are uncertain. In simplistic terms, earthquakes are caused by vibrations of the Earth's surface due to a spasm of ground shaking caused by a sudden release of energy in its interior. Seismologists have carried out

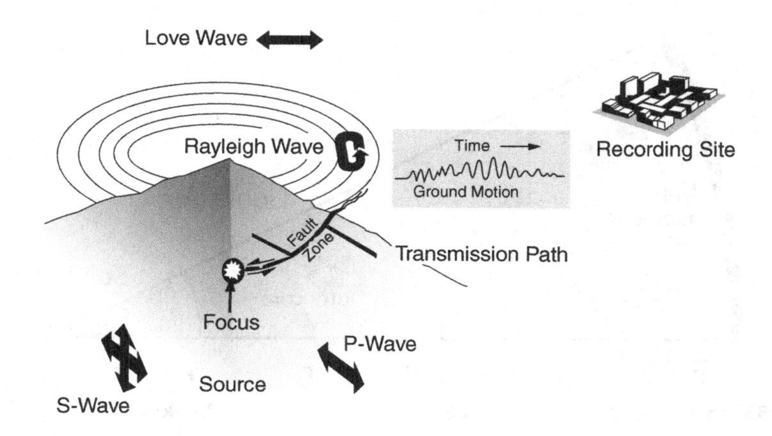

Figure 1.10 A schematic of the elements that affect ground motion.

studies of the geological aspects and changes that that have occurred inside the Earth over millions of years from which a broad picture of how earthquakes originate has emerged.

It is perhaps necessary to mention that tremors may be caused by other events, such as nuclear explosions or volcanic activity. However this book is concerned with earthquakes caused by disturbances deep inside the crust.

A schematic of the elements that affect ground motion is shown in Figure 1.10.

In the following sections some of the features of seismicity, the elements that affect ground motion, recording devices and evolution of the most important concepts of geology are explored.

1.3 Plate Tectonics

A cross section of the Earth is shown in Figure 1.11, illustrating the inner and outer core, mantle and crust, the crust itself being only 25–40 km in depth.

Various observations led the German meteorologist, Alfred Wegener, to expound the theory of *continental drift* in 1915. It was a revolutionary new idea and one of the founding principles of modern-day plate tectonics.

According to Wegener's theory, some 200–300 million years ago, the continents had formed a single land mass which slowly broke up and drifted apart. Wegener noticed that large-scale geological features on separated continents often matched very closely and that coastlines (for example, those of South America and Africa) seemed to fit together. Wegener also theorized that when continental land masses drifted together, crumpling and folding as they met, spectacular 'fold mountain' ranges such as the Himalayas and Andes were formed.

However, his fellow scientists found Wegener's theory hard to accept. This was mainly because he could not explain at that time *how* the continents had moved. His theory was finally shown to be right almost 50 years later. Scientists found (aided by modern recording instruments) that the sea floor is spreading apart in some places where molten rock is spewing out between two continents.

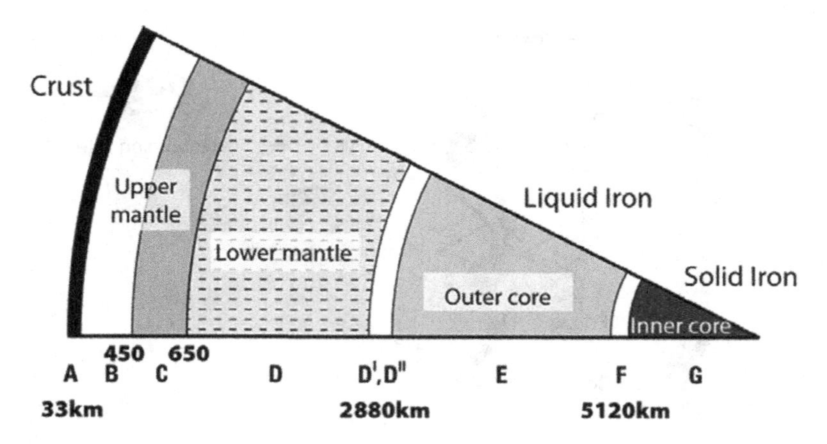

Figure 1.11 The Earth's structure, inner and outer core, mantle and crust. After Gubbins (1990) by permission of Cambridge University Press.

The theory thus gained acceptance by the scientific community and was hailed as one of the most revolutionary advances ever made in the field of geology from which the discipline of plate tectonics emerged. It is simple and elegant and has broad predictive power.

The basic principle of this discipline is that the Earth's crust consists of a number of plates that move with respect to each other floating on the underlying molten mantle. The movement is exceedingly slow on a geological scale but gives rise to enormous forces at the plate boundaries.

The movement of the plate boundaries is directly related to changes or reactions taking place inside the Earth. An accepted explanation of the source of plate movements is that the movement is governed by the requirement of thermo mechanical equilibrium of the mass inside the Earth. The upper portion of the mantle is in contact with the cooler crust whilst the lower portion is in contact with the hot outer core (Figure 1.12). A temperature gradient exists within the mantle, and leads to a situation where the cooler and dense crust rests on a less dense but warmer material. It gives rise to a situation where convection currents are set up within the mantle (Figure 1.12) which drives the plate movements. The cooler, denser material at the top begins to sink and the warmer less dense materials underneath rise. The sinking cooler material warms, becomes less dense, moves laterally and tries to rise, paving the way for subsequently cooled material to sink again; and this cycle would continue.

1.3.1 Types of Plate Boundaries

Plate tectonics is conceptually quite simple. The outer shell consists of 15 major plates about 100 km thick. The plates move relative to each other at very slow speeds (a few cm per year). The plates are considered rigid; there is little or no deformation within them. The deformation occurs at the boundaries. The outer strong shell forms the Earth's *lithosphere* and the movement takes place over the weaker layer called the *asthenosphere*. The types of movement at the plate boundaries are shown in Figure 1.13.

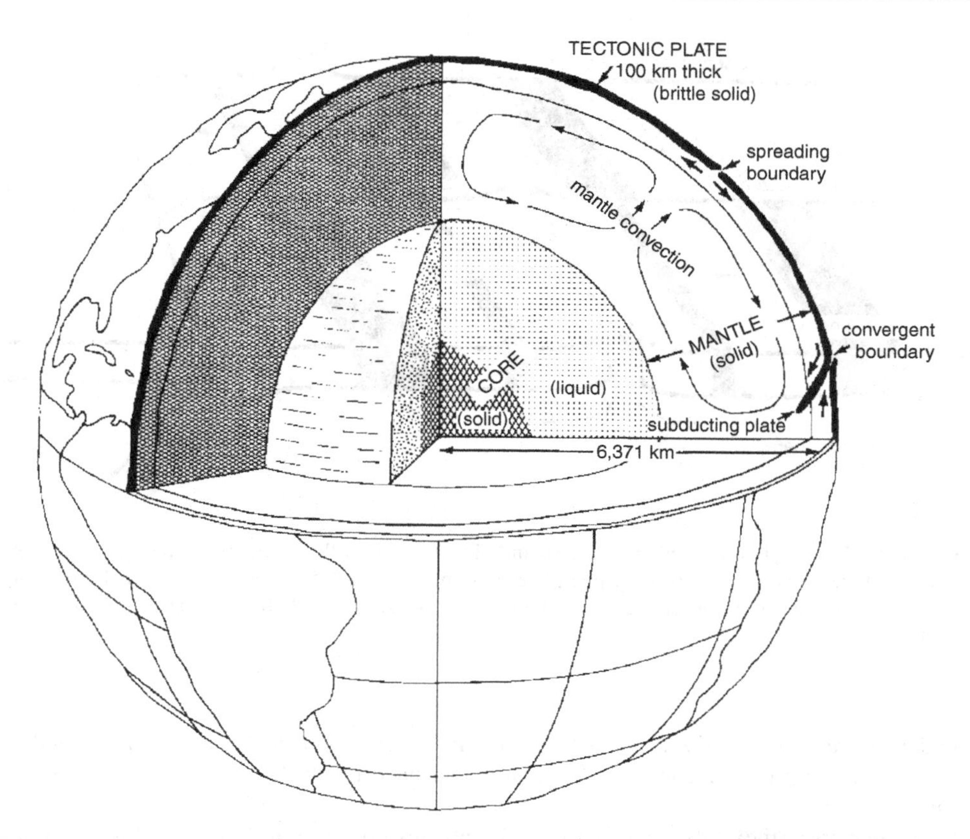

Figure 1.12 Convection currents in mantle. Near the bottom of the crust, horizontal components of convection currents impose shear stresses on bottom of crust, causing movement of plates on the Earth's surface. The movement causes the plates to move apart in some places and to converge in others. Reproduced from Noson *et al.* (1988) by permission of the Washington Division of Geology and Earth Resources.

1.3.2 Convergent and Divergent Boundaries

Three basic types of plate boundaries are as follows:

1. *Divergent* boundaries: where two plates are moving apart and new lithosphere is produced or old lithosphere is thinned. *Mid-oceanic ridges* (also known as *spreading centres*).
2. *Convergent* boundaries: where lithosphere is thickened or consumed by sinking into the mantle. *Subduction zones* and *alpine belts* are examples of convergent plate boundaries.
3. *Transcurrent* boundaries: where plates move past one another without either convergence or divergence. *Transform faults* and other strike-slip faults are examples of transcurrent boundaries.

At spreading centres both plates move away from the boundary. At subduction zones the subducting plates move away from the boundary. In general, divergent and transcurrent

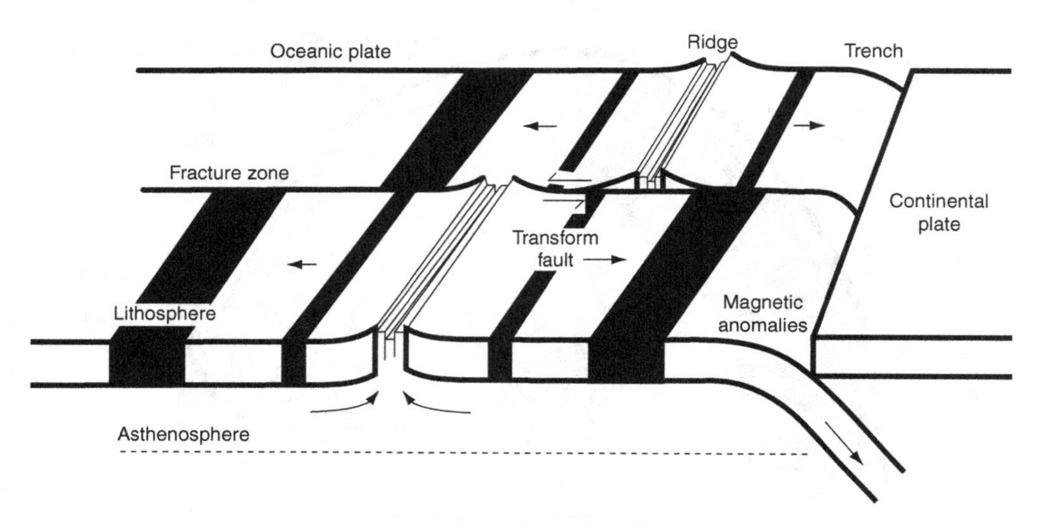

Figure 1.13 Types of movement at plate boundaries. Oceanic lithosphere is formed at ridges and subducted at trenches. At transform faults, plate motion is parallel to the boundaries. Each boundary has typical earthquakes. Reproduced from Stein and Wysession (2003) by permission of Blackwell Publishing.

boundaries are associated with shallow seismicity (focal depth less than 30 km). Subduction zones and regions of continental collision can have much deeper seismicity (Lay and Wallace, 1995).

Seismologists' study of plate tectonics points towards the following nature of plate behaviour and characteristics. Warm parts of the mantle material rise at *spreading centres* or mid-ocean ridges, and then cool. The cooling material forms strong plates of new oceanic lithosphere. The newly formed oceanic lithosphere while cooling down moves away from the ridges and eventually reaches the subduction zones or trenches, where it descends in downgoing slabs back into the mantle, reheating during the process. At a common point on the two boundaries, it is the direction of relative motion that will determine the nature of the boundary.

1.3.3 Seismicity and Plate Tectonics

Crustal deformations occur largely at plate boundaries; when the stresses at these boundaries exceed the strength of the plate boundary material, strain energy is released, causing a tremor. If strain energy accumulates within the plate the consequent release of energy results in a full-blown earthquake.

The theory of plate tectonics has established that there are movements at plate boundaries. The movement between the portions of the crust occurs at new or pre-existing discontinuities in the geological structure known as faults. These may vary in length from several metres to hundreds of kilometres. Generally the longer the fault the larger the earthquake it is likely to generate. Figure 1.14 shows the major tectonic plates, mid-ocean ridges, trenches, and transform faults of the earth.

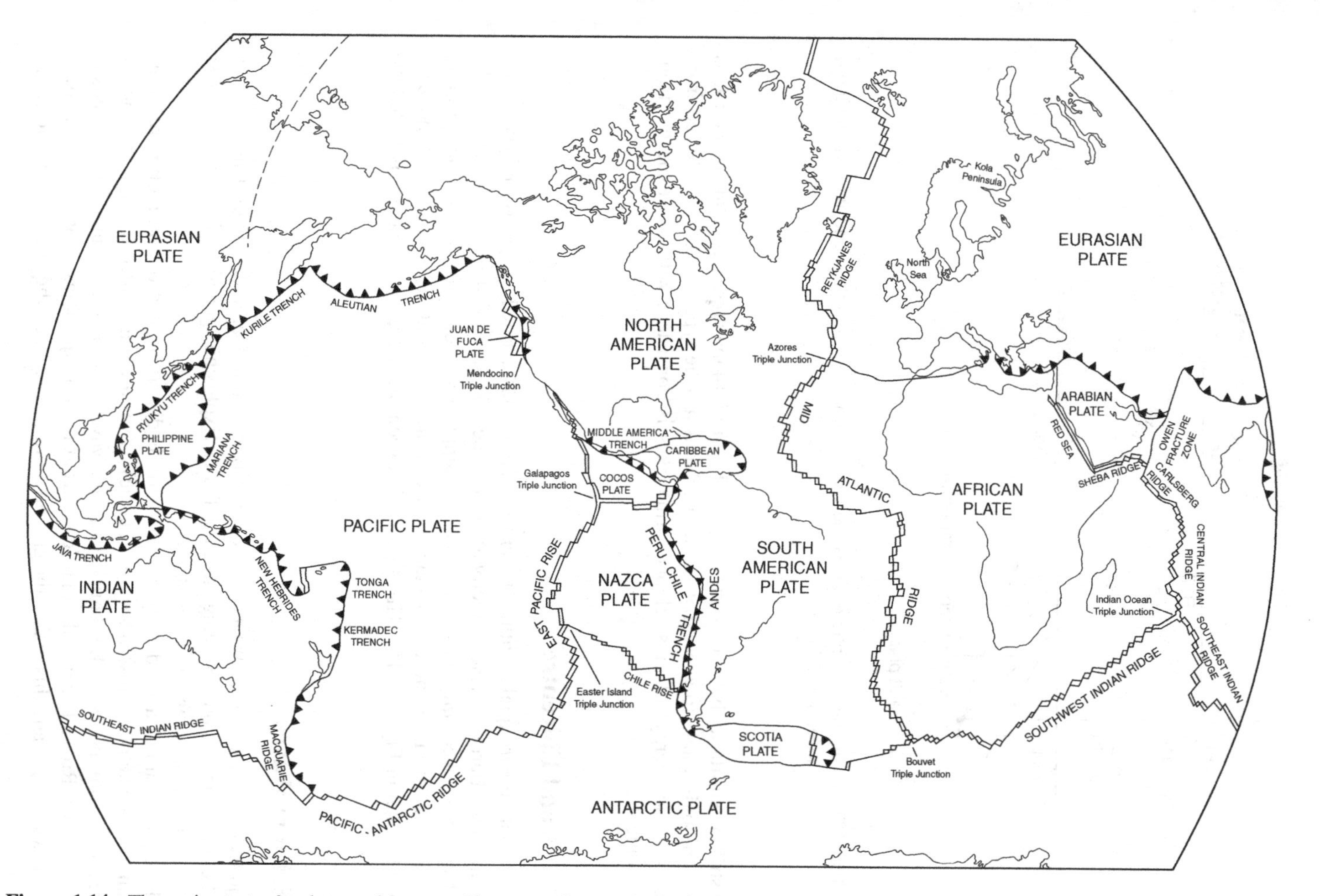

Figure 1.14 The major tectonic plates, mid-ocean ridges, trenches, and transform faults of the earth. Arrows indicate direction of plate movements. Reproduced from Fowler (1990) by permission of Cambridge University Press.

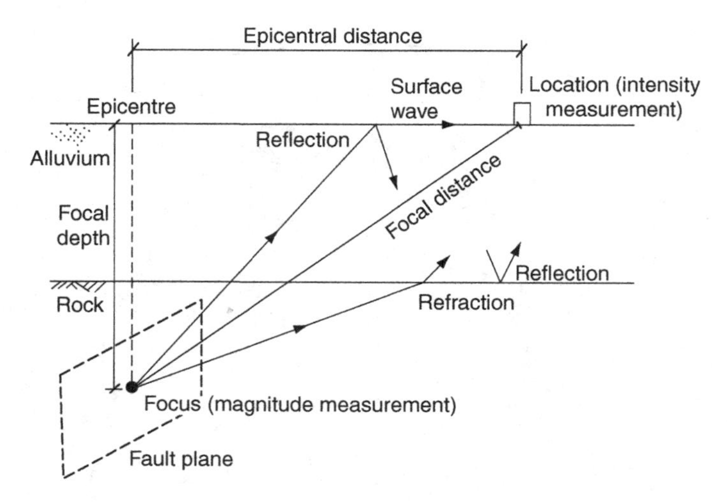

Figure 1.15 The focus and epicentre of an earthquake. Reproduced from Stein and Wysession (2003) by permission of Blackwell Publishing.

The striking similarity between Figure 1.5 and Figure 1.14 may be noted. Only in the 1960s were seismologists able to show that 'focal mechanism' (the type of faulting inferred from radiated seismic energy) of most global earthquakes is consistent with that expected from plate tectonic theory.

1.4 Focus and Epicentre

The source of an earthquake within the crust is commonly termed the *focus* or *hypocentre*. The point on the Earth's surface directly above the focus is the *epicentre*. The distance on the ground surface between *any site of interest* (for example a recording station) and the epicentre is the epicentral distance and the distance between the focus and the *site* is called the *focal distance* (shown in Figure 1.15). The distance between the focus and epicentre is called the *focal depth*.

1.5 Seismic Waves

Earthquakes generate elastic waves when one block of material slides against another, the break between the two blocks being called a 'fault'. Explosions generate elastic waves by an impulsive change in volume in the material.

If the equilibrium of a solid body like the earth is disturbed due to fault motion resulting from an earthquake or explosion seismic (elastic) waves are transmitted through the body in all directions from the focus. Earthquakes radiate waves with periods of tenths of seconds to several minutes. Rocks behave like elastic solids at these frequencies. Elastic solids allow a variety of wave types and this makes the ground motion after an earthquake or explosion quite complex.

1.5.1 Body Waves

Two categories of seismic body waves are produced during an earthquake:

- primary waves (P)
- secondary waves (S)

P and S waves travel through the interior of the earth from the focus to the surface. That is why they are called *body waves*. The velocities encountered depend upon the elastic constants and densities of the materials and other properties of the surrounding medium.

The first waves to arrive are the P ('pressure' or compression) waves (Figure 1.16a). The name P wave has its roots in the Latin *primus*. ('first'), since they are the first waves to arrive, having the highest velocity.

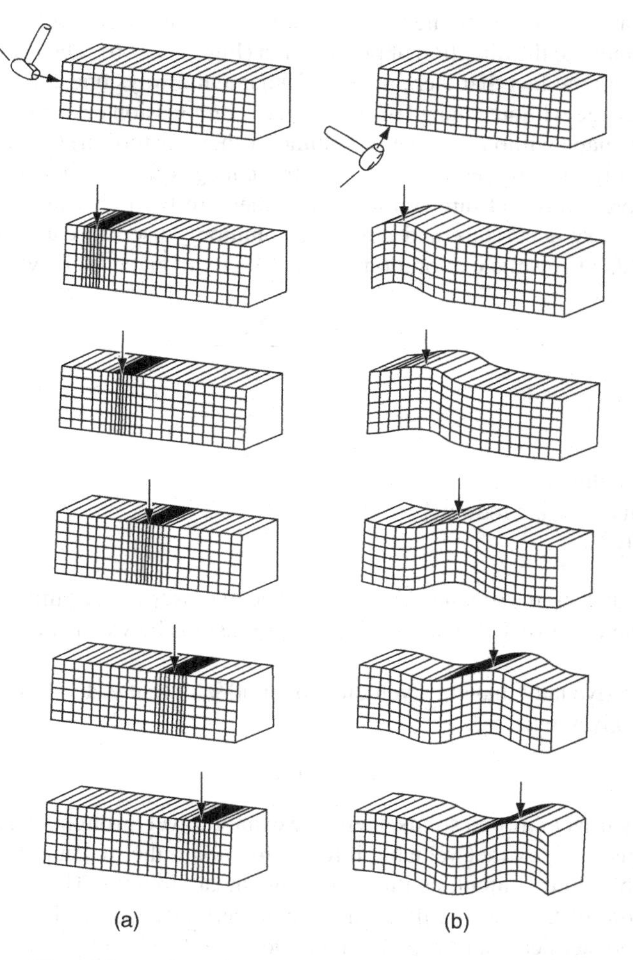

Figure 1.16 Primary (P) and Secondary (S) waves. Reproduced from Doyle (1996) by permission of John Wiley & Sons, Ltd.

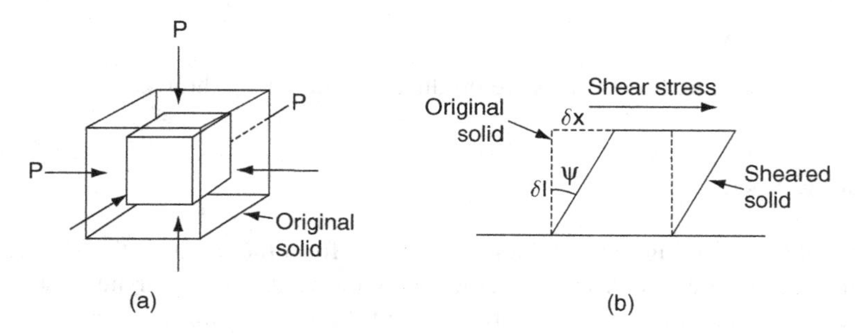

Figure 1.17 (a) Deformation by change of volume due to compression; (b) shear strain for case of block with lower side fixed. Reproduced from Doyle (1996) by permission of John Wiley & Sons, Ltd.

The second waves to arrive are the S (*secundus)* which have a transverse, shear vibration in a plane perpendicular to the direction of propagation (Figure 1.16b). The presence of two types of wave arises from the fact that there are two fundamental ways one can strain a solid body: (1) by volume change without change of shape (pure compression or expansion) Figure 1.17a; (2) by change of shape without change of volume (a shear distortion) (Figure 1.17b).

The P or compression waves transmit pressure changes through the Earth in a series of alternating compressions and rarefactions. Since they are the first waves to arrive, P waves can be recorded very accurately and are most commonly used in earthquake location and other related fields of seismic exploration. Their velocity is given by Doyle (1995):

$$V_p = \frac{(k + 4/3\mu)^{1/2}}{(\rho)}$$

Where

k is the bulk modulus
μ is the rigidity (Shear Modulus)
ρ is the density

(It may be noted that strains produced in the rock due to passage of seismic waves are normally very small (of the order of 10^{-6}), hence linear relationships between stress and strain may be assumed).

The S waves arrive later, having a velocity about 60 % of that of the P waves. The S (shear) wave velocity is given by:

$$V_s = (\mu/\rho)^{1/2}$$

S waves can only travel through solids, therefore cannot travel through the Earth's liquid outer core. Seismologists can extract valuable information about the makeup of the interior of the Earth and possible fluid content, as fluids have no shear strength. That is why it is believed that the outer core of the earth is fluid and almost certainly mainly liquid iron as has been deduced from geochemical and magnetic data (Doyle, 1995). P waves can travel through the core. An S_v wave is one in which the ground motion (vibration) is vertical and an S_h wave refers to one where the ground motion is horizontal (side to side).

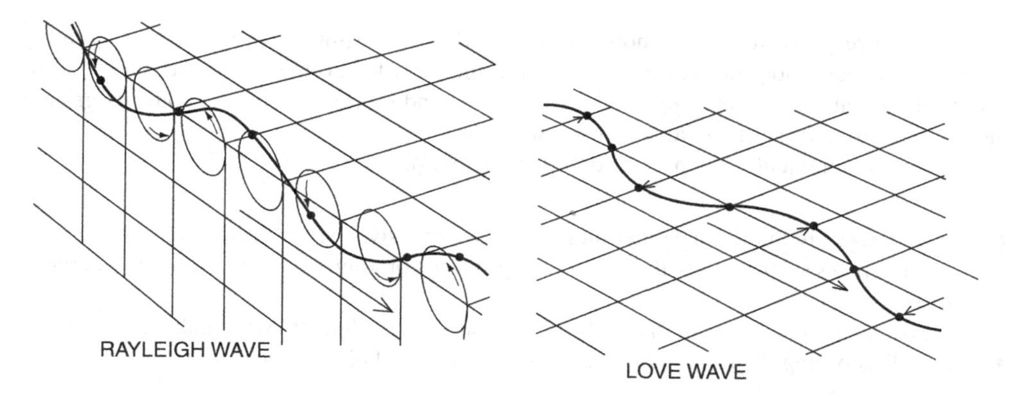

Figure 1.18 Surface waves: Love waves and Rayleigh waves. Reproduced from Doyle (1995) by permission of John Wiley & Sons, Ltd.

1.5.2 Surface Waves

When the two types of body wave reach the surface of the Earth, an interesting change occurs in the behaviour of the waves. The combination of the two types of wave in the presence of the surface leads to other types of waves, two of which are important for geophysics:

- Rayleigh waves
- Love waves

These are surface waves as distinct from body waves and produce large amplitude motions in the ground surface. They decay at a much slower rate than body waves and hence result in maximum damage. Surface waves are more destructive because of their low frequency, long duration and large amplitude.

The two types of surface waves are shown in Figure 1.18.

Rayleigh waves, also called 'ground roll', are the result of interaction between P and S_v waves and are analogous to ocean waves. The existence of these waves was first demonstrated by the English physicist, John William Strutt (Lord Rayleigh) in 1885. Rayleigh waves may be clearly visible in wide open spaces during an earthquake.

Love waves (Figure 1.18) are the result of interaction between P and S_h waves. Their existence was first deduced by the British mathematician, A.E.H. Love, in 1911 and they travel faster than Rayleigh waves.

1.6 Seismometers

Seismic waves are measured and recorded using the seismograph. The seismograph assists seismologists, geologists and scientists in the measurement and location of earthquakes. The instrumentation must above all be able to: (1) detect transient vibration within a moving reference frame (the pendulum of the instrument will be stationary as the Earth moves); (2) operate continuously with a detection device that is able to record accurately ground motion variation with time producing a seismogram; and (3) for instrument calibration purposes have a fully known linear response to ground motion; this would allow seismic recording to be accurately related to the amplitude and frequency content of the recorded ground motion. Such a recording system is called a *seismograph* and the actual ground motion sensor that converts ground motion into some form of signal is called a *seismometer*.

The basic components which make up most seismographs are:

- a frame lodged in the Earth, sometimes the most expensive part of the device;
- an inertial mass suspended in the frame, using springs or gravity to create a stable reference position;
- a damper system to prevent long-term fluctuation after reading an earthquake;
- a way of recording the motion or force of the mass, in relation to the frame.

1.6.1 Early Seismographs

John Milne is credited with the invention of the modern seismograph in 1880 while based at the Imperial College of Engineering in Japan. A surviving picture of Milne's device is shown below in Figure 1.19.

In the late nineteenth century most seismographs were developed by a team consisting of John Milne, T.A. Ewing and others working in Japan (1880–1895). These instruments consisted of a large stationary pendulum with a stylus. During an earthquake, as the ground

Figure 1.19 John Milne's device. http:/tremordeterra.blogspot.com 01/08/08

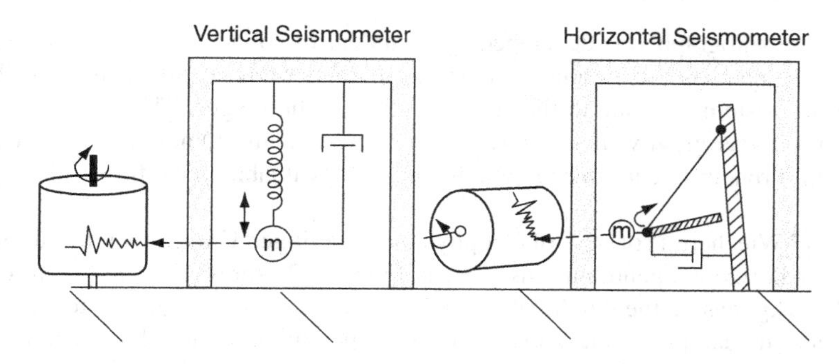

Figure 1.20 A schematic of vertical and horizontal seismometers. Actual ground motions displace the pendulums from their equilibrium positions, inducing relative motions of the pendulum masses. The dashpots represent a variety of possible damping mechanisms. Mechanical or optical recording systems with accurate clocks are used to produce the seisograms. Reproduced from Lay and Wallace (1995), Academic Press imprint by permission of Elsevier.

moved, the heavy mass of the pendulum remained stationary due to the inertia. The stylus at the bottom records markings which correspond to the Earth's movement.

Almost all seismographs are based on damped inertial-pendulum systems of one form or another. A schematic of simple vertical and horizontal instruments developed is shown in Figure 1.20. Passing seismic waves move the frame, while the suspended mass tends to stay in a fixed position. The seismometer measures the relative motion between the frame and the mass.

Two seismic instruments in the classic mould, developed around the turn of the earlier twentieth century are shown in Figure 1.21.

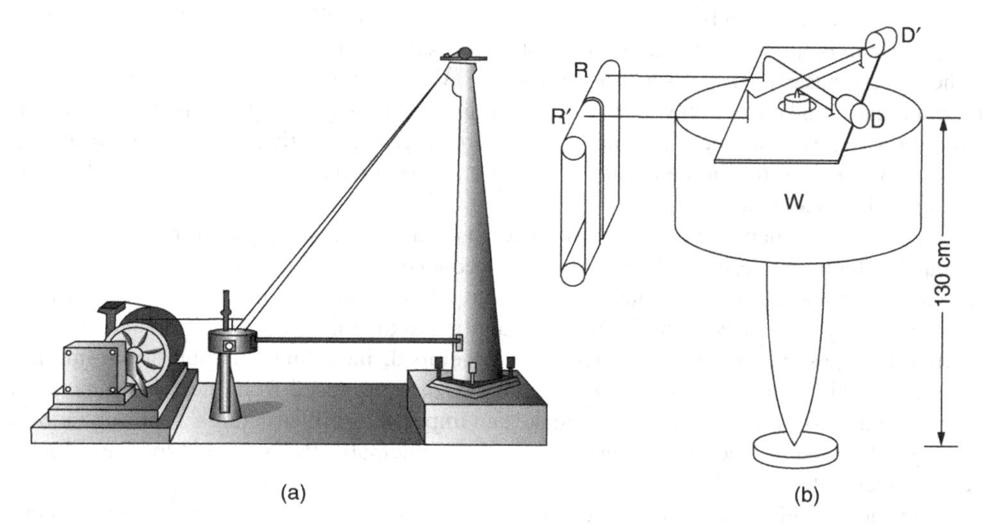

Figure 1.21 Early mechanical horizontal-motion seismographs: (a) the 1905 Omori 60-s horizontal-pendulum seismograph and (b) the 1904 1000-kg Wiechert inverted pendulum seismograph. Reproduced from Lay and Wallace (1995), Academic Press imprint by permission of Elsevier.

The Omori seismograph was developed by a student of John Milne in Japan. The instrument had direct response to ground displacement for periods of less than 60 s. The only damping in the system was due to the stylus friction in the hinges. The restoring force acting on the mass was gravity. It is interesting to note that the Omori instrument recorded the 1906 San Francisco earthquake which provided valuable data for further scientific research.

In 1898 Emil Wiechert, Professor of Geophysics at Gottingen University introduced viscous damping in his horizontal pendulum instrument. Figure 1.21 shows the version introduced in 1904 with 1000 kg mass at the top. Mechanical levers magnified the signal up to 200 times and pistons provide the damping. The Wiechert inverted-pendulum system has been in operation for more than 90 years.

Both instruments etched a record on smoked-paper recorders. Friction on the stylus provided the only damping in the Omori system, while air pistons (D and D′) damped the Wiechert instrument. Restoring springs connected to the mass, W, kept the inverted pendulum in equilibrium, with a special joint at the base of the mass permitting horizontal motion in any direction.

If the ground motion frequency is much higher than the natural frequency of the instrument ($\omega. >> \omega_0$) displacement on the seismometer is directly proportional to the ground displacement. Many of the early efforts in the development of seismometers focused on reducing ω_0 to yield displacement recordings for regional-distance seismographs.

1.6.2 Modern Developments

Earthquake design practices did not progress significantly until strong motion recording instruments were invented. Historical records can provide us with written descriptions of earthquake damage dating back 2000 years. However they are of little help in the development of appropriate methods of design or quantitative assessment of hazard.

There were basic forms of seismograph in use which were able to record the Great San Francisco earthquake in 1906 which was picked up as far away as Japan. The first measurements on a relatively modern seismograph of strong ground motion were recorded during the Long Beach (California) earthquake in 1933 and measurement techniques have advanced exponentially since then.

A recent development is the teleseismometer, a broadband instrument for registering seismographs which can record a broad range of frequencies. A servo feedback system keeps the mass motionless and the relative movement of the frame is measured by electronic displacement transducer with minimum force applied to the mass.

Different types of seismograph have been developed, including a digital strong-motion seismograph, also known as an *accelerograph*. The data produced from these instruments is essential to understanding how an earthquake can impact on built structures. A strong motion seismograph measures acceleration and this can be mathematically integrated to give velocity and displacement.

Strong-motion seismometers are not as sensitive to ground motion as teleseismometers but they stay on scale during the strongest seismic shaking.

A seismic recording centre is shown in Figure 1.22 and an example of a modern seismic recording in Figure 1.23.

Figure 1.22 A seismic recording centre at the University of Alaska. Geophysical Institute in Fairbanks, Alaska. The data is recorded digitally on a mainframe and on a PC and displayed on the recording drums. Photo by James Kocia, Courtesy of the US Geological Survey. The USGS home page is http://www.usgs.gov 18/07/2007.

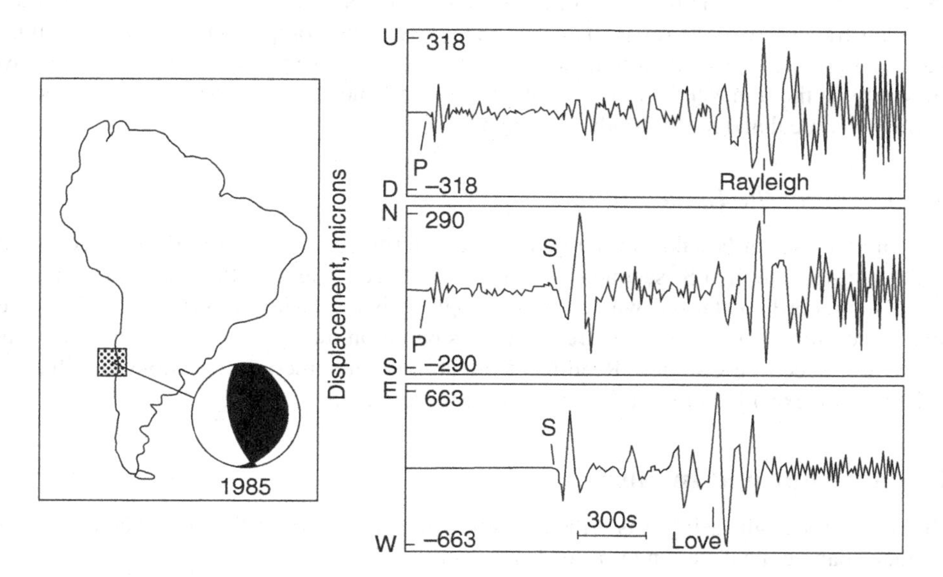

Figure 1.23 Recordings of the ground displacement history at station HRV (Harvard Massachusetts) produced by seismic waves from the 3 March 1985 Chilean earthquake, which had the location shown in the inset. Reproduced from Lay and Wallace (1995), Academic Press imprint by permission of Elsevier.

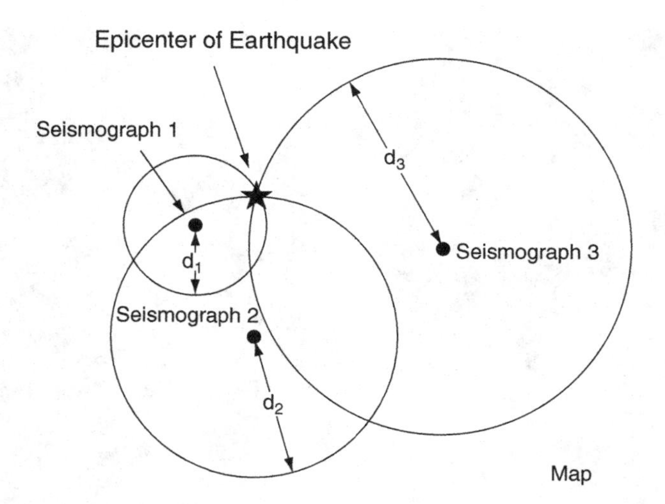

Figure 1.24 The intersection point between three seismographs, the location of the epicentre. Reproduced from http://www.tulane.edu/~sanelson 17/01/08 by permission of Professor Stephen A. Nelson, Department of Earth and Environmental Sciences, Tulane University, USA.

The three seismic traces correspond to vertical [U–D], north–south [N–S], and east–west [E–W] displacements. The direction to the source is almost due south; so all horizontal displacements transverse to the ray path appear on the east–west component. The first arrival is a *P* wave that produces ground motion along the direction of wave propagation. The Love wave occurs only on the transverse motions of the E–W component, and the Rayleigh wave occurs only on the vertical and north–south components.

1.6.3 Locating the Epicentre

The location of an earthquake is usually reported in terms of its epicentre. P (primary) waves travel at a faster rate than S (secondary) waves and are consequently the first waves to arrive at a seismic recording station (where the seismograph is located) followed by S waves. Preliminary identification of the epicentre location is based on the relative arrival times of P and S waves at the recording station. Readings from at least three recording stations are obtained to calculate the exact location of the epicentre (Figure 1.24).

1.7 Magnitude and Intensity

To be of use for engineering it is critical to know the size of an earthquake. The size of the earthquake can be measured in terms of *magnitude* and *intensity*.

- *Magnitude* is the amount of energy released from the source or focus of the earthquake.
- *Intensity* is the impact of ground-shaking on population, structures and the natural landscape; the impact will be greater nearer the site and less far away.

To measure these characteristics seismologists use two fundamentally different but equally important types of scale, the Magnitude scale and the Intensity scale.

1.7.1 Magnitude Scales

The magnitude of an earthquake is related to the amount of energy released by the geological rupture causing it and is therefore a measure of the absolute size of the earthquake without reference to the distance from the epicentre.

The best-known measure of earthquake magnitude was introduced by Charles Richter in the 1930s and became known as the *Richter Scale*, now referred to as the *local magnitude* (M_L). Richter was motivated by his desire to compile the first catalogue of Californian earthquakes and also saw the need for an objective size measurement to assess earthquakes' significance. Richter observed that the logarithm of maximum ground motion decayed with distance along parallel curves for many earthquakes. All the observations were from the same type of seismometer, a simple Wood-Anderson torsion instrument.

From his original recordings Richter expressed the magnitude scale as:

$$M_L = \log_{10} A(\Delta) - \log_{10} A_0(\Delta),$$

where A is the maximum trace amplitude for a given earthquake at a given distance as recorded by a Wood-Anderson instrument and A_0 is that for a particular earthquake selected as reference.

As may be seen it is on a logarithmic scale expressed in ordinary numbers and decimals. For example, the magnitude could be expressed as 4.3 on the Richter scale and the higher the number, the greater the damage.

Though M_L in its original form is rarely used today as Wood-Anderson torsion instruments are uncommon, it remains a very important magnitude scale because it was the first widely used 'size measure' and all other magnitude scales are tied to it. Further, M_L is useful for engineering. Many structures have natural periods close to that of a Wood-Anderson instrument (0.8 s) and the extent of earthquake damage is closely related to M_L (Lay and Wallace, 1995).

Although useful in establishing *local magnitude*, Wood-Anderson seismometers cease to be helpful for shocks at distances beyond 1000 km, hence the term *local magnitude*. It helps to distinguish it from magnitude measured in the same way, but from recordings on long-period instruments, which are suitable for more distant events. Beyond about 600 km the long-period seismographs of shallow earthquakes are dominated by surface waves (M_s) usually with a period of approximately 20 s. Surface wave amplitudes are strongly dependent on the source depth. Deep earthquakes ($>$ 45 km) do not generate much surface-wave amplitude and there is no appropriate correction for source depth.

Partly to overcome this problem and also to be applicable for shallow and deep earthquakes, Gutenberg proposed what he called the 'unified magnitude' denoted by m or m_b, which is dependent on body waves and is now generally referred to as *body wave magnitude* m_b. This magnitude scale is particularly appropriate for seismic events with a focal depth greater than 45 km.

The three magnitude scales (M_L, m_B *and* M_S) are interrelated by empirical formulae. Gutenberg and Richter (1956) reported the following empirical equations:

$$m_b = 0.63M_s + 2.5$$

$$M_s = 1.27(M_L - 1) - 0.016M_L^2.$$

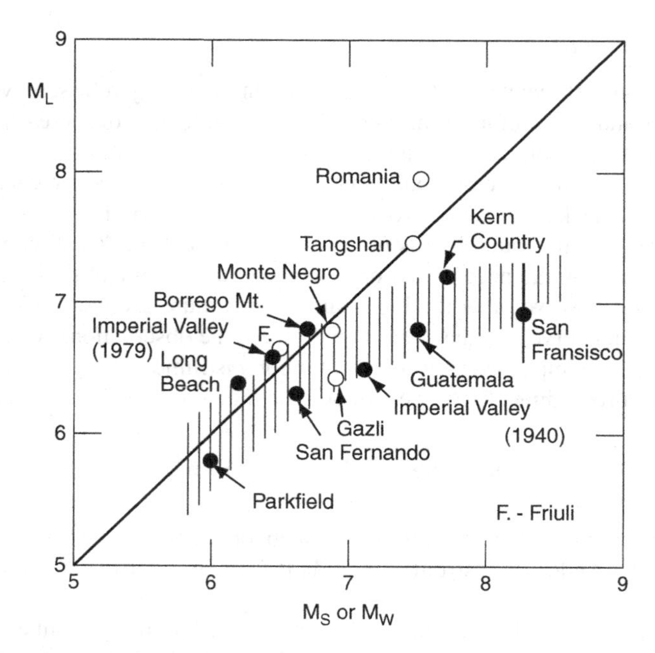

Figure 1.25 Effects of magnitude saturation for high frequency magnitude M_L versus the lower frequency magnitudes M_s or M_w (Note M_w is tied to M_s). Reproduced from Kanamori (1977), courtesy of the American Geophysical Union (AGU).

Since then several such equations have been published relating the different magnitudes (Utsu, 2002).

Both M_s and m_b were designed to be as compatible as possible with M_L; thus at times all three magnitudes give the same value for an earthquake. However this is rare. In all three cases we are making frequency dependent measurement of amplitudes at about 1.2, 1.0 and 0.05 Hz for M_L, m_b and M_s. Because of the frequency ω (or period) at which m_b is measured, earthquakes above a certain size will have a constant m_b (Lay and Wallace, 1995). This is referred to as *magnitude saturation*. Figure 1.25 shows measured values of M_L and M_s for several different earthquakes. It is clear that M_L begins to saturate at about magnitude 6.5, but for rare examples such as Tangshan and Romania, M_L does not saturate. A magnitude measure that does not suffer from this saturation deficiency is preferable.

1.7.2 Seismic Moment

Seismic moment, which characterizes the overall deformation at the fault, is considered by many researchers to be more accurate in calculating the earthquake's source strength. The advantage of seismic moment is that it can be interpreted simply in terms of the ground deformation. The moment is proportional to the product of the area of dislocation and the displacement across the fault. It is possible to relate seismic moment M_0, to seismic energy released. Also, surface wave magnitude M_s has been expressed (empirical relationship) in terms of seismic energy and thus M_0 and M_s are tied (Lay and Wallace, 1995).

The seismic moment, is estimated from the expression:

$$M_0 = G \cdot A \cdot d$$

where:

G is the shear modulus of the medium,
A is the area of the dislocation or fault surface,
d is the average displacement of slip on that surface.

It is possible at times to estimate d and A from field data and aftershock areas, but M_0 is usually estimated from the amplitudes of long period waves at large distances, corrected for attenuation, directional effects etc. (Doyle, 1995).

Hanks and Kanamori (1979) and Kanamori (1983) have, using certain assumptions, suggested a new magnitude scale, *moment magnitude* M_w, based on seismic moment:

$$M_w = \log_{10}(M_0/1.5) - 10.7$$

$$(M_0 \text{ in dyne-cm; } 10^5\text{dyne} = 1\text{N, thus} 10^7\text{dyne} - \text{cm} = 1\text{N} - \text{m}).$$

This scale has the important advantage that it does not saturate and the same formula can be used for shallow and deep earthquakes.

The range of seismic phenomenon in the earth is indicated in Figure 1.26.

From the above discussions it is clear how important it is to know the type of magnitude that is being used. In this book the local (i.e. Richter) magnitude (M_L) is used unless noted otherwise.

1.7.3 Intensity Scales

The intensity of an earthquake is a measurement of the observed damage at a particular location. This intensity will vary with distance from epicentre and depend on local ground conditions. It must be emphasized that intensity is a qualitative description of the effects of an earthquake at a particular site and this concept has proved very useful in interpreting historical data. Interpreting historical earthquake data can help in establishing the location, recurrence rates, and size of earthquakes in areas where no seismographs were installed at the time of the earthquake.

The Rossi-Forel (RF) scale of intensity (values ranging from I to X) was introduced in the late nineteenth century. The RF scale was in use for many years until its replacement in English-speaking countries by the Modified Mercalli Intensity (MMI) scale in 1931 (Table 1.1).

The MMI scale has twelve grades denoted by Roman numerals I–XII which describe the physical effects of earthquakes in a way similar to the Beaufort scale of wind strength.

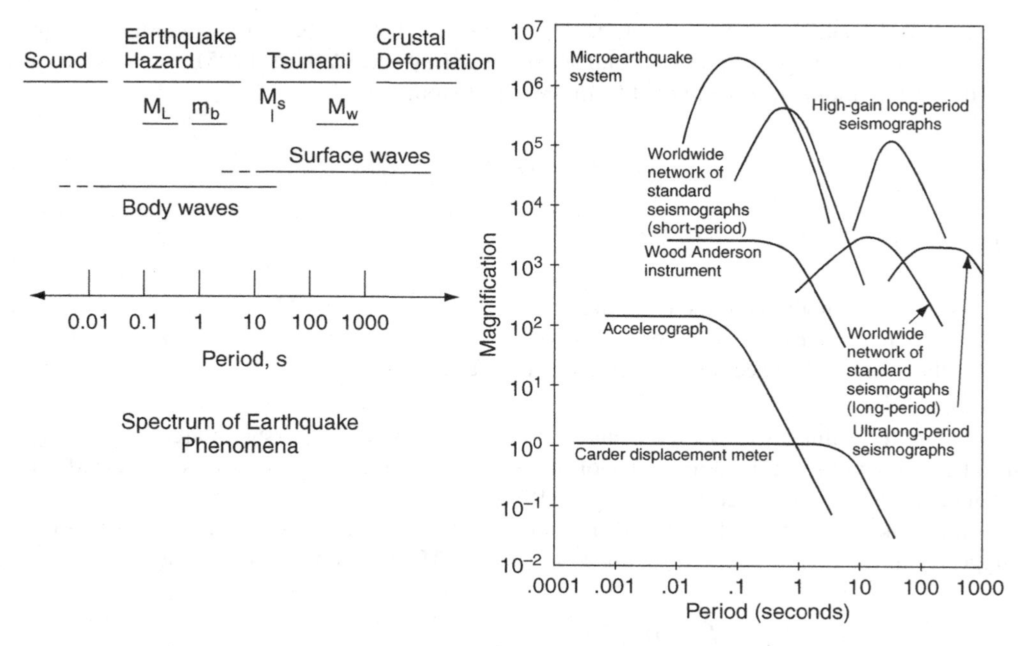

Figure 1.26 The range in period of seismic phenomenon in the Earth is shown on the left along with the characteristics: period of body waves; surface waves; and different seismic magnitude scales. On the right the amplitude responses of some major seismometer systems are shown. Each magnitude scale is associated with a particular instrument type; for example the Richter magnitude M_L is measured on the short period Wood-Anderson instrument. Reproduced from Kanamori (1988), Academic Press imprint by permission of Elsevier.

Table 1.1 Modified Mercalli Intensity Scale.

I. Not felt except by a very few under especially favourable conditions.
II. Felt only by a few persons at rest, especially on upper floors of buildings.
III. Felt quite noticeably by persons indoors, especially on upper floors of buildings. Many people do not recognize it as an earthquake. Standing motor cars may rock slightly. Vibrations similar to the passing of a truck. Duration estimated.
IV. Felt indoors by many, outdoors by few during the day. At night, some awakened. Dishes, windows, doors disturbed; walls make cracking sound. Sensation like heavy truck striking building. Standing motor cars rocked noticeably.
V. Felt by nearly everyone; many awakened. Some dishes, windows broken. Unstable objects overturned. Pendulum clocks may stop.
VI. Felt by all, many frightened. Some heavy furniture moved; a few instances of fallen plaster. Damage slight.
VII. Damage negligible in buildings of good design and construction; slight to moderate in well-built ordinary structures; considerable damage in poorly built or badly designed structures; some chimneys broken.

Table 1.1 (*continued*)

VIII. Damage slight in specially designed structures; considerable damage in ordinary substantial buildings with partial collapse. Damage great in poorly built structures. Fall of chimneys, factory stacks, columns, monuments, walls. Heavy furniture overturned.
IX. Damage considerable in specially designed structures; well-designed frame structures thrown out of plumb. Damage great in substantial buildings, with partial collapse. Buildings shifted off foundations.
X. Some well-built wooden structures destroyed; most masonry and frame structures destroyed with foundations. Rails bent.
XI. Few, if any (masonry) structures remain standing. Bridges destroyed. Rails bent greatly.
XII. Damage total. Lines of sight and level are distorted. Objects thrown into the air.

Source: Courtesy of the U.S. Geological Survey. The USGS home page is http://www.usgs.gov 18/07/2007.

1.8 Reid's Elastic Rebound Theory

As mentioned earlier in the chapter a key development in the study of earthquakes is Reid's elastic rebound theory first advanced by Henry Fielding Reid and based on his observations following the Great 1906 San Francisco Earthquake. Following an in-depth study he believed *elastic rebound* to be the cause of earthquakes and his simple theory has been confirmed over the years.

Reid examined ground displacement around the San Andreas fault which led him to conclude that the San Francisco earthquake was due to the sudden release of elastic energy accumulated in rocks on both sides of the fault. Elastic rebound theory may be explained in the simplest possible terms by thinking about what happens when an elastic band is stretched then either broken or cut – the energy stored in the band during stretching is released in a sudden 'elastic rebound'. Like the rubber band, the more the fault is strained the more energy is stored in the rocks. When the fault ruptures, this elastic energy is released. The energy is dissipated partly as heat, partly in cracking under ground rocks and partly as elastic waves. These waves cause the actual earthquake. The mechanism is illustrated in Figure 1.27.

In his studies of the San Andreas fault Reid discovered that several metres of relative motion occurred along several hundred kilometres of the fault (Figure 1.28). The Pacific plate slid as much as 4.7 metres in a northerly direction past the adjacent North American plate. Reid concluded that due to the plates sliding past each other, the rocks at the fault zone were bending and storing up elastic energy. When the rocks released this elastic energy and returned to their original form, the result was the Great 1906 San Francisco Earthquake.

1.9 Significant Milestones in Earthquake Engineering

Historically, earthquakes have shown the shortcomings of contemporaneous design methodologies and construction practices, resulting in structural failures and loss of life. Post-earthquake investigations have led to improvements in engineering analysis, design and construction practices. A summary of the historical development of earthquake engineering practice, showing how earthquake engineers have learned from past failures, is shown in Table 1.2.

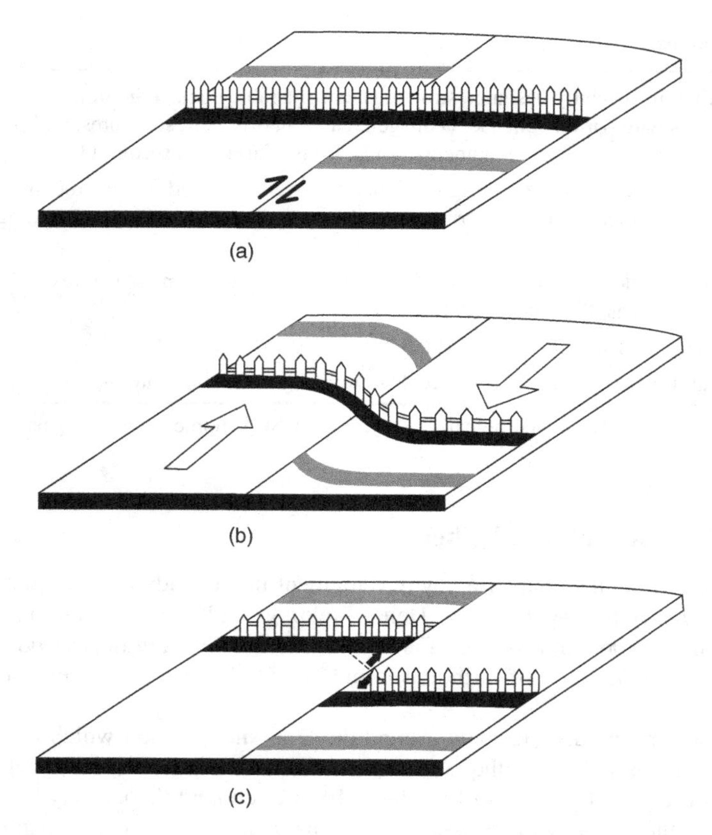

Figure 1.27 In his studies of the San Andreas fault, Reid discovered that several metres of relative motion occurred along several hundred kilometres of the San Andreas fault.

1.10 Seismic Tomography

Also known as seismic imaging, travel times along ray paths from different directions crossing the region of interest (Figure 1.29) are studied by seismologists to produce three-dimensional models of the interior, applying a process known as *mathematical inversion*. Recordings of ground motions as a function of time (Figure 1.23) provide the basic data that seismologists use to study elastic waves as they spread throughout the planet. Seismology is an observation-based science that tries to address the make-up of the Earth's interior by applying elastodynamic theory to interpret seismograms. There is also the physical constraint of being able to record seismic wave motions only at (or very near) the surface of the Earth. Hence seismologists draw heavily from mathematical methodologies for solving a system of equations commonly referred to as *inverse theory*. A good summary of state-of-the-art 'inverse theories' is provided by Romanowicz (2002).

The quest for mapping the interior of the Earth began from the time seismograms became available. Two eventful discoveries are of interest (a historical digression).

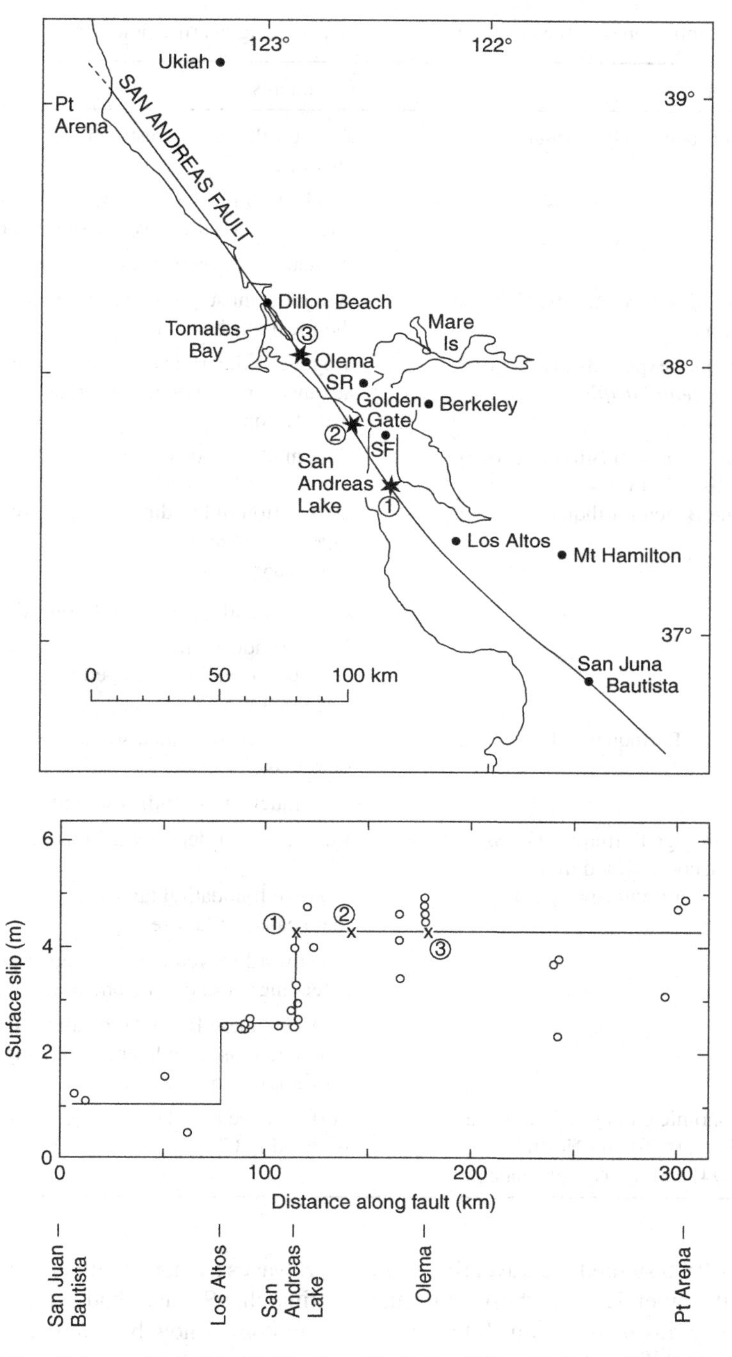

Figure 1.28 Map of the portion of the San Andreas fault that slipped in the 1906 San Francisco earthquake (top) and the amount of surface slip reported at various points along it (bottom) The slip is the distance by which the earthquake displaced originally adjacent features on the opposite side of the fault. Reproduced from Boore (1997), (c) Seismological Society of America (SSA).

Table 1.2 Key milestones in the development of earthquake engineering practice.

Year	Event	Comments
1906	San Francisco Earthquake	State Earthquake Investigation Commission appointed. Realization for the first time of the knock-on effects earthquakes have on our planet. A wealth of scientific knowledge derived.
1911	Reid introduces Elastic Rebound Theory	A significant step towards our understanding of the focal mechanism.
1915	Wegener expounds the theory of '*continental drift*'	Wegener's theory is one of the most significant advances in the twentieth century in the field of plate tectonics.
1927	UBC (Uniform Building Code) published in USA.	First modern code of practice.
1933	Long Beach Earthquake (USA).	Destruction of buildings, damage to schools especially severe. UBC 1927 revised. Field Act and Riley Act introduced. First earthquake for which records were obtained from the recently developed strong motion seismometers.
1964	Niigata Earthquake (Japan).	Showed that soil can also be a major contributor of damage. Soil liquefaction studies started.
1994	Northridge Earthquake (USA) Steel connections failed in bridges.	Importance of ductility in construction realized.
1995	Kobe Earthquake (Japan) Kawashima.	Massive foundation failure. Soil effects were the main cause of failure. Downward movement of a slope (lateral spreading) is said to be one of the main causes. JRA (Japanese Road Association) Code 1996 modified (based on lateral spreading mechanism) for design of bridges.
1995	Economic damages from Loma Prieta (1989) and Northridge (1994) earthquakes evaluated.	Performance based earthquake engineering initiated in USA.

Oldham in 1906 studied the travel times of P and S waves as they travelled from one side of the Earth to the other. He hypothesized for the first time that S waves had penetrated a central core where they travelled at a much lower rate. Seismologists now believe that the outer part of the core is fluid through which S waves do not propagate.

Gutenberg, a seismologist working in Germany in the early 20th century, fixed the depth to the boundary of the separate core to be about 2900 km. This value has remained virtually unchanged since then.

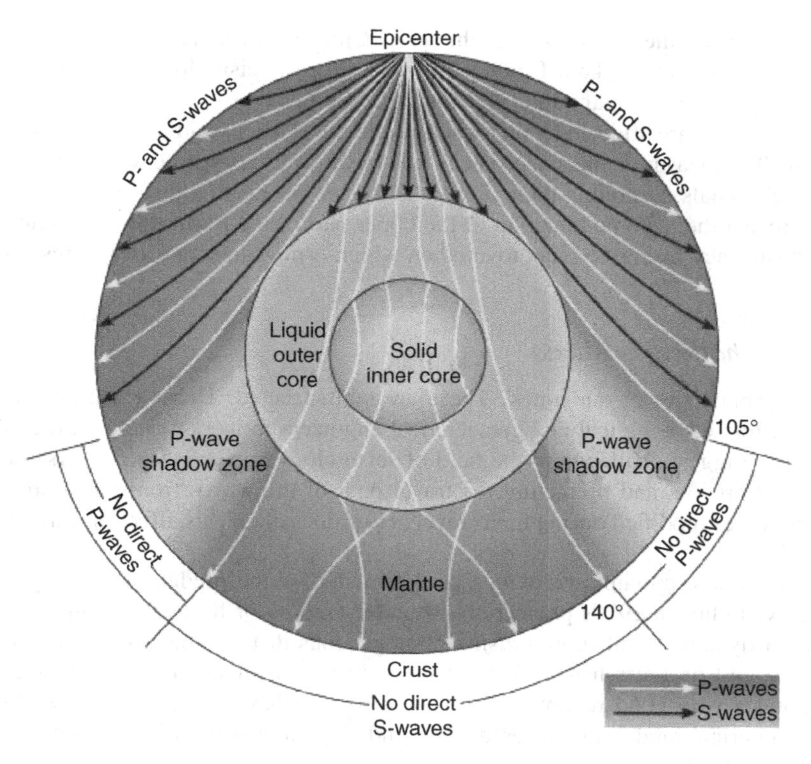

Figure 1.29 Travel paths of seismic waves [Note: S waves do not pass through the liquid outer core]. Levin, H.L. 2003, Reprinted by permission of John Wiley & Sons, Inc.

All high resolution methods for studying the Earth's interior are based on analysing the propagation of seismic body waves as they travel across the Earth. Recall that P waves are compressional waves and the nearest analogy is that of the sound waves in air. P waves can travel through solids and fluids. Shear waves (S waves) propagate only in solids.

Figure 1.29 shows the travel paths of seismic waves through the Earth's mantle from the source to seismographs at the surface. If travel times are available for enough ray paths in different directions crossing a small region, then it is possible to unscramble the recorded plots to reveal a three-dimensional wave speed distribution.

P and S waves penetrate though the Earth as X-rays do in the human body. The concept has an analogy from modern medicine. A common technique used by doctors to obtain internal images of the human body is a CAT (Computerized Axial Tomography) scan. By taking multiple X-rays, the variation in density of the body's tissues can be deduced mathematically and assembled into a three-dimensional image. Probing the interior of the Earth by P and S waves is known as *geophysical tomography*.

The data are treated in much the same way as in medical imaging. Usually a small block of different tectonic type (for example a block for tomography of UK) is probed. Numerical methods play a major role. The resolution of the block depends upon the number of rays passing through the block.

Figure 1.23 shows the recordings of the 1985 Chilean Earthquake. P, S, Love (L) and Rayleigh (R) waves are marked. From these recordings a seismologist can determine the location of the hypocentre, magnitude and source properties.

The surface waves are affected by the structure and the elasticity of the rocks through which they traverse. The measurement of the speeds and wave forms constitute what is referred to as 'tomographic' signals. Decoding these signals is the job of seismologists. Once decoded the signals can reveal the tectonic make-up of the upper part of the Earth. It is also important to establish consistency from one event to another, which is now an active area of research.

1.10.1 The Challenges Ahead

What are the challenges of our times? These are multi-faceted. There is a need, more than ever perhaps, for seismological and geotechnical engineers to work closely with earthquake engineers. The biggest challenge in the field of seismology appears to be understanding the fault rupturing process and predicting the travel path of the waves from the source to site. How are the waves modified during their travel? How does all of this affect the strong motion at the site?

Earthquake source dynamics provides key elements for the prediction of strong ground motion. Early studies in 1970s pioneered our understanding of friction and introduced simple models of dynamic fault rupture using homogeneous distributions of stress and friction parameters. Rapid progress has since been made in the understanding of dynamic rupture modelling and a very good review of the state-of-the-art has been provided by Madariaga (2006). A numerical model able to predict ground motion at a site is a development which would be of great value.

Understanding of soil behaviour and changes thereof during an earthquake continues to elude us. Sometimes, when liquefaction does not manifest itself on the ground surface, it is still very difficult to know; what changes have taken place beneath the surface and more importantly how the soil will respond in the future?

Perhaps the most exciting area of seismological research concerns movements deep inside the earth. Many of these movements can now be described but we know very little about the driving forces behind them.

At the other end of the seismic design spectrum, we note that structures need to be designed to withstand seismic forces. A key requirement for modelling the mechanical response of a structure is to have realistic estimates of how the ground beneath that structure will move during an earthquake.

The potential payoff for being able to perform accurate nonlinear analysis of soil structure behaviour is high. Accurate numerical simulations can provide a more realistic picture of how a structure actually responds during strong ground motion. With this information, engineers can be significantly more confident that a structure will survive a large-magnitude earthquake.

1.11 References

Bolt, B.A. (2004) *Earthquakes*. W.H. Freeman & Company, New York.

Bolt, B.A. Image of subsidence of running track following earthquake. http://www.whfreeman.com/bolt/content/bt00/figure2.jpg 08/04/08

Boore, D.M. (1977) 'Strong-motion recordings of the California earthquake of April 18, 1906'. *Bull. Seism. Soc. Am.* **67**: 561–577.

Doyle, H. (1995) *Seismology*. John Wiley & Sons, Chichester, New York.

Esteva, L. and Villaverde, R. (1973) 'Seismic risk, design spectra and structural reliability'. Proceedings of the 5th World Conference on Earthquake Engineering, Rome, 2586–2596.

Fowler, C.M.R. (1990) *The Solid Earth. An Introduction to Global Geophysics*. Cambridge University Press, Cambridge.

GEER (Geo-Engineering Earthquake Reconnaissance) Organization. http://gees.usc.edu/GEER/Bhuj/image15.gif 21/07/08 Damage to Fatehgarh Dam during Bhuj Earthquake 2001.

Gubbins, D. (1990) *Seismology and Plate Tectonics*. Cambridge University Press, Cambridge.

Gutenberg, B. and Richter, C.F. (1956) 'Magnitude and energy of earthquakes'. *Ann. Geofis.*, **9**: 1–15.

Hanks, T.C. and Kanamori, H. (1979) 'A moment magnitude scale'. *J. Geophys. Res.* **84**: 2348–2350.

Kanamori, H. (1977) 'The energy release in great earthquakes'. *J. Geophys. Res.* **82**(20): 2981–2987.

Kanamori, H. (1983) 'Magnitude scales and the quantification of earthquakes'. *Tectonophysics* **93**: 185–199.

Kanamori, H. (1988) 'Importance of historical seismograms for geophysical research', In W.H.K. Lee, H. Meyers and K. Shimazaki (eds), *History of Seismograms and Earthquakes of the World*. Academic Press, San Diego.

Lay, T. and Wallace, T.C. (1995) *Modern Global Seismology*. Vol. 58, International Geophysics Series. Academic Press, San Diego.

Levin, H.L. (2003) *The Earth through Time*. John Wiley & Sons Inc., New York. http://www3.interscience.wiley.com:8100/legacy/college/levin/0470000201/chap_tutorial/ch05/chapter05-2.html 21/07/08

Madariaga, R. (2006) 'Earthquake dynamics and the prediction of strong ground motion'. Proceedings of the 1st European Conference on Earthquake Engineering and Seismology, Geneva, Keynote Address, Paper ID K1b, 14 pp.

Milne seismometer. http:/tremordeterra.blogspot.com 01/08/08

National Geophysical Data Center. Image of leaning apartment houses following Niigata Earthquake, Japan 1964. http://www.ngdc.noaa.gov/seg/hazard/slideset/1/1_25_slide.shtml 01/06/08.

NASA. http://www.nasa.gov 01/12/07. Image of Preliminary Determination of Epicenters.

Noson, L.L., Qamar, A.I., Thorsen, G.W. (1988) *Washington State Earthquake Hazards*: Washington Division of Geology and Earth Resources Information Circular 85, Olympia, Washington.

Pararas-Carayaniss, G. http://www.drgeorgepc.com/Tsunami1985Mexico.html 21/7/2008. Image of damage during Mexico City Earthquake 1985.

Romanowicz, B. (2002) 'Inversion of surface waves: A review'. In W.H.K. Lee, H. Kanamori, P.C. Jennings and C. Kisslinger (eds), *International Handbook of Earthquake and Engineering Seismology*. Part A International Geophysics Series. Academic Press. San Diego.

Steim, J.M. (1986) 'The very broadband seismograph', Ph. D. Thesis, Harvard University, Cambridge, M.A.

Stein, S. and Wysession, M. (2003) *An Introduction to Seismology, Earthquakes and Earth Structure*. Blackwell Publishing, Oxford.

Tulane University, Department of Earth and Environmental Sciences. http://www.tulane.edu/~sanelson 17/01/08. Image of intersection point between three seismographs.

US Geological Survey. http://www.usgs.gov 18/07/2007. Selected images.

US Geological Survey. http://www.usgs.gov 18/07/2007. Modified Mercalli Intensity Scale http://earthquake.usgs.gov/learning/topics/mercalli.php 21/07/08.

University of Bristol, Department of Civil Engineering. http://www.bristol.ac.uk/civilengineering/research/structures/eerc/realearthquakes/kobe-fire.gif 13/12/07. Image of damage during Kobe Earthquake, 1995.

Utsu, T. (2002) 'Relationships between magnitude scales'. In W.H.K. Lee, H. Kanamori, P.C. Jennings and C. Kisslinger (eds), *International Handbook of Earthquake and Engineering Seismology*. Part A International Geophysics Series. Academic Press. San Diego.

Virtual Museum of the City of San Francisco. http://www.sfmuseum.org. Selected images of damage following the Great 1906 San Francisco Earthquake. http://www.sfmuseum.org/hist/pix48.html 18/07/08.

http://www.sfmuseum.net/hist10/mktstfire.html 18/07/08.

2

Single Degree of Freedom Systems

2.1 Introduction

Dynamic response of a practical structure is complex, and therefore it is preferable to begin the study of dynamic behaviour using simple systems. By simple systems we mean a system with one degree of freedom. A single degree of freedom (SDOF) system is defined as that in which only one type of motion is possible, or in other words the position of the system at any instant of time can be defined in terms of a single coordinate. As a guide, the complexity of a dynamical system may be judged from the *number of degrees of freedom* possessed by the system. This number is equal to the number of independent coordinates required to specify completely the displacement of the system. Systems with one degree of freedom are discussed in this chapter. Systems with many degrees of freedom are dealt with in Chapter 3. The emphasis to study simple one-degree-of-freedom systems is underlined by the fact that, in many practical situations, a simple and approximate solution to a complex system is sought.

Consider the elevated water tank shown in Figure 2.1(a) or the simple framing system in Figure 2.1(b).

The cantilever tower in Figure 2.1(a) supporting the water tank provides the lateral stiffness k. The frame in Figure 2.1(b) has a lumped mass of m equal to the mass of the roof and its lateral stiffness k is equal to sum of the stiffness of the two columns. Both these systems may be modelled as a single degree of freedom system with a lumped mass m supported by a mass less structure with stiffness k.

It is common knowledge that if the roof of the frame or the top of the water tank is pulled by a rope and suddenly released vibration in these structures will begin. The oscillations will continue with decreasing amplitude before coming to rest. The system comes to a rest because of the damping in the system.

Fundamentals of Seismic Loading on Structures Tapan Sen

(a)

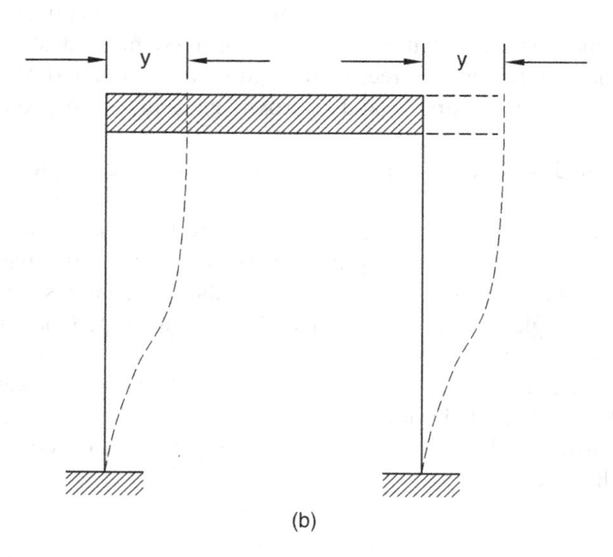

(b)

Figure 2.1 (a) Elevated water tank – an example of an SDOF system. http://www.walksydneystreets.net/photos/acacia-gardens-water-tank-w.jpg 29/07/08 (A Waddell). (b) A plane frame with lumped mass.

The basic concept of motion is derived from Newton's second law of motion – '*when a mass is acted upon by a force, the rate of* change *of momentum of the mass in the direction of the force is equal to the force*'. In other words

$$\frac{d}{dt}(mv) = F$$

$$m\frac{d^2y}{dt^2} = F.$$

We see that the force acting on the mass gives rise to acceleration. To extend the concept further, the schematic of a conventional representation of a single degree of freedom spring-mass system is shown in Figure 2.2(a) and the free-body diagram in Figure 2.2(b). Thus summing up the forces (Figure 2.2(b)) gives

$$m\ddot{y} + c\dot{y} + ky = F(t). \tag{2.1}$$

Note: In Equation (2.1) the dots denote differentiation with respect to time and this convention will be followed throughout the book unless stated otherwise.

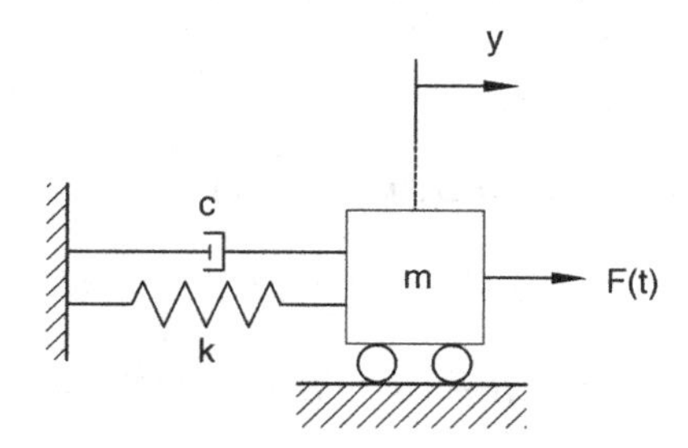

Single Degree of Freedom System

(a)

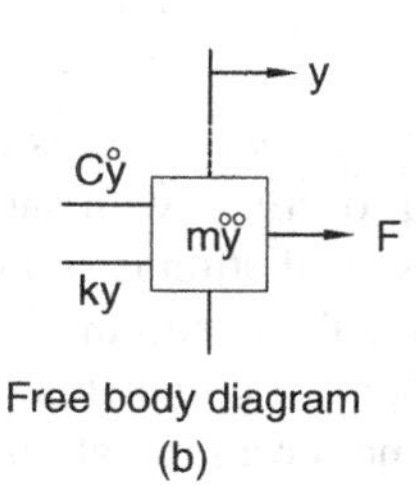

Free body diagram

(b)

Figure 2.2 (a) Schematic of a conventional representation of a single degree of freedom spring-mass system; (b) free-body diagram.

2.2 Free Vibration

This is the simplest form of dynamic response that may exist, i.e. in the absence of applied excitation (($F(t) = 0$). The equation motion for the undamped case is

$$m\ddot{y} + ky = 0. \tag{2.2}$$

If no damping is present then

$$\begin{Bmatrix} n_1 \\ n_2 \end{Bmatrix} = \pm i\sqrt{\frac{k}{m}}. \tag{2.3}$$

The solution may be expressed as

$$y = A_1 e^{n_1 t} + A_2 e^{n_2 t}. \tag{2.4}$$

From the relationship between exponential and trigonometric functions we have

$$e^{ix} = \cos x + i \sin x. \tag{2.5}$$

The solution (Equation 2.2) can also be written as

$$y = B_1 \sin \omega t + B_2 \cos \omega t \tag{2.6a}$$

or

$$y = B \sin(\omega t + \psi) \tag{2.6b}$$

where

$$B = (B_1^2 + B_2^2)^{1/2} \tag{2.6c}$$

$$\tan \psi = B_2 / B_1 \tag{2.6d}$$

$$\omega = \sqrt{k/m} \tag{2.6e}$$

and is known as the circular natural frequency of the system.

As the system is not damped, it will oscillate with same frequency ω. The units of ω are radians/per time (radians/second). This oscillation is shown in Figure 2.3.

It is seen that the amplitude, defined as the maximum displacement from the mean position, is B; the period T is the time required for one complete cycle of vibrations, and the frequency f is the number of vibrations in unit time. From Equations (2.6c, d) and the above definitions we have

$$\omega T = 2\pi \quad \text{and} \quad f = 1/T,$$

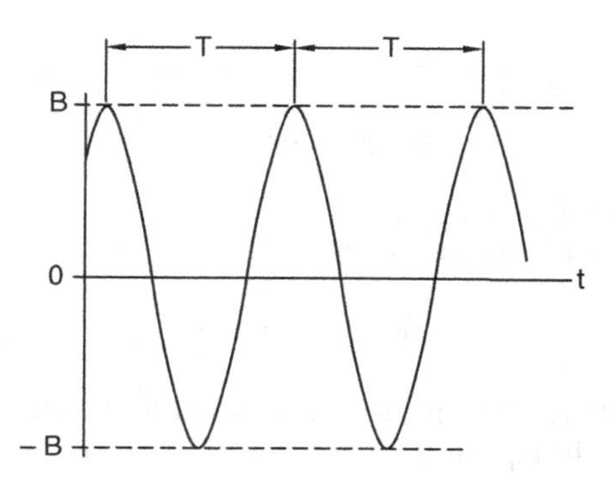

Figure 2.3 Free undamped oscillation.

i.e.

$$f = \omega/2\pi = 1/T.$$

Thus ω is the circular natural frequency of the system, ω_0 being measured in radians per second if the natural frequency f is in Hertz. The angle ψ in Equation (2.6d) is a phase angle. The constants B and ψ in Equations (2.6c, d) or B_1 and B_2 in Equation (2.6a) are determined from the initial conditions. For example, suppose that at time $t = 0$ the mass m is given a displacement y_0 and a velocity V. Then substituting in Equation (2.6a):

$$y_0 = B_2.$$

Differentiating (2.6a) with respect to t, we have

$$\overset{\circ}{y} = B_1\omega \cos \omega t - B_2\omega \sin \omega t$$

and substituting

$$\overset{\circ}{y} = V \text{ at } t = 0$$
$$B_1\omega = V.$$

Thus,

$$y = \frac{V}{\omega} \sin \omega t + y_0 \cos \omega t$$

or

$$y = \left[(\frac{V}{\omega})^2 + y_0^2 \right]^{1/2} \sin (\omega t + \psi)$$

with

$$\tan\psi = \omega y_0/V.$$

2.2.1 Equations of Motion with Damping

$$m\overset{\circ\circ}{y} + c\overset{\circ}{y} + ky = 0 \tag{2.7}$$

This is a second order differential equation of standard form. Methods of solution may be found in many textbooks. The equation readily yields a solution if we assume

$$y = Ae^{nt} \tag{2.8}$$

where both A and n may be complex numbers. Substitution of this expression into Equation (2.7) yields a quadratic in n:

$$n^2 + \frac{c}{m}n + \frac{k}{m} = 0 \tag{2.9}$$

and therefore, this equation has two roots n_1 and n_2:

$$\begin{Bmatrix} n_1 \\ n_2 \end{Bmatrix} = -\frac{c}{2m} \pm \sqrt{\frac{c^2}{4m^2} - \frac{k}{m}}. \tag{2.10}$$

The solution to Equation (2.7) is therefore:

$$y = A_1 e^{n_1 t} + A_2 e^{n_2 t} \tag{2.11}$$

where A_1 and A_2 are constants that depend upon initial conditions. Judging from the nature of the roots the solution is very much dependent on the amount of viscous damping in the system.

With damping present, there are two distinct types of behaviour depending on the magnitude of c. There are two possible situations, i.e.

1. If $c > 2\sqrt{km}$
   then the roots of Equation (2.7) are both real and negative. With

   $$y = A_1 e^{n_1 t} + A_2 e^{n_2 t},$$

   we see that no oscillation will occur. The motion will decay exponentially with time. The system is identified as being *over-damped.*

2. If $c < 2\sqrt{km}$,
   the roots are complex conjugates.

   $$n = -\frac{c}{2m} \pm i\sqrt{\frac{k}{m} - \frac{c^2}{4m^2}}. \tag{2.12}$$

We note, that the solution can be expressed as

$$y = e^{-ct/2m}\left[(A_1 + A_2)\cos\omega_d t + i(A_1 - A_2)\sin\omega_d t\right]. \tag{2.13}$$

Since A_1 and A_2 are arbitrary constants dependent on initial conditions, we can simplify the Equation (2.13) and write

$$y = e^{-ct/2m}\left[B_1\cos\omega_d t + B_2\sin\omega_d t\right]. \tag{2.14}$$

The solution will decay exponentially according to the exponential $e^{-ct/2m}$. In this case, the system is identified as being *under-damped*.

The transition between the over-damped and under-damped occurs when

$$c = c_c = 2\sqrt{mk} = 2m\omega. \tag{2.15}$$

The damping coefficient c_c is known as *critical damping*. A structure is said to be critically damped when, given a small initial displacement with no velocity, it just manages to return to rest without oscillatory motion. The relationship between c_c and the damping coefficient, is given by Equation (2.15).

2.2.2 Damping Ratio

The damping ratio ξ is defined as

$$\xi = c/c_c.$$

The damping in a system is defined in terms of ξ or as a percentage of critical damping. The damping ratio is a useful non-dimensional measure of damping and is usually expressed as a percentage. In most structures, there is no more than few percent of critical damping. In welded steel structures the damping is usually 2 % of critical.

The natural frequency of the system with damping less than *critical* may be expressed as:

$$\omega_d = \sqrt{\frac{k}{m} - \frac{c^2}{4m^2}} = \omega\sqrt{1 - \left(\frac{c}{c_c}\right)^2} = \omega\sqrt{1 - \xi^2} \tag{2.16}$$

where ω_d is the damped natural frequency. Thus the natural frequency of a damped system is less than an undamped system.

Also note that in Equation (2.12) the damped natural frequency will tend towards zero as the damping approaches the *critical* value c_c.

Consider a system with 10 % damping or $\xi = 0.1$. The damped frequency is then

$$\omega_d = \omega\,\sqrt{0.99} = 0.995\omega$$

which is only slightly different from the undamped natural frequency. Hence, the decrease in natural frequency due to damping may for practical purposes be ignored.

Usually ξ is smaller than 10 % (no more than 2–3 % in steel structures).

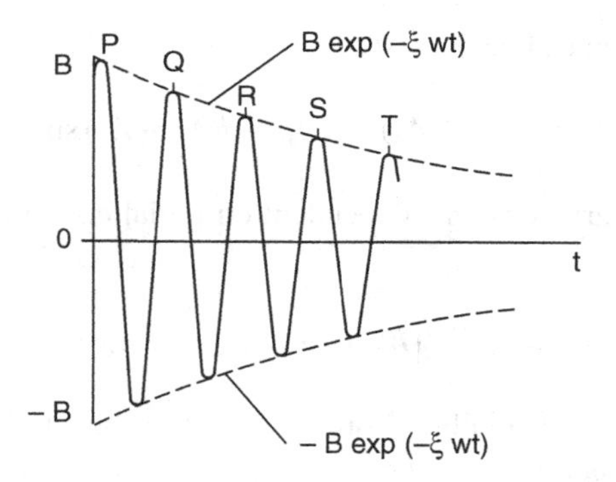

Figure 2.4 Free damped oscillation.

2.2.3 Treatment of Initial Conditions

Under-damped System

If the system is given an initial velocity $\overset{\circ}{y}$ and an initial displacement y_0 at time $t = 0$ we have from simple substitution in Equation (2.14):

$$y_0 = B_1 \text{ and}$$
$$\overset{\circ}{y}_0 = \omega B_2 - \frac{c}{2m} B_1$$

from which we obtain:

$$B_1 = y_0$$
$$B_2 = \frac{1}{\omega}\left(\overset{\circ}{y}_0 + \frac{c}{2m} y_0\right)$$

and the complete solution, with the above initial conditions is:

$$y = e^{-\xi \omega t}\left[y_0 \cos \omega_d t + \frac{(\overset{\circ}{y}_0 + \xi \omega y_0)}{\omega_d} \sin \omega_d t\right]. \tag{2.17}$$

The solution is a decaying oscillation as shown in Figure 2.4. Thus with an initial displacement and an initial velocity a dynamic response will ensue.

The initial conditions are of great importance in dynamics.

2.3 Periodic Forcing Function

Let us consider now the forcing function $F(t)$ in Equation (2.1). If the single mass system is under a sinusoidal excitation $F \sin \omega t$, free vibrations will be generated initially at the onset of periodic forcing. These decay rapidly due to damping and we may obtain a solution when the transient initial conditions have ceased to matter.

The equation of motion is:

$$m\overset{\circ\circ}{y} + c\overset{\circ}{y} + ky = F \sin \Omega t. \tag{2.18a}$$

Here, Ω is the frequency of the forcing function and the '*frequency ratio*' β is defined as:

$$\beta = \frac{\Omega}{\omega}$$

For the solution of this equation assume:

$$y = D_1 \sin \Omega t + D_2 \cos \Omega t. \tag{2.18b}$$

Substituting Equation (2.18b) into Equation (2.18a) and collecting together coefficients of sin Ωt and cos Ωt gives two equations in D_1 and D_2:

$$-mD_1\Omega^2 - cD_2\Omega + kD_1 = F \tag{2.18c}$$

$$-mD_2\Omega^2 + cD_1\Omega + kD_2 = 0. \tag{2.18d}$$

Solving for D_1 and D_2, from Equation (2.18c):

$$D_1(k - m\Omega^2) - c\Omega D_2 = F \tag{2.18e}$$

and from Equation (2.18d)

$$D_2(k - m\Omega^2) + c\Omega D_1 = 0 \tag{2.18f}$$

from which

$$D_2 = \frac{-cD_1\Omega}{(k - m\Omega^2)}$$

therefore from Equation (2.18e):

$$D_1(k - m\Omega^2) - c\Omega \cdot \frac{-cD_1\Omega}{(k - m\Omega^2)} = F.$$

Therefore:

$$D_1 = \frac{(k - m\Omega^2)F}{(k - m\Omega^2)^2 + c^2\Omega^2}. \tag{2.18g}$$

Substituting Equation (2.18g) into Equation (2.18f) we have

$$D_2(k - m\Omega^2) + c\Omega \cdot \frac{(k - m\Omega^2)F}{(k - m\Omega^2)^2 + c^2\Omega^2} = 0$$

from which

$$D_2 = -\frac{c\Omega F}{(k - m\Omega^2)^2 + c^2\Omega^2}. \tag{2.18h}$$

The derivation may be found in Chopra (2001). The solution may be rewritten in the form:

$$y = D_3 \sin(\Omega t - \psi) \tag{2.19}$$

and we have

$$y = \frac{F}{[(k - m\Omega^2)^2 + c^2\Omega^2]^{1/2}} \sin(\Omega t - \psi) \tag{2.20a}$$

and

$$\tan\psi = \frac{c\Omega}{(k - m\Omega^2)}. \tag{2.20b}$$

If we recall that

$$\omega = \sqrt{\frac{k}{m}}$$

and

$$c_c = 2m\omega$$

then

$$y = \frac{(F/k)\sin(\Omega t - \psi)}{[(1 - \frac{\Omega^2}{\omega^2})^2 + 4\frac{\Omega^2}{\omega^2}\frac{c^2}{c_c^2}]^{1/2}} \tag{2.21a}$$

$$\tan\psi = \frac{2\frac{\Omega}{\omega}\frac{c}{c_c}}{1 - \frac{\Omega^2}{\omega^2}}. \tag{2.21b}$$

The full solution due to a periodic forcing function should include terms for free vibrations.

A useful fact to be deduced from Equations (2.21a, b) is that the response lags behind the forcing function, as can be seen in Figure 2.5. When the frequency of the forcing function is small compared with the natural frequency of the system, the dynamic response follows the forcing function in phase and the displacement is not amplified. As the frequency of the forcing function increases, the dynamic response is amplified and also begins to lag.

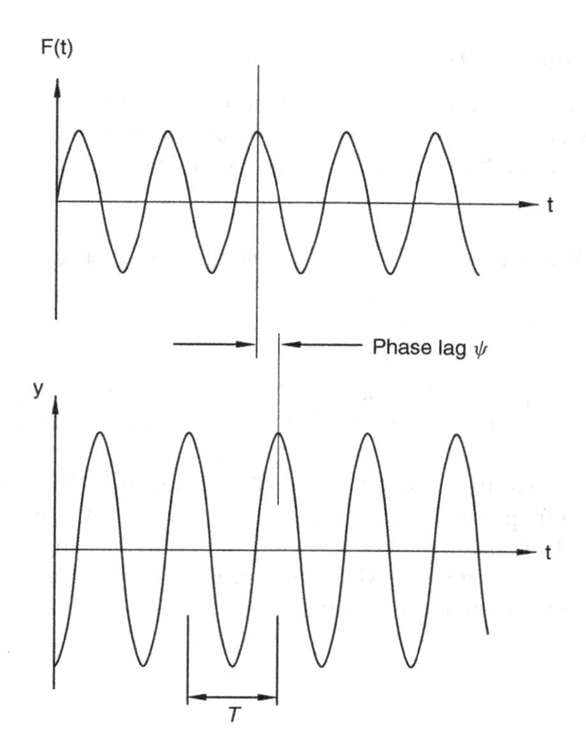

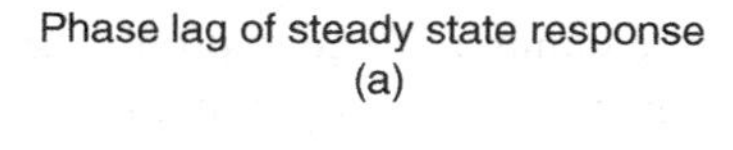

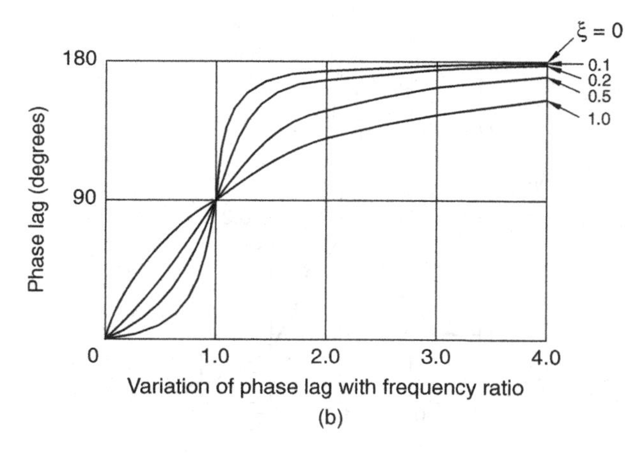

Figure 2.5 Phase lag of steady state response and variation of phase lag with frequency ratio. (a) Phase lag of steady state response. (b) Variation of phase lag with frequency ratio.

2.3.1 *Magnification Factors*

Referring to Equation (2.21a), it may be noted that the static defection is given by F/k and hence we can investigate to see how the deflection y varies with the frequency of excitation Ω. The dynamic magnification factor, also known as '*dynamic load factor*' (DLF) is expressed as $|y|/(F/k)$.

$$\begin{aligned}\text{DLF} &= \text{max. dynamic deflection/max. static deflection} \\ &= \frac{1}{[(1-\beta^2)^2 + (2\xi\beta)^2]^{1/2}}.\end{aligned} \tag{2.22}$$

The DLF may now be plotted (Equation (2.22)) to show how $|y|/(F/k)$ varies with the frequency of excitation. The plot of $|y|/(F/k)$ against various values of ξ is shown in Figure 2.6.

This is also referred to as the *resonance curve*. The curves show that the magnification has a maximum value at a frequency ratio just below 1.0. It is apparent from Figure 2.6 that even with small damping, theoretically infinite amplitudes (as in the undamped case) do not occur at resonance. However, the condition, where, ω is near the forcing frequency, Ω needs to be avoided during the design of dynamic systems.

2.3.2 *Damping*

Viscous Damping

It is noted from Equation (2.7) that the damping force is proportional to velocity. This kind of velocity-proportional damping is known as viscous damping as it is observed in viscous fluids. Damping in real life is rarely so simple but this is a good assumption. The hallmark of this system is that it is simple and has general usefulness.

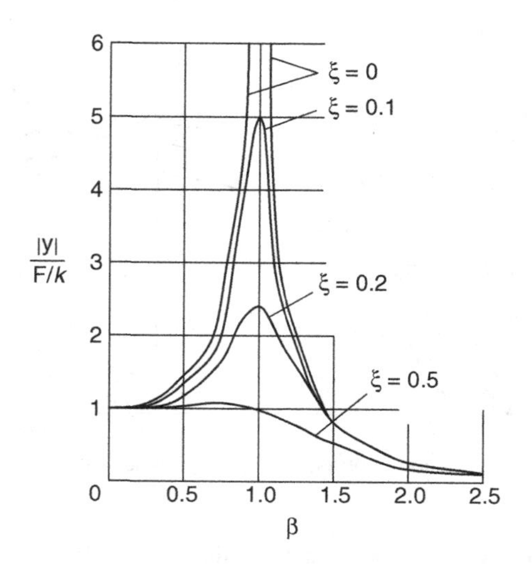

Figure 2.6 Response to forced vibration. Variation of dynamic magnification factor $|y|/(F/k)$ with frequency ratio $\beta = \Omega/\omega$.

Logarithmic Damping

However, another widely used measure of damping is the *logarithmic damping*. It is defined as

$$\delta = \log_e \ [(\text{amplitude at cycle n})/(\text{amplitude at cycle } (n+1))]$$

Therefore, referring to Figure 2.4, the ratio of peak amplitudes of two successive cycles is given by

$$\frac{y_n}{y_{n+1}} = \frac{y_0 e^{-\xi\omega t}}{y_0 e^{-\xi\omega(t+T)}} = e^{\xi\omega T} \tag{2.23}$$

and it follows that the logarithmic and the damping ratio are related:

$$\delta = \xi\omega T = 2\pi\xi\omega/\omega_d \approx 2\pi\xi. \tag{2.24}$$

Coulomb Damping

Damping in actual structures is not strictly due to viscosity, but mostly caused by friction at interfaces. A mathematically convenient approach is to idealize them through an equivalent viscous damping model. Although this equivalent viscous damping model is sufficiently accurate for most practical analysis, it may not be appropriate when special friction devices have been introduced to reduce vibration in a structure. Currently, there is much interest in such devices and several of these are discussed in Chapter 9.

Friction opposes the motion. If it is denoted by F_R then the equation of motion takes the form:

$$m\overset{\circ\circ}{y} + F_R + ky = F(t)\overset{\circ}{y} \ - \text{positive} \tag{2.25a}$$

$$m\overset{\circ\circ}{y} - F_R + ky = F(t)\overset{\circ}{y} \ - \text{negative.} \tag{2.25b}$$

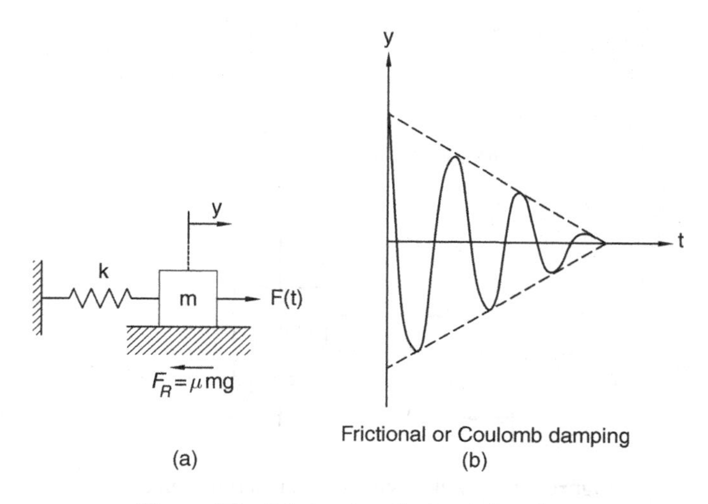

Figure 2.7 Frictional or Coulomb damping.

A simple model of frictional, or Coulomb damping is shown in Figure 2.7.

The solution to these equations in the case of free vibrations, $F(t) = 0$, reveals that there is linear decay (Figure 2.7).

Damping in real structures must consist partly of Coulomb friction, since only this mechanism can stop motion in free vibration. The various damping mechanisms that exist in real structures are rarely modelled individually, unless special frictional devices have been included in the structure.

2.3.3 Support Motion

An important class of problems arises from vibration due to movement of supports. While other sources such as passing traffic or pile driving can cause vibration of supports, it is extremely important in relation to earthquakes. Consider the simple system shown in Figure 2.8.

The ground/support movement is denoted by $y_g\ (t)$. The equation of motion is

$$m\overset{\circ\circ}{y} + c\,(\overset{\circ}{y} - \overset{\circ}{y}_g) + k(y - y_g) = 0. \tag{2.26}$$

For earthquake engineering applications a different approach will be more helpful. As the velocity of the system and the spring force are affected we can introduce the following variables

$$y_r = y - y_g;$$
$$\overset{\circ}{y}_r = \overset{\circ}{y} - \overset{\circ}{y}_g;$$
$$\overset{\circ\circ}{y}_r = \overset{\circ\circ}{y} - \overset{\circ\circ}{y}_g;$$

hence we have:

$$m(\overset{\circ\circ}{y}_r + \overset{\circ\circ}{y}_g) + c\overset{\circ}{y}_r + ky_r = 0 \tag{2.27}$$

or

$$m\overset{\circ\circ}{y}_r + c\overset{\circ}{y}_r + ky_r = -m\overset{\circ\circ}{y}_g. \tag{2.28}$$

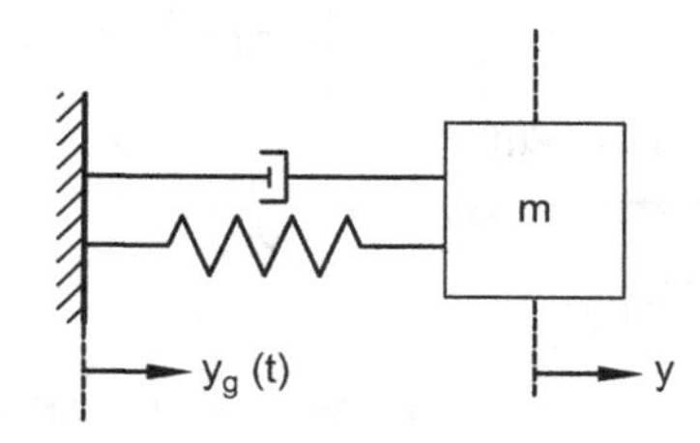

Figure 2.8 SDOF system with support movement.

This is similar to Equation (2.18a), except that the time dependent external forcing function $F \sin \Omega(t)$ has been replaced by $-m\overset{\circ\circ}{y}_g$ and the absolute displacement of the mass by the relative displacement of the spring. This formulation is helpful for earthquake analysis as seismograms record ground acceleration. The ground acceleration may be treated as arbitrary loading. A numerical procedure for analysis of an SDOF system subjected to arbitrary loading is discussed in Section 2.4 below.

2.4 Arbitrary Forcing Function

2.4.1 Duhamel Integral

In most practical applications the loading function is not sinusoidal, but rather quite arbitrary. Acceleration arising from earthquake records is a good example. Of special interest in our case of earthquake loading is the fact that the method to be outlined here will be useful for development of the response spectra discussed in Chapter 6.

One of the ways of obtaining a solution to this kind of problem is to utilize the concept of *'impulse'*, which is defined as the area under the load-time curve. Consider a short impulse as shown in Figure 2.9.

If the duration of the impulse is short, then it can be reasoned that there is no significant change of displacement during that short time interval but only a change in velocity $\Delta \overset{\circ}{y}$. The change in velocity can be estimated from Newton's second law of motion – '*the rate of change of momentum of a mass is equal to the force*':

$$d(m\Delta \overset{\circ}{y}) = F(\tau)d\tau \tag{2.29}$$

$$\Delta \overset{\circ}{y} = \frac{F(\tau)d\tau}{m}. \tag{2.30}$$

This is equivalent to imparting an initial velocity $\overset{\circ}{y}$ applied for a short duration and removed.

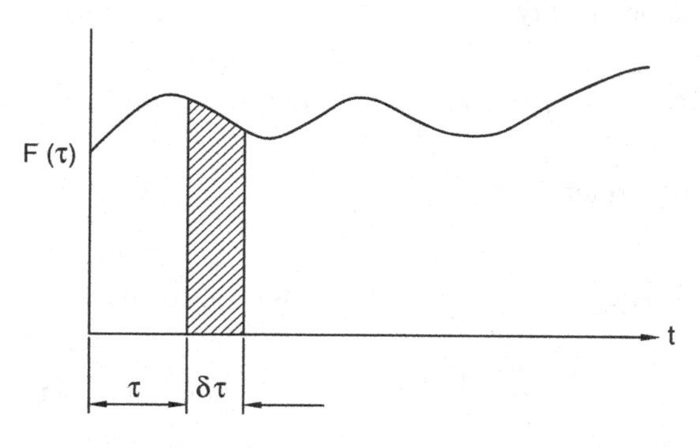

Figure 2.9 Arbitrary forcing function.

The initial conditions are:

$$y_0 = 0 \tag{2.31a}$$

$$\overset{\circ}{y} = \int \frac{F(\tau)d\tau}{m}. \tag{2.31b}$$

The vibration at a later time may be obtained from Equation (2.17), which gives the motion of a single degree of freedom system with initial velocity and displacement. In the equation the time t must be replaced by $(t - \tau)$ as shown in Figure 2.9.

With the initial conditions, the displacement at time t caused by an impulse at τ is given by:

$$y(t) = \frac{1}{m\omega_d} \int_0^t F(\tau)e^{-\xi\omega(t-\tau)} \sin \omega_d (t - \tau)\, d\tau \tag{2.32}$$

This is identified as the damped response equivalent of the *Duhamel Integral*. Each impulse in Figure 2.9 will produce a response of this form.

For an undamped system this result simplifies to

$$y(t) = \frac{1}{m\omega} \int_0^t F(\tau) \sin [\omega (t - \tau)]\, d\tau \tag{2.33}$$

In an arbitrary loading pattern, the basic assumption is that it is composed of a number of short duration impulses. Thus if we have a summation procedure for the individual responses (principle of superposition will hold) by way of numerical integration, we can then evaluate the responses due to any arbitrary loading.

2.4.2 Numerical Evaluation

It is not easy to manipulate Equations (2.32 or 2.33) for numerical integration. It requires some substitutions and algebraic manipulations to make the summation process a little easier.

Using the trigonometric identity

$$sin(\omega_d t - \omega_d \tau) = (sin\ \omega_d t \cdot cos\ \omega_d \tau - cos\ \omega_d t \cdot sin\ \omega_d \tau)$$

Equation (2.32) may be written

$$y(t) = \frac{1}{m\omega_d} \int_0^t e^{-\xi\omega(t-\tau)} F(\tau). \sin \omega_d t. \cos \omega_d \tau d\tau$$

$$- \frac{1}{m\omega_d} \int_0^t e^{-\xi\omega(t-\tau)} F(\tau). \cos \omega_d t \sin \omega_d \tau d\tau \tag{2.34}$$

$$= \{A(t) \sin \omega_d t - B(t) \cos \omega_d t\} \tag{2.35}$$

where

$$A(t) = \frac{1}{m\omega_d} \int_0^t F(\tau) \frac{e^{(\xi\omega\tau)}}{e^{(\xi\omega t)}} \cos\omega_d\tau d\tau \tag{2.36a}$$

$$B(t) = \frac{1}{m\omega_d} \int_0^t F(\tau) \frac{e^{(\xi\omega\tau)}}{e^{(\xi\omega t)}} \sin\omega_d\tau d\tau. \tag{2.36b}$$

The new integral equations above, it may be seen, account for the exponential decay behaviour due to damping. The integral Equations (2.36a, b) can be calculated efficiently by using different numerical procedures (simple summation, trapezoidal rule, etc.).

Numerical Scheme

We begin by evaluating the function, $y(\tau) \equiv F(\tau)\cos\omega_d\tau$. The computations are usually carried out at equal intervals of time as explained in Figure 2.10.

With reference to Figure 2.10 the response is obtained by summing the individual responses at the end of each time interval $\Delta\tau$ [Note: $(\Delta\tau = t - \tau)$]. The summation process will depend upon the numerical scheme adopted and the approximate recursive forms would be as follows.

Simple Summation:

$$A_n \cong A_{n-1} \cdot e^{-(\xi\omega\Delta\tau)} + \frac{\Delta\tau}{m\omega_d} \cdot y_{n-1} \cdot e^{-(\xi\omega\Delta\tau)} \quad n = 1, 2, 3.. \tag{2.37}$$

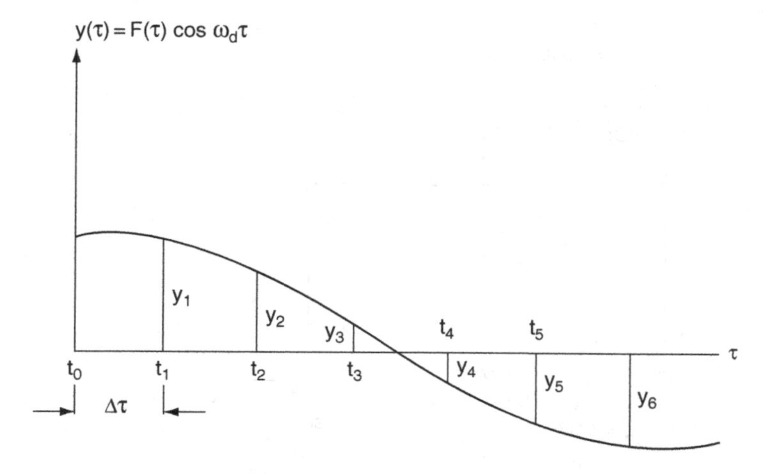

Figure 2.10 Numerical summation process (Duhamel Integral).

Trapezoidal Rule:

$$A_n \cong A_{n-1} \cdot e^{-(\xi\omega\Delta\tau)} + \frac{\Delta\tau}{2m\omega_d} \cdot y_{n-1} \cdot e^{-(\xi\omega\Delta\tau)} + y_n \quad n = 1, 2, 3.. \tag{2.38}$$

A FORTRAN routine may easily be developed to carry out the computations.

2.4.3 Worked Example – Duhamel Integral

As an example of the Duhamel integration process, consider the portal frame shown in Figure 2.11a subjected to a ground motion shown in Figure 2.11b. The impulse loading from a blast reaches a peak of $4.0\,\text{m/sec}^2$ in 0.04 second and decreases to zero in another 0.04 sec.

The portal frame as mentioned earlier is assumed to consist of massless columns with a lumped mass on the roof. The frame is modelled as a single degree of freedom system.

Calculate 'f' natural frequency

$$f = \frac{1}{2\pi}\sqrt{\frac{2 \times 10^6}{2000}} = 5.035\,\text{Hz}.$$

$$\text{Damping ratio } \xi = 6.32\%$$

$$\text{Period (tp)} = \frac{1}{f} = 0.199 \qquad \omega = \frac{2\pi}{tp} = 31.5738\,(\omega \approx \omega_d)$$

A recursive relation may now be set up to calculate $A(t)$ and $B(t)$ as outlined in Equations 2.37 or 2.38.

The response may be obtained from Equation (2.35):

The SDOF response for the loading shown in Figure 2.11(b) using the simple summation rule is shown in Figure 2.12.

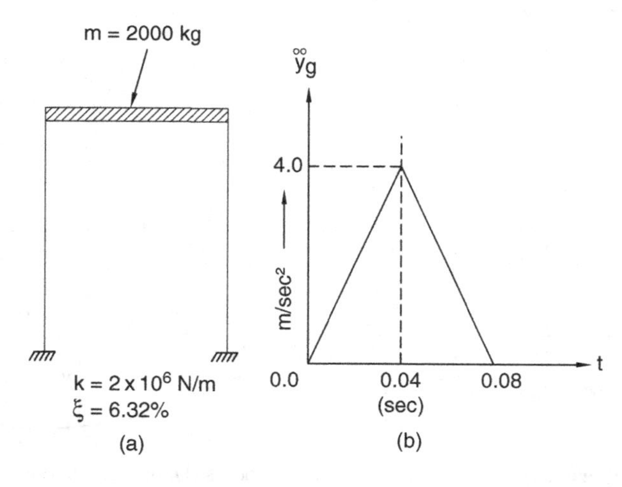

Figure 2.11 Worked example – SDOF system subjected to ground motion from a blast.

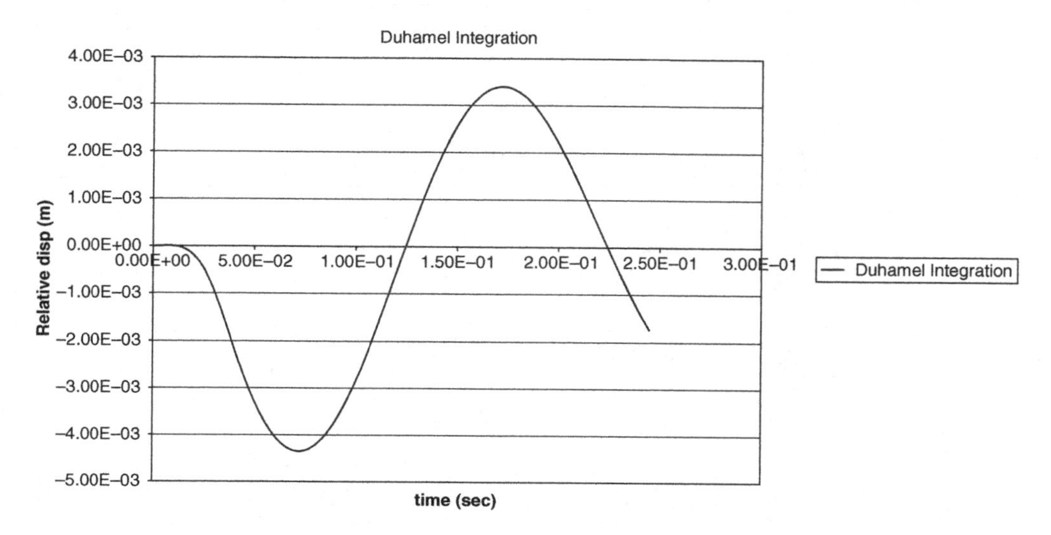

Figure 2.12 SDOF response for the impulsive loading (Figure 2.11(b))

2.5 References

Chopra, A.K. (2001), *Dynamics of Structures: Theory and Applications to Earthquake Engineering*. Prentice Hall, New Jersey.

Waddell, A. Image of elevated water tank. http://www.walksydneystreets.net/photos/acacia-gardens-water-tank-w.jpg 29/07/08.

3

Systems with Many Degrees of Freedom

3.1 Introduction

In the previous chapter single degree of freedom systems were discussed. It was stated that a single degree of freedom system could be defined in terms of a single coordinate. Most practical structures present a complex dynamical system. Usually they cannot be represented by a single degree of freedom system. The complexity of the system depends upon the number of independent degrees of freedom possessed by the system. The number is equal to the number of independent coordinates required to specify completely the displacement of the system, i.e. the motion must be represented by more than one degree of freedom. It is possible to extend the methods of single degree of freedom systems to multi degrees of freedom systems, making a more refined analysis possible. However, it is convenient to model a system with continuous variation of deflection as a system consisting of a limited number of degrees of freedom, each subjected to its own deflection. Acceptable results are usually obtained by considering only a few degrees of freedom.

General analysis of the type mentioned can only be handled conveniently in matrix form and in this chapter the matrix method will be adopted. Attention will be focused on the 'lumped parameter model'. Systems with distributed mass and stiffness have been adequately described in other texts, including Clough and Penzien (1993), Smith (1988), and Chopra (2001), to name a few. Discussion topics have been restricted to topics required later in Chapter 6.

3.2 Lumped Parameter Systems with Two Degrees of Freedom

A system has two degrees of freedom if its configuration, at any time, can be described by two independent coordinates. Two examples are shown in Figures 3.1 and 3.2.

Figure 3.1 depicts a system that is represented by the two independent coordinates, y_1 and y_2, the absolute displacements of m_1 and m_2 from their positions of static equilibrium.

Fundamentals of Seismic Loading on Structures Tapan Sen

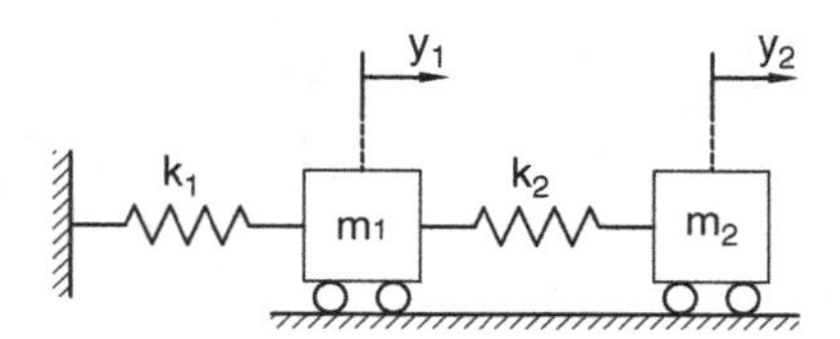

Figure 3.1 Lumped mass system with two degrees of freedom.

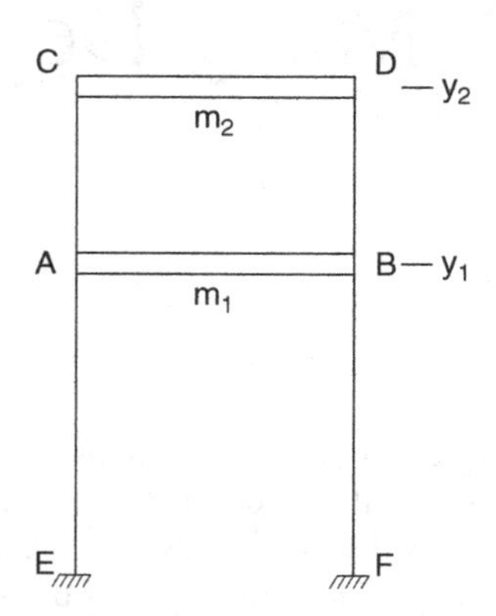

Figure 3.2 A system represented by the two independent coordinates, y_1 and y_2, the absolute displacements of m_1 and m_2 from their positions of static equilibrium.

Figure 3.2 represents a two-storey building frame consisting of two rigid horizontal members AB and CD and four light, elastic members AE, FB, AC and BD. This frame is an example of a 'shear building' with two storeys. It is assumed that the floors are rigid and the joints cannot rotate and the masses are concentrated at the floors. This concept is very useful for simplifying the analysis of tall buildings. For a shear building with 'n' floors the model is reduced to one with 'n' degrees of freedom.

Only the swaying of the frame shown in Figure 3.2 is permitted. The absolute displacements of the members AB and CD in the horizontal direction are denoted as y_1 and y_2. The lateral resistance is provided by the lateral stiffness of the columns. Let k_1 represent the equivalent stiffness of EA and FB, and k_2 that of AC and BD.

Considering the dynamic equilibrium of each mass in turn leads to the following equation of motion:

$$m_1 \ddot{y}_1 + k_1 y_1 + k_2(y_1 - y_2) = 0 \tag{3.1a}$$

$$m_2 \ddot{y}_2 + k_2(y_2 - y_1) = 0. \tag{3.1b}$$

3.3 Lumped Parameter Systems with more than Two Degrees of Freedom

Most real life structures would require more than two degrees of freedom for modelling. What is required is a systematic procedure for lumped masses and springs that at the same time is general enough to accommodate different types of structural members. The finite element method is the most versatile one in this respect. At the heart of the method is the philosophy that an irregular structure may be subdivided in to a large number of small regular elements for which the force-displacement relationships are relatively simple. The individual elements are assembled in an organized manner and the interactions between the elements analysed by formal mathematical procedures employing matrix algebra. This formalized approach lends itself to computer implementation. Though not covered in this text, many good publications exist on the subject, including Zienkiewicz (1971), Bathe and Wilson (1976), and Desai and Abel (1972).

Now consider a generalized multi degree of freedom system with masses, springs and dashpots as shown in Figure 3.3.

The equations of equilibrium of the combined system, taking each mass in turn, may be written as

$$m_1\ddot{y}_1 + c_1\dot{y}_1 + k_1y_1 + c_2(\dot{y}_1 - \dot{y}_2) + k_2(y_1 - y_2) = F_1(t), \tag{3.2a}$$

$$m_2\ddot{y}_2 + c_2(\dot{y}_2 - \dot{y}_1) + k_2(y_2 - y_1) + c_3(\dot{y}_2 - \dot{y}_3) + k_3(y_2 - y_3) = F_2(t), \tag{3.2b}$$

$$m_N\ddot{y}_N + c_N(\dot{y}_N - \dot{y}_{N-1}) + k_N(y_N - y_{N-1}) = F_N(t). \tag{3.2c}$$

It will be convenient to refer to equations of motion for a large system in compact matrix notation. Therefore, we write the equation of motion in compact form as follows:

$$[M]_D\left\{\ddot{y}\right\} + [C]\left\{\dot{y}\right\} + [K]\{y\} = \left\{F(t)\right\}. \tag{3.2d}$$

$[M]_D$ is the diagonal mass matrix containing the masses $m_1,\ m_2\ldots.m_n$:

$$[M]_D = \begin{bmatrix} m_1\ldots..0\ldots\ldots..0\ldots\ldots.0 \\ 0\ldots\ldots\ldots m_2\ldots.0\ldots\ldots.0 \\ 0\ldots\ldots\ldots 0\ldots\ldots..m_3\ldots 0 \\ . \\ 0\ldots\ldots\ldots 0\ldots\ldots..0\ldots\ldots m_n \end{bmatrix}. \tag{3.2e}$$

The matrix containing the damping coefficients $c_1,\ c_2\ldots..c_N$, is called the damping matrix. The matrix containing the coefficients $k_{11},\ k_{12}$ etc. is the linear elastic stiffness matrix.

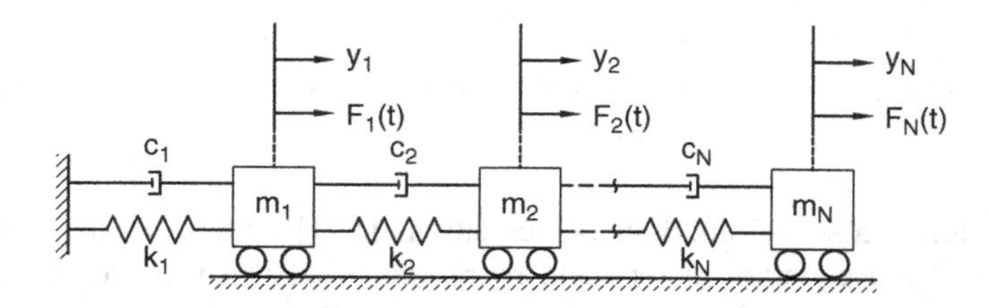

Figure 3.3 Lumped mass multi degree of freedom system.

A stiffness matrix for individual members may be developed which makes it possible to accommodate different member properties. The assemblage of the individual matrices to form the overall stiffness matrix is carried out in a finite element (FE) program.

The damping and the stiffness matrices are banded matrices.

Equation (3.2d) represents n simultaneous linear coupled differential equations. Methods of uncoupling the equations are discussed later in the section.

3.3.1 Free Vibration

Equation (3.2d) can be modified to represent the equation of motion for an undamped system undergoing free vibration in any of the possible modes of vibration. Hence we may write:

$$[M]_D \left\{ \overset{\circ\circ}{y} \right\} + [K]\{y\} = \{0\}, \tag{3.3}$$

where $[M]_D$ and K are matrices as explained above and the right hand is a column vector containing only zeros. We know from the free vibration response of an SDOF system that simple harmonic motion is of the form

$$\overset{\circ\circ}{y} = -\omega^2 y. \tag{3.4}$$

It may be assumed that a similar solution will be valid for the free vibration response of MDOF systems. Therefore substituting (3.4) into (3.3) we have

$$-\omega^2 [M]_D \{y\} + [K]\{y\} = \{0\} \tag{3.5}$$

or

$$\left| [K] - \omega^2 [M]_D \right| \{y\} = 0. \tag{3.6}$$

In expanded form this becomes

$$\begin{bmatrix} (k_{11} - m_1\omega^2) \ldots\ldots\ldots\ldots k_{12} \ldots\ldots\ldots\ldots\ldots\ldots\ldots\ldots\ldots\ldots k_{1n} \\ k_{21} \ldots\ldots\ldots\ldots\ldots\ldots (k_{22} - m_2\omega^2) \ldots\ldots\ldots\ldots\ldots\ldots\ldots k_{2n} \\ \cdot \\ \cdot \\ k_{n1} \ldots\ldots\ldots\ldots\ldots\ldots k_{n2} \ldots\ldots\ldots\ldots\ldots\ldots\ldots\ldots (k_{nn} - m_n\omega^2) \end{bmatrix} \begin{bmatrix} y_1 \\ y_2 \\ \\ \\ y_n \end{bmatrix} = 0 \tag{3.7}$$

Equation (3.6) forms a set of simultaneous equations in $\{y\}$ with ω^2 as the unknowns. The right-hand-side of the equation is zero and represents what is known as an '*Eigen value*' problem in matrix algebra. Cramer's rule, which may be used to solve Equation (3.6), requires that for a non-trivial solution the *determinant* of the coefficients of $\{y\}$ be zero, or we may write:

$$|D| = \begin{bmatrix} (k_{11} - m_1\omega^2) \ldots\ldots\ldots\ldots k_{12} \ldots\ldots\ldots\ldots\ldots\ldots\ldots\ldots\ldots k_{1n} \\ k_{21} \ldots\ldots\ldots\ldots\ldots (k_{22} - m_2\omega^2) \ldots\ldots\ldots\ldots\ldots\ldots\ldots k_{2n} \\ \cdot \\ \cdot \\ k_{n1} \ldots\ldots\ldots\ldots\ldots k_{n2} \ldots\ldots\ldots\ldots\ldots\ldots (k_{nn} - m_n\omega^2) \end{bmatrix} = 0. \tag{3.8}$$

The matrix $[D]$ is known as the frequency matrix. We see that if the determinant $|D|$ becomes zero, then the amplitudes will be infinite. This situation will arise when the system is vibrating at a frequency ω that is equal to one of the natural frequencies of free vibration of the system. The natural frequencies of free vibration can be calculated by finding those frequencies at which

$$|D| = 0$$

By expanding the determinant into its co-factors an equation of nth degree in ω^2 is obtained, n being the number of degrees of freedom. For simple systems, as we shall see, it is feasible to evaluate the determinant algebraically.

The values of ω_1, $\omega_2 \ldots\ldots\ldots \omega_n$ which satisfy this equation are called '*Eigen*' values and correspond to the frequencies at which the system can vibrate. The smallest frequency is usually identified as ω_1 and referred to as the fundamental frequency.

We can appreciate that, as the number of degrees of freedom increases, the determination of the roots by polynominal expansion becomes impractical. Many procedures have been developed to obtain Eigen values of a large system and these have been covered in many standard texts (Bathe and Wilson, 1976, Clough and Penzien, 1993, Chopra, 2001).

Mode Shapes

Once we have obtained the frequencies of vibration, we can define the mode shapes. The mode shapes will be given by the Eigen vectors $\{y\}$. In trying to obtain the values of $\{y\}$ from (3.6) we see that the actual values of the displacement are *indeterminate*. However, it is possible to determine the shape of a mode by assigning an arbitrary value to a particular displacement and solving all other displacements relative to the assigned arbitrary value.

If we represent the ith Eigen vector by φ_i for a system with n degrees of freedom, then the shape of the ith mode is given by

$$\varphi_i = \begin{Bmatrix} \varphi_{1i} \\ \varphi_{2i} \\ \cdot \\ \phi_{ni} \end{Bmatrix}. \tag{3.9}$$

Since there will be n modes, the shapes of which are given by $\varphi_1,\ \varphi_2,\ \varphi_3 \ldots\ldots.\varphi_n$, hence the size of the mode shape matrix will be $n \times n$ i.e.

$$\varphi = \begin{bmatrix} \varphi_{11} \ldots \varphi_{12} \ldots \varphi_{13} \ldots\ldots \phi_{1n} \\ \varphi_{21} \ldots \varphi_{22} \ldots \phi_{23} \ldots\ldots \varphi_{2n} \\ \varphi_{31} \ldots \varphi_{32} \ldots \varphi_{33} \ldots\ldots \varphi_{3n} \\ . \\ . \\ \varphi_{n1} \ldots \varphi_{n2} \ldots\ldots \varphi_{n3} \ldots \varphi_{nn} \end{bmatrix}. \tag{3.10}$$

A worked example for a two degrees of freedom system follows.

3.3.2 A Worked Example (Two degrees of Freedom System)

A 20 m tall cantilever structure is shown in Figure 3.4. It is treated as a two degrees of freedom system with the masses lumped at nodes 1 and 2. The properties are also shown. It is required to compute the Eigen values and mode shapes.

The mass matrix is given by

$$M = \begin{bmatrix} 3503 \ldots\ldots.. 0 \\ 0 \ldots\ldots\ldots 7006 \end{bmatrix} \text{kg.}$$

The stiffness matrix obtained from an FE program is:

$$K = \begin{bmatrix} 0.1206 \times 10^7 \ldots\ldots\ldots - 0.3015 \times 10^7 \\ -0.3015 \times 10^7 \ldots\ldots\ldots 0.9647 \times 10^7 \end{bmatrix} \text{N/m.}$$

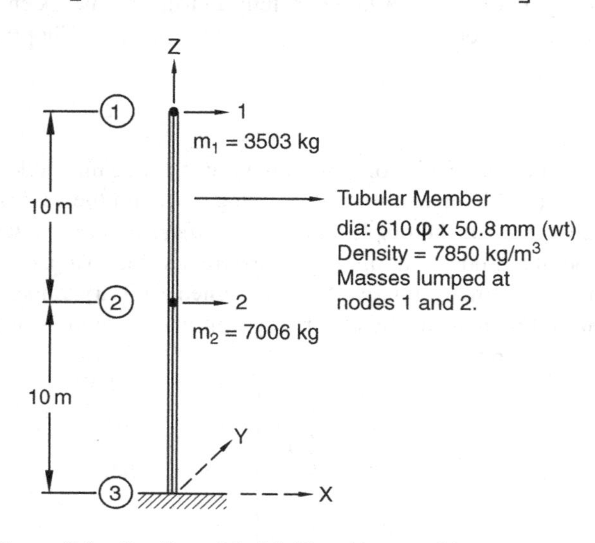

Figure 3.4 Cantilever Model (Two degrees of freedom system).

From Equation (3.8) we have

$$|D| = \begin{bmatrix} (k_{11} - m_1\omega^2) \ldots\ldots k_{12} \ldots\ldots k_{1n} \\ k_{21} \ldots\ldots (k_{22} - m_2\omega^2) \ldots\ldots k_{2n} \\ \cdot \\ \cdot \\ k_{n1} \ldots\ldots k_{n2} \ldots\ldots (k_{nn} - m_n\omega^2) \end{bmatrix} = 0.$$

Hence the determinant for the natural frequencies is

$$D = \begin{vmatrix} (0.1206 \times 10^7 - 3503\omega^2) \ldots\ldots - 0.3015 \times 10^7 \\ -0.3015 \times 10^7 \ldots\ldots (0.9647 \times 10^7 - 7006\omega^2) \end{vmatrix} = 0. \quad (3.11)$$

When expanded, it will lead to a quadratic equation with ω^2 as the unknown. Thus

$$2.45x10^7\omega^4 - 4224x10^7\omega^2 + 0.02539x10^{14} = 0 \quad (3.12)$$

$$\omega^2 = \frac{-b \pm \sqrt{b^2 - 4ac}}{2a}$$

$$\omega_1 = 7.896$$

$$\omega_2 = 40.7$$

Mode Shapes

Mode 1

$$n = 1$$

The mode shapes can be determined by assuming a normalized shape for each mode. Assuming $\varphi_{1n} = 1$ for each mode gives

$$\varphi_1 = \begin{Bmatrix} 1 \\ \varphi_{2n} \end{Bmatrix}.$$

Noting that $\omega_1 = 7.896$, from Equation (3.11) we may write:

$$\begin{bmatrix} 0.98777 \times 10^6 \ldots\ldots - 0.3015 \times 10^7 \\ -0.3015 \times 10^7 \ldots\ldots 0.92105 \times 10^7 \end{bmatrix} \begin{Bmatrix} 1 \\ \varphi_{2n} \end{Bmatrix} = \begin{Bmatrix} 0 \\ 0 \end{Bmatrix}. \quad (3.13a)$$

The solution of this equation gives

$$\varphi_{2n} = 0.328$$

and hence

$$\varphi_1 = \begin{Bmatrix} \varphi_{11} \\ \varphi_{21} \end{Bmatrix} = \begin{Bmatrix} 1 \\ 0.328 \end{Bmatrix} \tag{3.13b}$$

Mode 2

$$n = 2$$

We note that $\omega_2 = 40.7$ and a similar exercise gives

$$\varphi_2 = \begin{Bmatrix} \varphi_{12} \\ \varphi_{22} \end{Bmatrix} = \begin{Bmatrix} 1 \\ -1.529 \end{Bmatrix} \tag{3.14}$$

The mode shapes are shown in Figure 3.5.

3.3.3 Normalization of Mode Shapes

It was noted earlier that the vibration mode amplitudes from an Eigen value solution are arbitrary. Assigning any arbitrary value to a particular displacement will satisfy the frequency Equation (3.6) and only the resulting shapes are uniquely defined. In the example worked out above the amplitude of degree of freedom 1 (Figure 3.5) was set to unity, and other displacement determined relative to this reference value.

Other normalizing procedures are also in use. For example, in many computer programs the shapes are normalized with respect to the largest value. Thus the largest value in each modal

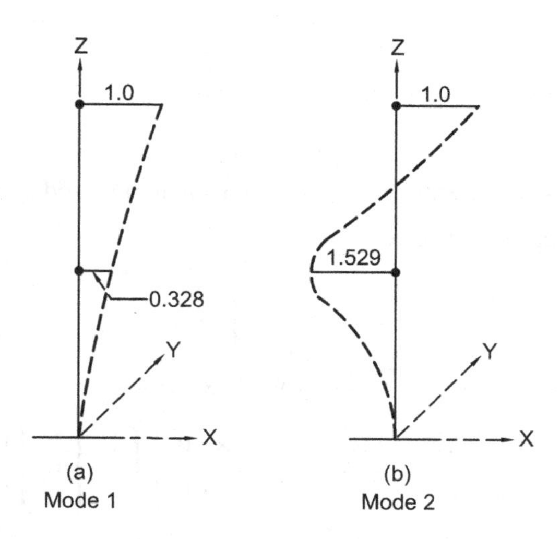

Figure 3.5 Mode shapes.

vector is unity. The normalizing procedure implemented in most computer programs involves adjusting the modal amplitudes to $\overset{\approx}{\varphi}_n$ (say) such that

$$\overset{\approx}{\varphi}_n^T M \overset{\approx}{\varphi}_n = 1. \tag{3.15}$$

For the worked example above this normalizing procedure yields

$$\overset{\approx}{\varphi}_1 = \begin{Bmatrix} \overset{\approx}{\varphi}_{11} \\ \overset{\approx}{\varphi}_{21} \end{Bmatrix} = \begin{Bmatrix} 0.0153 \\ 0.005019 \end{Bmatrix}$$

$$\overset{\approx}{\varphi}_2 = \begin{Bmatrix} \overset{\approx}{\varphi}_{12} \\ \overset{\approx}{\varphi}_{22} \end{Bmatrix} = \begin{Bmatrix} 0.007098 \\ -0.01084 \end{Bmatrix}.$$

It may be seen that Equation (3.7) is satisfied.

3.3.4 Orthogonality of Mode Shapes

A very important property of free vibration is that the modes are orthogonal to one another. This is extremely useful for further dynamic analysis procedures. In order to demonstrate the orthogonal properties take any two modes vibrating at frequencies ω_i and ω_j. Then from Equation (3.5) and noting that the ith Eigen vector is represented by $\{\varphi\}_i$, we have

$$\omega_i^2 [M]_D \{\varphi_i\} = [K] \{\varphi_i\} \tag{3.16}$$

and

$$\omega_j^2 [M]_D \{\varphi_j\} = [K] \{\varphi_j\} \tag{3.17}$$

Noting that the transpose of matrix $[A].[B]$ is $[B]^T.[A]^T$, we take the transpose of Equation (3.16) and post multiply by $\{\varphi_j\}$. Therefore

$$\omega_i^2 \{\varphi_i\}^T [M]_D^T \{\varphi_j\} = \{\varphi_i\}^T [K]^T \{\varphi_j\}. \tag{3.18}$$

Pre-multiplying the second Equation (3.17) by $\{\varphi_i\}^T$ the result is

$$\omega_j^2 \{\varphi_i\}^T [M]_D \{\varphi_j\} = \{\varphi_i\}^T [K] \{\varphi_j\}. \tag{3.19}$$

Since both $[M]_D$ and $[K]$ are symmetric

$$[M]_D = [M]_D^T$$
$$[K] = [K]^T.$$

Now, subtracting Equation (3.19) from (3.18) we have

$$\left(\omega_i^2 - \omega_j^2\right) \{\varphi_i\}^T [M]_D \{\varphi_j\} = 0 \tag{3.20}$$

Since $\omega_i \neq \omega_j$

$$\{\varphi_i\}^T [M]_D \{\varphi_j\} = 0 \text{ and } \{\varphi_i\}^T [K] \{\varphi_j\} = 0 \tag{3.21}$$

Equation (3.21) proves the condition of orthogonality.

3.3.5 Worked Example – Orthogonality Check

Referring to the worked example, of the two noded cantilever column, in Section 3.3.2. we have

$$\overset{\approx}{\varphi}_1 = \begin{Bmatrix} \overset{\approx}{\varphi}_{11} \\ \overset{\approx}{\varphi}_{21} \end{Bmatrix} = \begin{Bmatrix} 0.0153 \\ 0.005019 \end{Bmatrix}$$

$$\overset{\approx}{\varphi}_2 = \begin{Bmatrix} \overset{\approx}{\varphi}_{12} \\ \overset{\approx}{\varphi}_{22} \end{Bmatrix} = \begin{Bmatrix} 0.007098 \\ -0.01084 \end{Bmatrix}$$

$$M = \begin{bmatrix} 3503 \ldots\ldots\ldots 0 \\ 0 \ldots\ldots\ldots 7006 \end{bmatrix}$$

Thus

$$\begin{bmatrix} 0.0153 \ldots\ldots\ldots 0.005019 \end{bmatrix} \begin{bmatrix} 3503 \ldots\ldots\ldots\ldots 0 \\ 0 \ldots\ldots\ldots\ldots 7006 \end{bmatrix} \cdot \begin{bmatrix} 0.007098 \\ -0.01084 \end{bmatrix} = 0$$

This is easily verified on simple multiplication.

Modal Mass and Stiffness

When $\omega_i = \omega_j$ we have the following condition from Equation (3.20):

$$\{\varphi_i\}^T [M]_D \{\varphi_i\} \neq 0$$

Upon expanding

$$\{\varphi_i\}^T [M]_D \{\varphi_i\} = \varphi_{1i}^2 m_1 + \varphi_{2i}^2 m_2 + \varphi_3^2 m_{3i} + \ldots\ldots\ldots\ldots \varphi_{ni}^2 m_n \tag{3.22a}$$

$$= \sum_{r=1}^{n} m_r \varphi_{ri}^2 \tag{3.22b}$$

$$= M_i \tag{3.22c}$$

The quantity M_i is known as the modal mass in the ith mode.

If we pre-multiply Equation (3.16) by $\{\varphi_i\}^T$ and from Equation (3.22c) we get

$$\{\varphi_i\}^T [K] \{\varphi_i\} = \omega_i^2 M_i \tag{3.23}$$

This is the equation for modal stiffness.

3.4 Mode Superposition

3.4.1 Use of Normal or Generalized Coordinates

The formulation of the equations of motion in matrix form was presented in Section 3.3, Equation (3.2d) which took the form:

$$[M]_D \left\{ \overset{\circ\circ}{y} \right\} + [C] \left\{ \overset{\circ}{y} \right\} + [K]\{y\} = \{F(t)\}$$

Since the stiffness matrix which is square and symmetric, has off-diagonal terms, the equations of motion are coupled.

Determining mode shapes for free vibration was discussed in Section 3.3.1. Solving for coupled equations is cumbersome and it is often advantageous to obtain the same response from the superposition of free vibration mode shapes. The mode shapes, in this respect serve the same purpose as do the trigonometric functions in Fourier series. Also, the response we are looking for, may often be obtained considering only a few modes. The principle of mode superposition is shown in Figure 3.6.

$$y = \varphi A \quad y_1 = \varphi_1 A_1 \quad y_2 = \varphi_2 A_2 \quad y_3 = \varphi_3 A_3$$

Note: If vector y is time dependent the co-ordinates A_n will also be time dependent

Referring to Equation (3.2d), the response y can be found in terms a linear combination of the mode shapes. The basis for being able to do so is the property of orthogonality of Eigen functions.

We may write

$$\{y\} = \{\varphi_1\}A_1(t) + \{\varphi_2\}A_2(t) + \ldots\ldots + \{\varphi_n\}A_n(t) \tag{3.24}$$

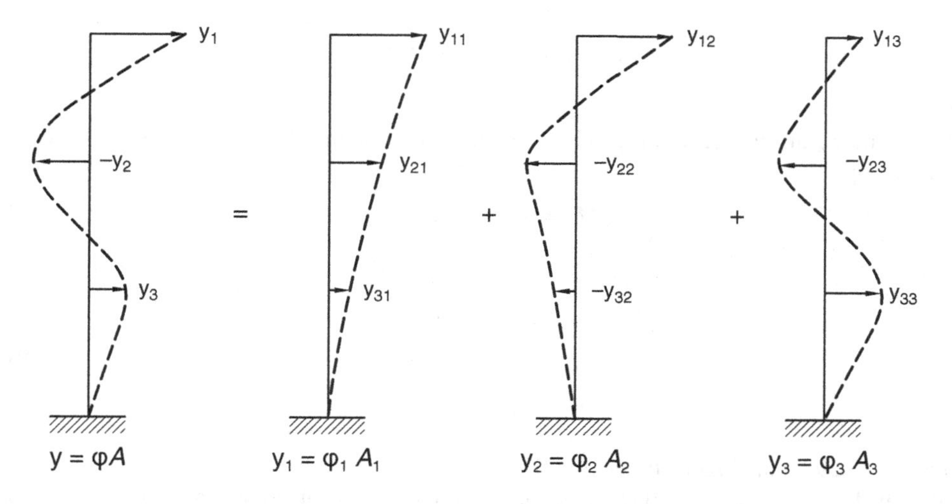

Figure 3.6 Deflection as a sum of mode shapes.

where

$\varphi_1, \varphi_2, \ldots\ldots\ldots \varphi_n$ represent the Eigen vectors of the 1, 2, ..., n[th] mode of free vibration. As noted earlier, for an nDOF system there will be n independent mode shapes.

A_1, A_2, are the modal amplitudes, the time-varying coefficients which determine how the normal modes must be added to determine the total response of the system. Thus,

$$\{y\} = \left[\{\varphi_1\}\{\varphi_2\}\ldots\ldots\ldots\{\varphi_n\}\right]\begin{Bmatrix} A_1 \\ A_2 \\ \cdot \\ A_n \end{Bmatrix} \tag{3.25a}$$

$$= [\varphi_i]\{A_i\} \tag{3.25b}$$

This defines the transformation from geometric coordinates to normal coordinates A.

Substituting into the equations of motion, we get

$$[M]_D[\varphi]\left\{\overset{\circ\circ}{A}\right\} + [K][\varphi]\{A\} = F(t). \tag{3.26}$$

Pre-multiplying by the transpose of $[\varphi]$ we have

$$[\varphi]^T[M]_D[\varphi]\left\{\overset{\circ\circ}{A}\right\} + [\varphi]^T[K][\varphi]\{A\} = [\varphi]^T F(t) \tag{3.27}$$

By using the orthogonality condition

$$\{\varphi_i\}^T[M]_D\{\varphi_j\} = 0 \qquad (i \neq j) \tag{3.28a}$$

$$= M_i \qquad (i = j) \tag{3.28b}$$

We can write Equation (3.27) as:

$$M_n\overset{\circ\circ}{A}_n + K_nA_n = F_n(t) \tag{3.29}$$

where

M_n, K_n and F_n are the generalized mass, stiffness and loading for mode n, respectively and are given by

$$M_n = \{\varphi_n\}^T[M]_D\{\varphi_n\}; \tag{3.30a}$$

where

$$n = \text{mode number}$$

$$K_n = \{\varphi_n\}^T[K]\{\varphi_n\}; \tag{3.30b}$$

$$F_n = \{\varphi_n\}^T\{F(t)\} \tag{3.30c}$$

Equation (3.29) is analogous to the SDOF system.

It may be seen that the coupled equations set up in local coordinates may be uncoupled by the use of normal or generalized coordinates. The transformation is defined by the relationship in Equations (3.25a, b). The orthogonality of the damping matrix is discussed next.

3.5 Damping Orthogonality

It is of definite interest to review the conditions under which the normal coordinate transformation will also uncouple the damped equations of motion.

Equation (3.29) may be extended to include damping. Thus

$$M_n \overset{\circ\circ}{A}_n(t) + C_n \overset{\circ}{A}_n(t) + K_n A_n(t) = F_n(t) \tag{3.31}$$

where

$$C_n = \{\varphi_n\}^T [C] \{\varphi_n\} \tag{3.32}$$

This is known as modal damping.

In keeping with earlier discussions in Chapter 2 we can assume

$$C_n = 2\xi_n \omega_n M_n .. \text{ or } \xi_n = \frac{C_n}{2\omega_n M_n} \tag{3.33}$$

Substituting this in Equation 3.31 and dividing by the modal mass, the expression for modal equation of motion may be expressed as:

$$\overset{\circ\circ}{A}_n(t) + 2\xi_n \omega_n \overset{\circ}{A}_n(t) + \omega_n^2 A_n(t) = \frac{F_n(t)}{M_n} \tag{3.34}$$

where ξ_n represents the modal viscous damping ratio.

'In general the damping matrix $[C]$ cannot be constructed from element damping matrices such as the mass and stiffness matrices of the element assemblage and its purpose is to approximate the overall energy dissipation during the system response'. (Bathe and Wilson, 1976).

It is recommended that the damping of an MDOF system be defined using the damping ratio for each mode, rather than evaluating the coefficients of the damping matrix $[C]$ because the modal damping ratios ξ_n can be determined experimentally or estimated with precision in many cases (Clough and Penzien, 1993). However, we often encounter linear systems having nonproportional damping. In such situations an appropriate damping matrix can be constructed from appropriate proportional damping matrices derived by other means. The way forward is to assume damping to be proportional to mass and stiffness damping[1]:

$$C = \alpha M + \beta K \tag{3.35}$$

It is of interest to note that in Equation (3.35) when damping matrix ($C = \alpha M$), i.e. the damping ratio, is inversely proportional to the frequency of vibration then the higher modes

[1] This is also known as Rayleigh damping after Lord Rayleigh who first proposed it.

of a structure will have very little damping. Similarly, when the damping is proportional to the stiffness matrix ($C = \beta K$) i.e. $\alpha = 0$, the damping ratio is directly proportional to frequency; the higher modes of the structure will be heavily damped (Clough and Penzien, 1975).

If the damping does not have this form the modes are coupled. This means that vibration in one mode may excite vibration in another mode and the breakdown of structural responses into separate modal responses is not possible. It is shown that that the damping ratio ξ_n can be expressed as (Clough and Penzien, 1993):

$$\xi_n = \frac{\alpha}{2\omega_n} + \frac{\beta\omega_n}{2} \tag{3.36}$$

The two damping factors α and β can be evaluated by the solution of two simultaneous equations if the damping ratios ξ_i and ξ_j for two specific modes ω_i and ω_j are known.

3.6 Non-linear Dynamic Analysis

3.6.1 Introduction

The dynamic analysis procedures discussed so far assume that the structures are linearly elastic. The principle of mode superposition will be valid. Consider a structure subjected to a strong motion earthquake. Very few structures could withstand such strong motion without some plastic deformation. As discussed in Chapter 6, it would be uneconomical to design a structure so that it remains elastic. As a result, structures are designed to be able to dissipate energy by means of ductile hysteretic behaviour (see Chapter 9). The mode superposition technique will no longer be applicable. Resorting to non-linear analysis becomes necessary.

Direct step-by-step integration techniques reflecting the non-linear response of the system is usually resorted to. In the step-by-step procedure, the equilibrium of the structure with regard to internal and external forces is established at time t (refer to Chapter 9 on non-linear design concepts). The next equilibrium is sought at time $t + \Delta t$. The stiffness properties are not constant and hence the nonlinear properties need to be updated at each time increment. The solution is sought as a series of incremental time steps where the properties have modified at each time step. The displacements, which have been obtained at time t become the starting values for the next time step at $t + \Delta t$. A numerical solution process of this kind has to be implemented and executed on a high speed digital computer.

3.6.2 Incremental Integration Process

Consider the SDOF system shown in Figure 3.7a.

Referring to Figure 3.7b, during an interval of time Δt the change in the applied force $\Delta F(t)$ will be balanced by corresponding changes in the restoring forces, namely, inertia, damping (Figure 3.7c) and stiffness (Figure 3.7d). The mass is assumed to remain constant, and $c(t)$ and $k(t)$ are damping and stiffness during the time interval Δt. The increase in displacement during the time interval is Δy. Acceleration and velocity are represented by $\Delta \overset{\circ\circ}{y}$ and $\Delta \overset{\circ}{y}$. The incremental equation of equilibrium may be written as:

$$m\Delta \overset{\circ\circ}{y} + c(t)\Delta \overset{\circ}{y} + k(t)\Delta y = \Delta F(t) \tag{3.37}$$

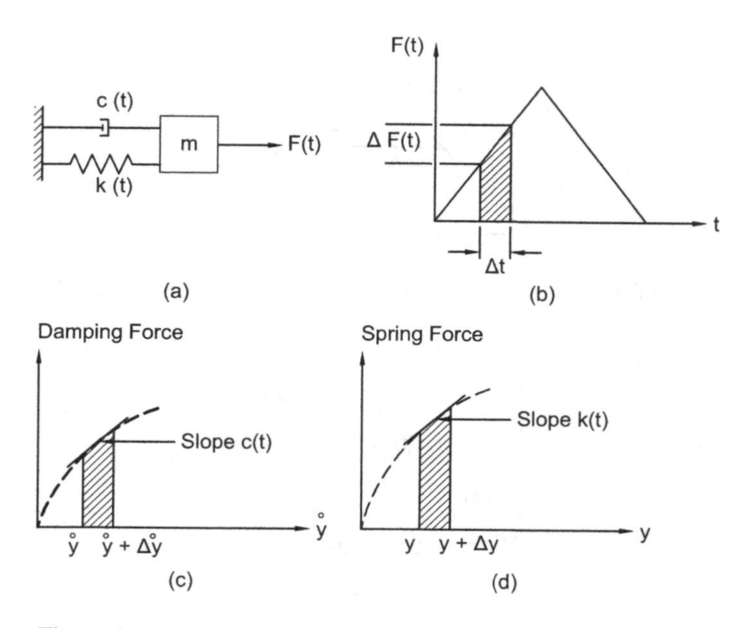

Figure 3.7 An arbitrary non-linear single degree of freedom system.

In terms of modelling non-linearity, it may be seen that it can be very general. Any kind of non-linearity in terms of damping and stiffness (stiffness reduction due to plasticity in the members etc.) may be introduced in the model. The values of $c(t)$ and $k(t)$ which correspond to a particular increment may be calculated.

3.6.3 Numerical Procedures for Integration

Finite difference methods are well established as solutions for differential equations in general. The solution of Equation (2.18a) (Chapter 2) i.e.

$$m\ \overset{\circ\circ}{y} + c\ \overset{\circ}{y} + ky = F\ \sin\ \Omega t$$

may be solved by finite differences.

Referring to Equation (2.18a), the derivative of y with respect to t is defined as the limit of the ratio of the increment in y to the increment in t, as the increments approach zero i.e.

$$\frac{dy}{dt} = \lim_{\Delta t \to 0} \left(\frac{\Delta y}{\Delta t} \right) \tag{3.38a}$$

The relationship with small increments in time is shown in Figure 3.8.

We can now approximate expressions for the slope of the curve and the rate of change of slope. The representation is shown in Figure 3.9. The slope $\left(\frac{dy}{dt}\right)$ at y_0 may be approximated as:

$$\frac{y_{t+\Delta t} - y_{t-\Delta t}}{2\Delta t} \tag{3.38b}$$

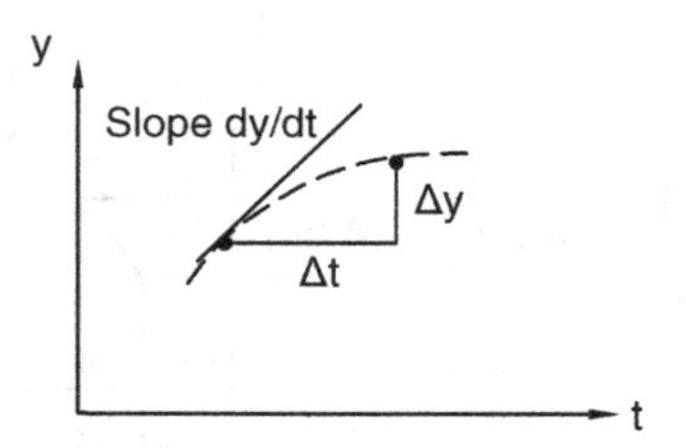

Figure 3.8 Finite difference representation of slope dy/dt approximated with $\Delta y/\Delta t$.

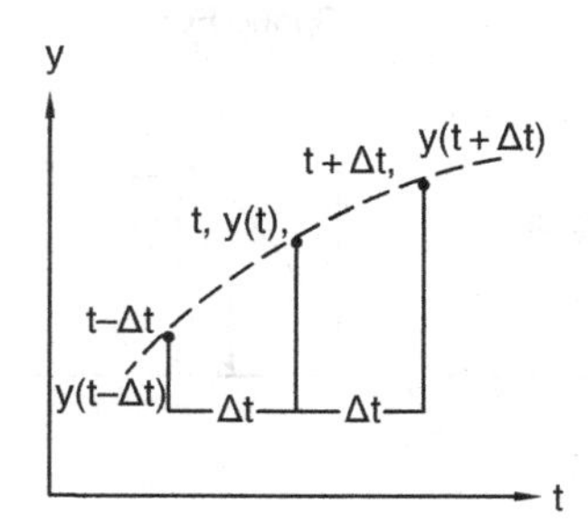

Figure 3.9 Finite difference approximation for acceleration.

and similarly $\left(\dfrac{d^2y}{dt^2}\right)$ as

$$\frac{y_{t+\Delta t} - 2y_t + y_{t-\Delta t}}{\Delta t^2} \tag{3.38c}$$

Note: The finite difference is with respect to time.

Figure 3.9 Finite difference approximation for Equation (2.18a) may now be written as

$$m\left(\frac{y_1 - 2y_0 + y_{-1}}{\Delta t^2}\right) + c\left(\frac{y_1 - y_{-1}}{2\Delta t}\right) + ky_0 = F_0 \tag{3.39}$$

Where

y_1 represents $y_{t+\Delta t}$
y_0 represents y_t
y_{-1} represents $y_{t-\Delta t}$
F_0 represents F_t

Solving for y_1 we have

$$y_1 = \left[\frac{1}{\left(\frac{m}{(\Delta t)^2} + \frac{c}{2\Delta t}\right)}\right]\left[\left(\frac{2m}{(\Delta t)^2} - k\right)y_0 + \left(\frac{c}{2\Delta t} - \frac{m}{(\Delta t)^2}\right)y_{-1} + F_0\right] \tag{3.40a}$$

For an MDOF system we may write:

$$\{Y\}_1 = \left[\frac{1}{\left(\frac{[M]_D}{(\Delta t)^2} + \frac{[C]}{2\Delta t}\right)}\right]\left[\left(\frac{2[M]_D}{(\Delta t)^2} - [K]\right)y_0 + \left(\frac{[C]}{2\Delta t} - \frac{[M]_D}{(\Delta t)^2}\right)y_{-1} + \{F\}_0\right] \quad (3.40b)$$

The computational scheme discussed above assumed the function

$$y = f(y_0, y_{-1}, y_1)$$

This is known as the '*central difference scheme*' as the function utilizes y_0, y_1 and y_{-1} and the point y_0 is central to y_{-1} and y_1.

The function y can also be expressed as

$$y = f(y_0, y_1, y_2, y_3 \ldots\ldots) \quad (3.41a)$$

Or

$$y = f(y_0, y_{-1}, y_{-2}, y_{-3} \ldots\ldots) \quad (3.41b)$$

The first function (3.41a) is a '*forward difference scheme*' and the second (3.41b) is a '*backward difference scheme*'. It is seen that the forward difference scheme will not be useful as we do not know the value of the displacements ahead of our calculations. In the backward difference scheme, the value at the next time step is expressed in terms of values at the preceding time step, which are known.

However, reverting back to Equation (3.40a), it may be deduced that it is possible to calculate the displacement of the mass at y_1 if the previous history of displacements y_0 and y_{-1} is known. Knowing the value at y_1, the same process may be repeated to calculate y_2 by simply changing the indices. It is extremely simple in concept and with repeated application can yield the complete time-history of the behaviour of the system. This process of finding new displacements y_2, with the knowledge of y_0 and y_{-1} and F_0 is known as the direct step-by-step integration process.

This approach, as may be seen, is general and not restricted to SDOF systems; it may also be applied to systems with many degrees of freedom. It is also applicable to nonlinear systems. It is evident that these techniques are most useful where no analytical solutions are possible.

3.6.4 *Estimate of Errors*

Intuitively we can assess that the process of representing the derivatives of Equations (3.38b, c) with such simple expressions will involve some errors. There is an error involved in each of the derivatives. The size of the errors may be investigated by Taylor's series expansion of the function.

The computational molecule for the first and second derivative and the error associated with it is shown in Fig 3.10.

The term $0(h^2)$ stands for terms of order h^2 and smaller.

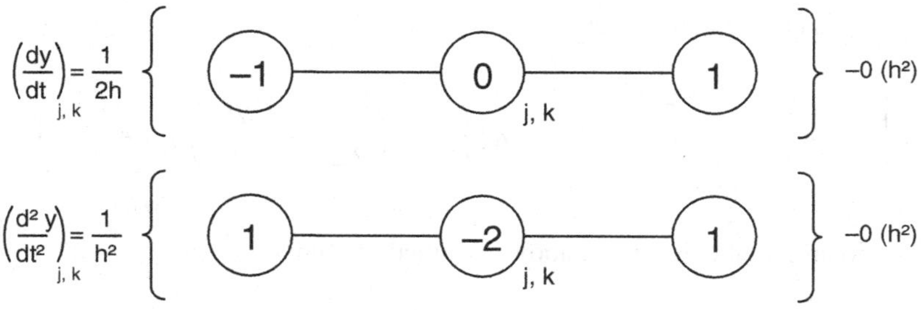

$$\left(\frac{dy}{dt}\right)_{j,k} = \frac{1}{2h}\left\{ \;(-1) \text{——} (0)_{j,k} \text{——} (1)\; \right\} - 0\,(h^2)$$

$$\left(\frac{d^2y}{dt^2}\right)_{j,k} = \frac{1}{h^2}\left\{ \;(1) \text{——} (-2)_{j,k} \text{——} (1)\; \right\} - 0\,(h^2)$$

Figure 3.10 Computational molecule for common two dimensional operators.

3.6.5 Houbolt's Method

Houbolt's method utilizes a backward scheme for the solution involving y_{-2}, y_{-1}, y_0 and y_1. Houbolt utilizes the higher order approximation from Taylor's series expansion of the function to obtain the expression for velocity and acceleration:

$$\dot{y}_1 = \frac{1}{6\Delta t}(11y_1 - 18y_0 + 9y_{-1} - 2y_{-2}) \tag{3.42}$$

$$\ddot{y}_1 = \frac{1}{(\Delta t)^2}(2y_1 - 5y_0 + 4y_{-1} - y_{-2}) \tag{3.43}$$

When these are substituted into the governing differential equation

$$m\ddot{y} + c\dot{y} + ky = F \qquad \text{we have} \tag{3.44}$$

$$\left(\frac{2m}{(\Delta t)^2} + \frac{11c}{6\Delta t} + k\right) y_1 = \left(\frac{5m}{(\Delta t)^2} + \frac{3c}{\Delta t}\right) y_0 - \left(\frac{4m}{(\Delta t)^2} + \frac{3c}{2\Delta t}\right) y_{-1} + \left(\frac{m}{(\Delta t)^2} + \frac{c}{3\Delta t}\right) y_{-2} + F_1 \tag{3.45a}$$

From this equation y_1 may be obtained, once the past displacements y_0, y_{-1} and y_{-2} are known. It is recommended that special starting procedures are used to calculate y_{-1} and y_{-2} (Bathe and Wilson, 1976).

For an MDOF System

$$\left(\frac{2[M]_D}{(\Delta t)^2} + \frac{11[C]}{6\Delta t} + [K]\right)\{Y\}_1 = \left(\frac{5[M]_D}{(\Delta t)^2} + \frac{3[C]}{\Delta t}\right)\{Y\}_0 - \left(\frac{4[M]_D}{(\Delta t)^2} + \frac{3[C]}{2\Delta t}\right)\{Y\}_{-1} + \left(\frac{[M]_D}{(\Delta t)^2} + \frac{[C]}{3\Delta t}\right)\{Y\}_{-2} + \{F\}_1 \tag{3.45b}$$

3.6.6 Explicit and Implicit Scheme

The expressions developed for the central difference method and the Houbolt's method are equally valid for multi degree of freedom (MDOF) systems. With reference to Equation (3.40a), it should be emphasized that the solution y_1 representing $y_{t+\Delta t}$ is thus based on equilibrium conditions at time t. For this reason the integration procedure is called an *'explicit integration method'*. For systems with many degrees of freedom (Equation (3.40b)), the *'explicit integration method' does not require a factorization of the (effective) stiffness matrix* K in the step-by-step solution. On the other hand Houbolt's method, uses the equilibrium conditions at time $t+\Delta t$ and for this reason is called an '*implicit integration scheme*'. It should be mentioned that there are other implicit schemes (Newmark beta, Wilson θ, etc.), which have been discussed in other texts (Clough and Penzien, 1993).

A basic difference between Houbolt's method and the central difference scheme is the appearance of the stiffness K as a factor to the required displacement Y_1. The term $K\{Y\}_1$ (Equation 3.45b) appears because the equilibrium is considered at time $t+\Delta t$ and not at time t as in the central difference method.

3.6.7 Minimum Time Step Δt (Explicit Integration Scheme)

A second, very important point to consider in the use of the explicit integration scheme is that the scheme requires that the time step Δt *is smaller than the critical value* Δt_{cr}, which can be calculated from mass and stiffness properties. It is shown that to obtain a valid solution

$$\Delta t \leq \Delta t_{cr} = \frac{T_n}{\pi} \tag{3.46}$$

where T_n is the smallest period of the finite element assemblage and n is the order of the element system, respectively. 'Integration schemes, that require the use of a time step Δt smaller than a critical time step Δt_{cr}, such as the central difference method, are said to be conditionally stable' (Bathe and Wilson, 1976).

'For 'implicit schemes' there is no critical time-step limit and Δt can in general be selected much larger than given in Equation (3.46) for the central difference method' (Bathe and Wilson, 1976).

3.7 References

Bathe, K.J., and Wilson, E.L. (1976) Numerical methods in finite element analysis. Prentice Hall, New Jersey.

Chopra, A.K. (2001) *Dynamics of structures: Theory and Applications to Earthquake Engineering*. Prentice Hall, New Jersey.

Clough, R.W. and Penzien, J. (1975) *Dynamics of Structures*. McGraw-Hill Inc., Kogakusha, Ltd., Tokyo, Japan

Clough, R.W. and Penzien, J. (1993) *Dynamics of Structures*. McGraw Hill Inc., New York.

Desai, C.S. and Abel, J.F. (1972) *Introduction to Finite Element Method*. Van Nostrand Reinhold Co., New York.

Smith, J.W. (1988) *Vibration of Structures*, Chapman & Hall, London.

Zienkiewicz, O.C. (1971) *The Finite Element Method in Engineering Science*. McGraw Hill Inc., New York.

3.7 References

4

Basics of Random Vibrations

4.1 Introduction

The occurrence of random vibrations is a feature in our daily lives. Many of the natural phenomena on our planet provide examples of random vibrations. The vibrations could be that of the swaying of a tree or a small plant in wind. The tree or the plant is moving backwards and forwards with time and is in other words, subject to random excitations. The displacements, say at the top, vary randomly with time and the precise value at a chosen time cannot be predicted. However, the rate and the amount of movement will depend on the mass, stiffness, internal damping within the system and the intensity of wind excitations.

Since random vibrations are intimately associated with many of the natural phenomena like wind, waves or earthquakes, their study has become a key topic for engineers. As an example of the time-varying fluctuations in engineering practice, consider the deflections at a joint in an offshore structure subjected to random wave loading. The random nature of the deflections is shown in Figure 4.1.

Therefore, there is a need to understand the nature of these deflections in order to ensure a reliable design. Familiarity with the concepts and methods involved has become a necessity in

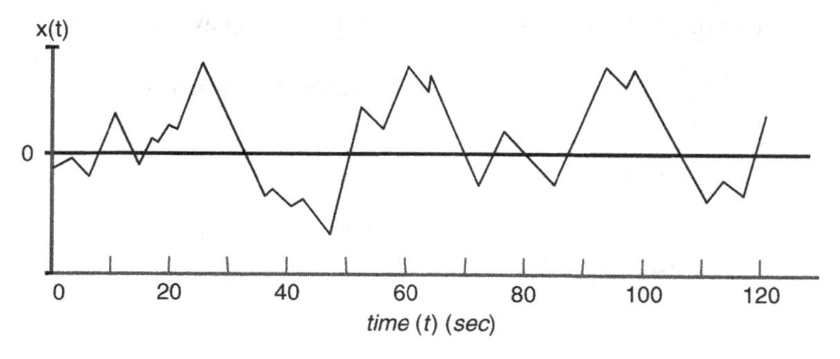

Figure 4.1 Fluctuations at a joint in an offshore structure subjected to random wave loading (deflections).

Fundamentals of Seismic Loading on Structures Tapan Sen

any engineering course or degree. Moreover methods today extend far beyond the traditional fields of engineering. These theories are now used extensively in telecommunications and control systems and other non-traditional, related fields of economic theory requiring random data processing.

However in practice, as one might imagine, the emphasis depends upon the type of problem being dealt with.

4.2 Concepts of Probability

Probability theory is the branch of mathematics that is concerned with random (or chance) phenomena. It has applications in many areas within physical, biological and social sciences as well as in engineering, and the business world.

There are phenomena we come across, where repeated observations under a specified set of conditions do not always lead to the same outcome. A familiar example is that of tossing a coin. If a coin is tossed 1000 times the occurrences of heads or tails alternate in an apparently unpredictable and erratic manner. It is such a phenomenon that we think of as being random and which will briefly be discussed in this section.

4.2.1 Random Variable Space

In the study of random variables we are always dealing with physical quantities such as voltage, torque, stress and so on, which can be measured in physical units. In these cases the chance occurrences are related to real numbers and not just things like heads or tails. This brings us to the notion of the random variable. Let us suppose we have a conceptual experiment for which we have defined a suitable sample space, an appropriate set of events and a probability assignment for the set of events. A *random variable* is simply a function that maps every point in the same space (things) on to the real line (numbers). A simple example of the mapping referred to is shown in Figure 4.2. Each face of the hypothetical coin shows symbols, but no number; the assignment of numbers is up to us and we have assigned the numbers to equate to the number of symbols in this case. However it could have been different.

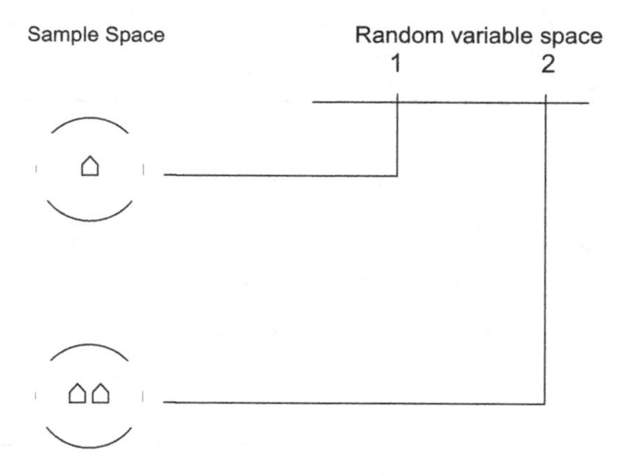

Figure 4.2 Mapping of a hypothetical coin.

So in the probability space, probabilities have been assigned to the events in the sample space that were presumably based on basic axioms of probability. Associated with each event in the original sample space, there will be a corresponding event in the random variable space. To illustrate the concept, consider another example with coins.

This time, in an experiment with coins, a person decides to toss three coins.

The following points have been assigned for the possible outcomes:

$$\begin{aligned} \mathrm{H}: &\quad 1 \\ \mathrm{T}: &\quad 0 \end{aligned}$$

We are interested in the points for H. Let variable X be the number of HEADS in 3 tosses. X takes on values of 0, 1, 2, 3 respectively. If, for example, the outcome is H H H then X will be 3. Table 4.1 lists the values of X (points) corresponding to each of the possible outcomes. If 'p' is the probability of outcome H then '$(1 - p)$' is the probability of having outcome T. The last column in the table gives the probabilities associated with the eight possible outcomes.

From Table 4.1, X takes on values of 0, 1, 2, 3 respectively, with probabilities:

$$(1-p)^3, 3p(1-p)^2, 3p^2(1-p) \text{ and } p^3.$$

Probability Distribution

In the examples above, the sample space consisted of a finite number of elements, and in this situation probability assignment can be made directly on the sample space elements according to what we feel to be reasonable in terms of likelihood of occurrence. This then defines the probabilities for all events defined in the sample space. These probabilities, in turn, transfer directly to equivalent events defined in the sample space. The allowable realizations (i.e. real numbers) in the random-variable space are elementary equivalent events themselves, so the result is a probability associated with each allowable realization in the random variable space. The sum of these probabilities must be unity.

A very good description and discussion on discrete random variables may be found in (Hoel *et al.*, 1971).

Table 4.1 The values of X (points) corresponding to each of the possible outcomes.

	α		$X(\alpha)$	$P(\alpha)$
H	**H**	**H**	**3**	p^3
H	**H**	**T**	**2**	$p^2(1-p)$
H	**T**	**H**	**2**	$p^2(1-p)$
H	**T**	**T**	**1**	$p(1-p)^2$
T	**T**	**T**	**0**	$(1-p)^3$
T	**H**	**T**	**1**	$p(1-p)^2$
T	**T**	**H**	**1**	$p(1-p)^2$
T	**H**	**H**	**2**	$p^2(1-p)$

Continuous Random Variable

In many situations, we encounter continuous random variables. As an illustration, consider the decay of radioactive particles, or the usual electronic noise (voltage etc.) encountered in a wide variety of applications. The corresponding sample space, in the case of continuous variables, contains an infinite number of points, so we cannot assign probabilities directly on the points of the sample space; but it must be done on the defined events.

Let us now consider the situation in a game of roulette. A player spins a pointer mounted on a circular board. The pointer is free to turn around the centre. A simplified sketch is shown in Figure 4.3. We wish to carry out an experiment and define the outcome as the location on the periphery at which the pointer stops. The sample space then consists of an infinite number of points along a circle. It would help if each point in the sample space be identified in terms of an angular coordinate measured in radians. The functional mapping that maps all points on the circle to corresponding points on the real line between 0 and 2π would then define an appropriate random space.

Let X denote a continuous random variable to the angular position of the pointer at which it comes to a stop. Presumably, this would be any angle between 0 and 2π radian, therefore the probability of any particular position is infinitesimal or stated differently the probability of X taking up any *specific value x* would be zero. A formal mathematical expression is available in other texts including Hoel *et al.* (1971).

Consider another illustration – that of a probabilistic model for decay times of a finite number of radioactive particles. Let T be the random variable denoting the time until the first particle decays. Then T would be the continuous random variable. The probability would be zero that the first decay occurs at any specific time (e.g. $T = 2.000000\ldots$ seconds).

Referring back to the pointer example, we can assign probability to the event that the pointer stops within a certain continuous range of values, say, between 0 and θ radians. If all positions are equally likely, we may assign probabilities as follows:

$$P(0 \leq X \leq \theta) = \begin{cases} \dfrac{1}{2\pi}\theta, & 0 \leq \theta \leq 2\pi \end{cases}$$

The function is shown in Figure 4.4. It may be noted that the probability assignment is a function of the parameter θ and the function is shown in Figure 4.4. We have chosen to adopt the 'equal chance of occurrence' assumption, hence the portion of the curve between 0 and 2π is linear.

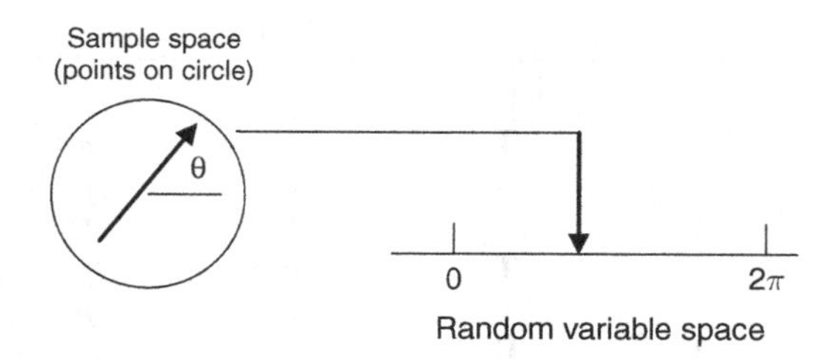

Figure 4.3 Mapping for the roulette game (no markings shown).

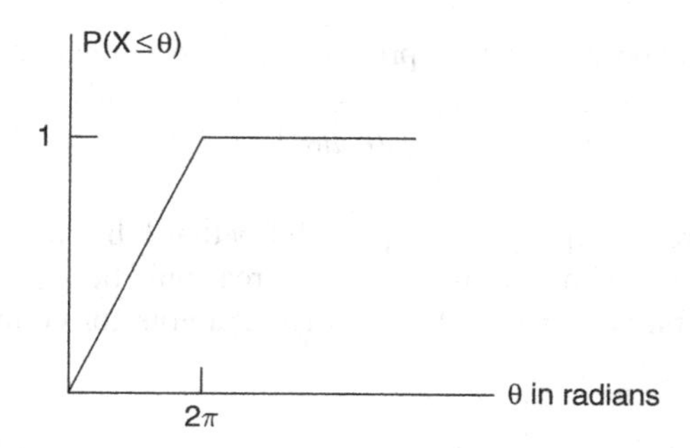

Figure 4.4 Probability distribution function for pointer example.

The function sketched in Figure 4.4 is known as the *cumulative distribution function*. It describes the probability assignment as it maps on to equivalent events in the random variable space. Mathematically, the probability distribution function associated with the random variable X is defined as

$$F_X(\theta) = P(X \leq \theta)$$

The following properties of the *cumulative distribution function* may be noted:

(a) $F_X(\theta) \to 0 \quad \text{as} \quad \theta \to -\infty$

(b) $F_X(\theta) \to 1 \quad \text{as} \quad \theta \to \infty$

(c) $F_X(\theta)$ is a non-decreasing function of θ.

It is possible to present the information contained in the distribution shown in Figure 4.4 in a derivative form provided the distribution is absolutely continuous. Let $f_X(\theta)$ be defined as

$$f_X(\theta) = \frac{d}{d\theta} F_X(\theta)$$

The function $f_X(\theta)$ is known as the *probability density function* associated with X, the random variable with which we have been dealing. The probability density function for the roulette game with the pointer is shown in Figure 4.5.

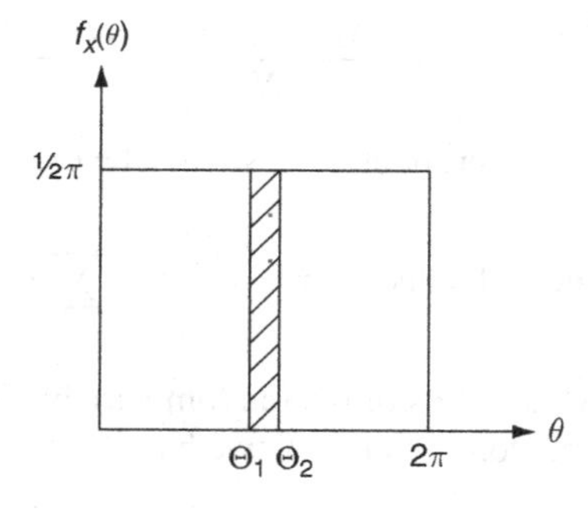

Figure 4.5 Probability density function for pointer example.

From Figure 4.5, it may be seen that the probability density function satisfies the following:

$$\int_{-\infty}^{\infty} f_X(\theta)d\theta = 1 \tag{4.1}$$

The shaded area in Figure 4.5 represents the probability that X lies between θ_1 and θ_2. If θ_1 and θ_2 are separated by an infinitesimal amount, $\Delta\theta$, the area could be approximated by $f_X(\theta_1) \cdot \Delta\theta$ which defines the probability element. A numerical example for computation of probability follows later.

Notation

Usually the cumulative distribution function is denoted by an upper case symbol and the corresponding lower case symbol is used for the density function. The subscript in each case indicates the random variable being considered. The argument function is a dummy which may be anything.

Expectation and Averages

We have all come across the process of averaging at some stage or another. It is worthwhile, however, to put together some ideas about averaging, especially relating to probability. In the simplest case the 'sample average' or the 'sample mean' of a random variable 'X' would be:

$$\overline{X} = \frac{X_1 + X_2 + X_3 + \ldots\ldots X_N}{N} \tag{4.2}$$

Where $\overline{X}$ denotes the mean and $X_1,\ X_2 \ldots X_N$ are sample realizations obtained from repeated trials of the chance situation. We should note that the averaging is for a finite number of trials. Associated with these trials we can also think along the lines of a conceptual average that would occur after an infinite number of trials. This hypothetical average has been termed '*expected value*'. It simply refers to the value one would expect in a statistically similar situation.

Imagine a random variable whose 'n' possible realizations are $x_1,\ x_2,\ x_3, \ldots . x_n$ Let the corresponding probabilities be $p_1,\ p_2,\ p_3 \ldots\ldots . p_n$. If we have N trials and if N is large, we would expect p_1N number of x_1 s, p_2N number of x_2s, etc... The sample average in this case would be

$$\overline{X} \approx \frac{(p_1N)x_1\ +\ (p_2N)x_2\ +\ (p_3N)x_3 \ldots\ldots (p_nN)x_n}{N}$$

The expected value for the discrete probability case may be expressed as

$$\text{Expected value of } X = E(X) = \sum_i^n p_i x_i \tag{4.3}$$

where n is the number of allowed values of the random variable X.

Similarly, for a continuous random variable X, we have

$$\text{Expected value of } X = E(X) = \int_{-\infty}^{\infty} x f_x(x)dx \tag{4.4}$$

Using the same arguments, we can define the expectation of a function of X as well as for X. Thus we have the following:

Discrete case:

$$E(g(X)) = \sum_{i}^{n} p_i g(x) \tag{4.5}$$

Continuous case

$$E(g(X)) = \int_{-\infty}^{\infty} g(x) f_x(x) dx \tag{4.6}$$

If the function $g(x)$ happens to be X^k we may extend Equation (4.6) to find the kth moment of X. Thus

$$E(X^k) = \int_{-\infty}^{\infty} x^k f_x(x) dx \tag{4.7}$$

Variance

The second moment of X, about the mean, is of interest because it leads to the expression for '*variance*' – a quantity frequently encountered. Thus

$$E(X^2) = \int_{-\infty}^{\infty} x^2 f_x(x) dx \tag{4.8}$$

The variance is defined as

$$\text{Variance of } X = E[(X - E(X))^2] \tag{4.9}$$

The variance is a measure of dispersion of X about its mean. If the mean is zero the variance is identical to the second moment. Equation (4.9) may be expanded to yield a more convenient form for computation. Expanding (4.9)

$$\begin{aligned} \text{Var } X &= E[X^2 - 2X \cdot E(X) + (E(X))^2] \\ &= E(X^2) - (E(X))^2 \end{aligned} \tag{4.10}$$

Standard Deviation

The square root of *variance* is known as the '*standard deviation*'

$$\text{Standard Deviation of } X = \sqrt{\text{var } \mathit{of}\ X} \tag{4.11}$$

4.2.2 Gaussian or Normal Distribution

There are several commonly occurring probability distributions which can be expressed in a compact mathematical form by means of a probability density function (p.d.f). The most frequently encountered distribution (at least as an acceptable approximation to many naturally occurring phenomenon) is that formulated by Gauss (1777–1855), the eighteenth-century

German mathematician who first invented it, and is known as the Gaussian distribution. It is also known as 'normal' distribution.

A signal $x(t)$ is said to be Gaussian if its probability density function has the form

$$f_x(x) = \frac{1}{\sigma\sqrt{2\pi}} \exp\left[-\frac{1}{2\sigma^2}(x - m_x)^2\right] \tag{4.12}$$

We may note that this density function contains two parameters m_x and σ^2. These are the mean and variance of the random variable. Thus, for the f_x specified by Equation (4.12)

$$\int_{-\infty}^{\infty} x f_x\,(x)\,dx = m_x \tag{4.13}$$

and

$$\int_{-\infty}^{\infty} (x - m_x)^2 f_x(x) dx = \sigma^2 \tag{4.14}$$

Note, that the normal density function is completely specified by assigning numerical values to the mean and variance.

It is possible to confirm that (4.12) satisfies

$$\int_{-\infty}^{\infty} f_x(x) dx = 1 \tag{4.15}$$

The normal density function and distribution function are shown in Figure 4.6.

Note that the density function is symmetric and peaks at its mean. Qualitatively this would indicate the mean to be the most likely value, with values on either side of the mean gradually becoming less and less likely as the distance from the mean increases. Many naturally occurring phenomenon conform to this distribution and hence its wide applicability in applied probability.

As discussed earlier, the quantity σ^2 is a measure of the dispersion of the random function and is known as the *variance*. Thus, a small value of σ corresponds to a sharp peaked density curve and large value of σ would correspond to a curve with a flat peak.

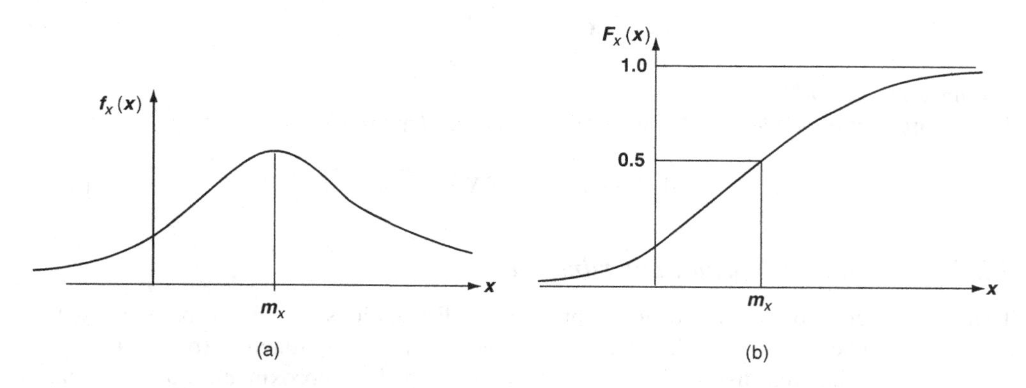

Figure 4.6 (a) Normal density function; (b) normal distribution function.

The normal distribution function is of course the integral of the density function (Equation 4.12):

$$F_x(x) = \int_{-x}^{x} f_x(u)du = \frac{1}{\sigma\sqrt{2\pi}} \int_{-x}^{x} \exp\left[-\frac{1}{2\sigma^2}(u - m_x)^2\right] du \tag{4.16}$$

According to Equation (4.16), the function $F_x(x)$ either becomes or approaches zero and unity with increasing negative and positive values of x, as shown in Figure 4.6.

Unfortunately, the integral above cannot be represented in closed form in terms of familiar functions. Clearly it can be found by numerical integration of the distribution function.

The cumulative density function for the Gaussian distribution is known as the 'error function'. Detailed discussions on other probability distributions may be found in many good standard texts on the subject (Ang and Tang, 1975; Newland, 1993).

Standard Normal Variable

If x is a normal random variable with mean m_x and variance σ_x^2, then $z = \frac{x-m_x}{\sigma_x}$ is a normal random variable with mean 0 and variance = 1. The random variable z is called a 'standard normal variable'.

The density function of the standardized random variable $z = \frac{X-m_x}{\sigma_x}$ is given by

$$p_Z(z) = \frac{1}{\sqrt{2\pi}} \exp\left(-\frac{z^2}{2}\right) \qquad -\infty < z < \infty \tag{4.17}$$

The areas under the standard normal curve are given in Table 4.2. The entries in Table 4.2 are the areas under the normalized curve between $z = 0$ and a value of z to the right of the mean (see Figure 4.7).

Table 4.2 Normal curve areas.

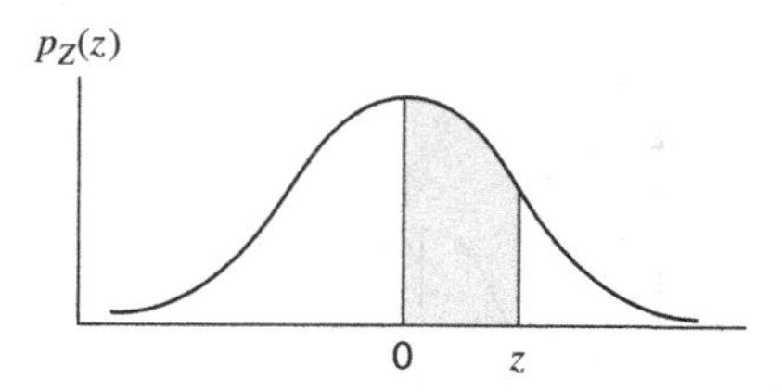

z	.00	.01	.02	.03	.04	.05	.06	.07	.08	.09
.0	.0000	.0040	.0080	.0120	.0160	.0199	.0239	.0279	.0319	.0359
.1	.0398	.0438	.0478	.0517	.0557	.0596	.0636	.0675	.0714	.0753
.2	.0793	.0832	.0871	.0910	.0948	.0987	.1026	.1064	.1103	.1141
.3	.1179	.1217	.1255	.1293	.1331	.1368	.1406	.1443	.1480	.1517
.4	.1554	.1591	.1628	.1664	.1700	.1736	.1772	.1808	.1844	.1879
.5	.1915	.1950	.1985	.2019	.2054	.2088	.2123	.2157	.2190	.2224

(*continued overleaf*)

Table 4.2 (*continued*)

z	.00	.01	.02	.03	.04	.05	.06	.07	.08	.09
.6	.2257	.2291	.2324	.2357	.2389	.2422	.2454	.2486	.2517	.2549
.7	.2580	.2611	.2642	.2673	.2704	.2734	.2764	.2794	.2823	.2852
.8	.2881	.2910	.2939	.2967	.2995	.3023	.3051	.3078	.3106	.3133
.9	.3159	.3186	.3212	.3238	.3264	.3289	.3315	.3340	.3365	.3389
1.0	.3413	.3438	.3461	.3485	.3508	.3531	.3554	.3577	.3599	.3621
1.1	.3643	.3665	.3686	.3708	.3729	.3749	.3770	.3790	.3810	.3830
1.2	.3849	.3869	.3888	.3907	.3925	.3944	.3962	.3980	.3997	.4015
1.3	.4032	.4049	.4066	.4082	.4099	.4115	.4131	.4147	.4162	.4177
1.4	.4192	.4207	.4222	.4236	.4251	.4265	.4279	.4292	.4306	.4319
1.5	.4332	.4345	.4357	.4370	.4382	.4394	.4406	.4418	.4429	.4441
1.6	.4452	.4463	.4474	.4484	.4495	.4505	.4515	.4525	.4535	.4545
1.7	.4554	.4564	.4573	.4582	.4591	.4599	.4608	.4616	.4625	.4633
1.8	.4641	.4649	.4656	.4664	.4671	.4678	.4686	.4693	.4699	.4706
1.9	.4713	.4719	.4726	.4732	.4738	.4744	.4750	.4756	.4761	.4767
2.0	.4772	.4778	.4783	.4788	.4793	.4798	.4803	.4808	.4812	.4817
2.1	.4821	.4826	.4830	.4834	.4838	.4842	.4846	.4850	.4854	.4857
2.2	.4861	.4864	.4868	.4871	.4875	.4878	.4881	.4884	.4887	.4890
2.3	.4893	.4896	.4898	.4901	.4904	.4906	.4909	.4911	.4913	.4916
2.4	.4918	.4920	.4922	.4925	.4927	.4929	.4931	.4932	.4934	.4936
2.5	.4938	.4940	.4941	.4943	.4945	.4946	.4948	.4949	.4951	.4952
2.6	.4953	.4955	.4956	.4957	.4959	.4960	.4961	.4962	.4963	.4964
2.7	.4965	.4966	.4967	.4968	.4969	.4970	.4971	.4972	.4973	.4974
2.8	.4974	.4975	.4976	.4977	.4977	.4978	.4979	.4979	.4980	.4981
2.9	.4981	.4982	.4982	.4983	.4984	.4984	.4985	.4985	.4986	.4986
3.0	.4987	.4987	.4987	.4988	.4988	.4989	.4989	.4989	.4990	.4990

Source: Abridged from Table 1 Hald, *Statistical Tables and Formu*las 1952. Reproduced by permission of John Wiley & Sons, Inc., New York.

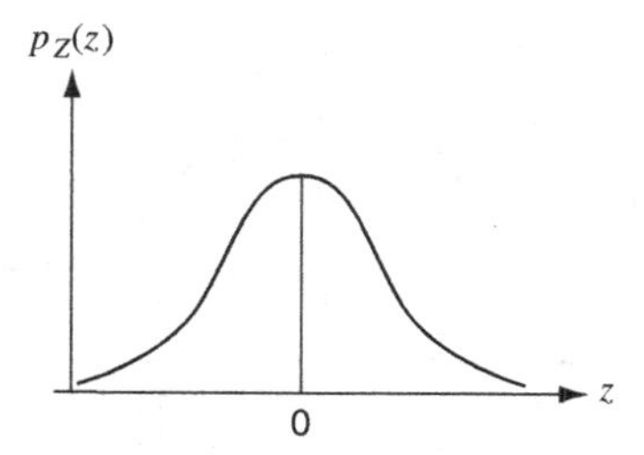

Figure 4.7 Normalized curve.

4.2.3 Worked Example with Standard Normal Variable

The use of tables of normalized distribution is made clearer by an example. In the case of zero mean, $m_x = 0$, and standard deviation $\sigma_x = 100\,\text{N/mm}^2$ (a situation of stress fluctuations), we might seek the probability of a sample value lying in the range of 149–151 N/mm^2. Treating this as a range $\Delta x = -2\,\text{N/mm}^2$ centred on $x = 150\,\text{N/mm}^2$, the normalized quantities are $\Delta z = \Delta x/\sigma_x = 0.02$ and central value $z = (150 - 0)/100 = 1.50$. Then

$$p_Z(z) = \left(1/\sqrt{2\pi}\right) \cdot \exp(-1.125) = 0.1295.$$

The sought solution is:

$$p_Z(z) \cdot \Delta z \quad or \quad 2.6 \times 10^{-3}.$$

4.3 Harmonic Analysis

4.3.1 Introduction

Procedures for harmonic analysis are well developed, but these procedures are not applicable to random processes. Over the years, the harmonic analysis procedures have been developed in such way as to be applicable to random signals, while retaining the basic features and concepts of harmonic analysis. Concepts of Fourier series, integrals and transforms are developed first. Concepts of power spectra are built upon these concepts, which pave the way for basic numerical procedures for computation of the frequency content of a random signal. All major software incorporates the Fast Fourier Transform (FFT) for processing random signals. This has not been covered, being beyond the scope of the present basic text. Discussions on FFT may be found in Newland (1993). In the discussions which follow, we shall assume as before, a random signal, which is both stationary and ergodic. This implies that a single random signal would be sufficiently representative of the random process under consideration.[1]

4.3.2 Fourier Series (Robson, 1963)

If $x(t)$ is periodic, with period $T = 2\pi/\omega_1$, it can be expanded as a series of harmonically varying quantities in the form:

$$x(t) = a_0 + \sum_{r=1}^{\infty} (a_r \cos r\omega_1 t + b_r \sin r\omega_1 t) \tag{4.18}$$

where

$$a_0 = \frac{1}{T}\int_{-T/2}^{T/2} x(t)dt, \tag{4.19a}$$

$$a_r = \frac{2}{T}\int_{-T/2}^{T/2} x(t)\cos r\omega_1 t dt \tag{4.19b}$$

$$b_r = \frac{2}{T}\int_{-T/2}^{T/2} x(t)\sin r\omega_1 t dt \tag{4.19c}$$

[1] There are many established texts on the subject (Crandall, 1958; Newland, 1993). The line of argument developed by Robson (1963) is outlined in the following sections and extracted from the text (reproduced from Robson (1963) by permission of Edinburgh University Press) for the benefit of readers.

On substituting the expression for $x(t)$ into Equation (4.19) and expanding the right-hand sides the above can be easily proved using the orthogonal properties of the modes.

The terms of the series are sinusoids whose frequencies are multiples of the fundamental frequency ω_1 and the amplitude of the various components can thus be plotted to give a discrete spectrum in which the lines have frequency spacing ω_1.

Expressing $< x^2(t) >$, the mean square value of $x(t)$, in terms of the coefficients a_r and b_r: squaring Equation (4.18) and integrating over the period $T = 2\pi/\omega_1$, we have

$$\int_{-T/2}^{T/2} x^2(t)dt = a_0^2 T + \sum_1^{\infty} \left(a_r^2 \frac{T}{2} + b_r^2 \frac{T}{2} \right)$$

all other product terms vanishing on integration over the period. Thus

$$< x^2(t) >= \frac{1}{T} \int_{-T/2}^{T/2} x^2(t)dt = a_0^2 + \frac{1}{2} \sum_1^{\infty} (a_r^2 + b_r^2) \tag{4.20}$$

This is a form of Parseval's Theorem discussed by Newland (1993).

We can also express a periodic function $x(t)$ as a series in complex form. Thus

$$x(t) = \sum_{-\infty}^{\infty} c_r e^{ir\omega_1 t} \tag{4.21}$$

where

$$c_r = \frac{1}{T} \int_{-T/2}^{T/2} x(t) e^{-ir\omega_1 t} dt \tag{4.22}$$

This can be proved by using our previous results from Equation 4.19. Thus from Equations (4.22) and (4.19)

$$c_r = \frac{1}{T} \int_{-T/2}^{T/2} x(t) \left(\cos r\omega_1 t - i \sin r\omega_1 t\right) dt = \frac{1}{2}(a_r - ib_r) \tag{4.23a}$$

$$c_{-r} = \frac{1}{T} \int_{-T/2}^{T/2} x(t) \left(\cos r\omega_1 t + i \sin r\omega_1 t\right) dt = \frac{1}{2}(a_r + ib_r) \tag{4.23b}$$

$$\text{and} \quad c_0 = \frac{1}{T} \int_{-T/2}^{T/2} x(t) dt = a_0 \tag{4.23c}$$

This gives

$$x(t) = c_0 + \sum_1^{\infty} c_r e^{ir\omega_1 t} + \sum_1^{\infty} c_{-r} e^{-ir\omega_1 t}$$

from Equation (4.21)

$$= a_0 + \sum_{1}^{\infty} \frac{1}{2}\{(a_r - ib_r)(\cos\ r\,\omega_1 t + i\sin\ r\,\omega_1 t) + (a_r + ib_r)(\cos\ r\,\omega_1 t - i\ \sin\ r\,\omega_1 t)\}$$

$$= a_0 + \sum_{1}^{\infty} [a_r\ \cos r\,\omega_1 t + b_r\ \sin r\,\omega_1 t]$$

as before.

Again the mean square value of $x(t)$ can be expressed in terms of the coefficients:

$$< x^2(t) > = \sum_{-\infty}^{\infty} |c_r|^2 \tag{4.24}$$

The result can be proved by expanding the right-hand side and using Equation (4.23):

$$\begin{aligned}
\sum_{-\infty}^{\infty} |c_r|^2 &= |c_0|^2 + \sum_{1}^{\infty} |c_r|^2 + \sum_{1}^{\infty} |c_{-r}|^2, \\
&= a_0^2 + \sum_{1}^{\infty} \frac{1}{4}(a_r^2 + b_r^2) + \sum_{1}^{\infty} \frac{1}{4}(a_r^2 + b_r^2), \\
&= a_0^2 + \frac{1}{2}\sum_{1}^{\infty} (a_r^2 + b_r^2), \\
&= < x^2(t) >,
\end{aligned}$$

by Equation (4.20).

4.3.3 Fourier Integrals (Robson, 1963, with permission)

A non-periodic function – a transient loading for example, can be expressed as a Fourier series if we consider it to be periodic, with infinite period. With this assumption our fundamental frequency is infinitesimal, so that the discrete spectrum closes up to give a continuous curve and the series becomes an integral.

Let us assume that the fundamental frequency ω_1 is very small and denote it by $\Delta\omega$. Then we write, combining Equations (4.21) and (4.22)

$$x(t) = \sum_{-\infty}^{\infty} \frac{\Delta\omega}{2\pi} \left\{ \int_{-T/2}^{T/2} x(t) \cdot e^{-ir\Delta\omega t} dt \right\} \cdot e^{ir\Delta\omega t}$$

If we now let $\Delta\omega$ become the infinitesimal $d\omega$, the period T will tend to infinity: the summation must now include a continuous sequence of frequencies, so that the summation becomes an integral and $r \cdot \Delta\omega$ simply becomes ω. We have, therefore, when period is infinite – where there is no periodicity – the result:

$$x(t) = \int_{-\infty}^{\infty} \left[\frac{d\omega}{2\pi} \left\{ \int_{-\infty}^{\infty} x(t)e^{-i\omega t} \cdot dt \right\} \cdot e^{i\omega t} \right]$$

This may be conveniently expressed as two equations, corresponding to Equations (4.21) and (4.22). Thus,

$$x(t) = \frac{1}{2\pi} \int_{-\infty}^{\infty} A(i\omega)e^{i\omega t}\, d\omega \tag{4.25}$$

With

$$A(i\omega) = \int_{-\infty}^{\infty} x(t)e^{-i\omega t} dt \tag{4.26}$$

Equations (4.25) and (4.26) give the *Fourier Integral* expression for $x(t)$. The quantity $A(i\omega)$ defined by equation (4.26) is called the *Fourier transform* of $x(t)$. It is a function of ω – and in general complex – and it shows how $x(t)$ may be considered to be *distributed over the frequency range*. The response of a linear system due to a transient loading can, in principle, always be determined by expressing the loading in the form (4.25), and calculating the separate responses corresponding to each infinitesimal strip $A(i\omega)d\omega$ of the Fourier transform spectrum (although this is not always the best way to solve such a problem). The function $x(t)$ is said to be the inverse transform of $A(i\omega)$: the two quantities $x(t)$ and $A(i\omega)$, related by Equations (4.25) and (4.26) are said to be a *Fourier Transform Pair*.

The near symmetry of Equations (4.25) and (4.26) is very striking and is often made more so by dividing the coefficient $1/2\pi$ into two factors $1/\sqrt{(2\pi)}$ and giving one to each equation; the equations then differ only in the sign of the exponent. The device is, however, unnecessary if we work in terms of the actual frequency f instead of the circular frequency ω. Putting $\omega = 2\pi f$ in Equations (4.25) and (4.26), we have

$$x(t) = \int_{-\infty}^{\infty} A(if)e^{i2\pi ft} df \tag{4.27}$$

with

$$A(if) = \int_{-\infty}^{\infty} x(t)e^{-i2\pi ft} dt \tag{4.28}$$

As there are good practical reasons for working in terms of f, we shall always use this form of the Fourier transform relationship.

Let us denote the complex conjugate of the Fourier Transform $A(if)$ by writing $A^*(if)$: clearly, by (4.28), this is given by

$$A^*(if) = \int_{-\infty}^{\infty} x(t)e^{i2\pi ft} dt \tag{4.29}$$

We also know that it is a property of complex conjugates

$$A(if) \cdot A^*(if) = |A(if)|^2 . \tag{4.30}$$

Thus

$$\begin{aligned}\int_{-\infty}^{\infty} x^2(t)\,dt &= \int_{-\infty}^{\infty} x(t)\cdot x(t)\,dt \\ &= \int_{-\infty}^{\infty} x(t)\left[\int_{-\infty}^{\infty} A(if)\,e^{i2\pi ft}df\right]dt \\ &= \int_{-\infty}^{\infty} A(if)\left[\int_{-\infty}^{\infty} x(t)\,e^{i2\pi ft}dt\right]df\end{aligned}$$

changing the order of integration,

$$\begin{aligned}&= \int_{-\infty}^{\infty} A(if)\cdot A^*(if)\,df, \quad \text{by (4.29)} \\ &= \int_{-\infty}^{\infty} |A(if)|^2\,df \qquad (4.31)\end{aligned}$$

by Equation (4.30).

This also proves Parseval's theorem (Newland, 1993).

It is often convenient in practical applications to consider only positive frequencies: in such cases Equation (4.31) can be rewritten in the form:

$$\int_{-\infty}^{\infty} x^2(t)dt = 2\int_{0}^{\infty} |A(if)|^2\,df \qquad (4.32)$$

because $|A(if)|^2$ is an even function of f.

The result of Equation (4.32) will prove valuable in the next section.

4.3.4 Spectral Density (Robson, 1963)

The results of the previous paragraphs are not applicable to random signals as they stand. A random signal is not periodic and thus cannot easily be expressed as a Fourier series. To have stationary properties a random signal must be assumed to continue over an infinite time, and in such a case neither the real part nor the imaginary part of the Fourier transform converges to a steady value.

Moreover, we are not looking for a description only of a single function of time $x(t)$, which a Fourier integral would provide. We must have a description equally applicable to any member function of the random process $\{x(t)\}$ that might instead have been generated by the physical system under consideration.

Definition of New Quantity: Spectral Density

From our results, however, it is possible to develop a new quantity, the '*spectral density*', which has no convergence difficulties and which is applicable to a whole class of similarly generated functions.

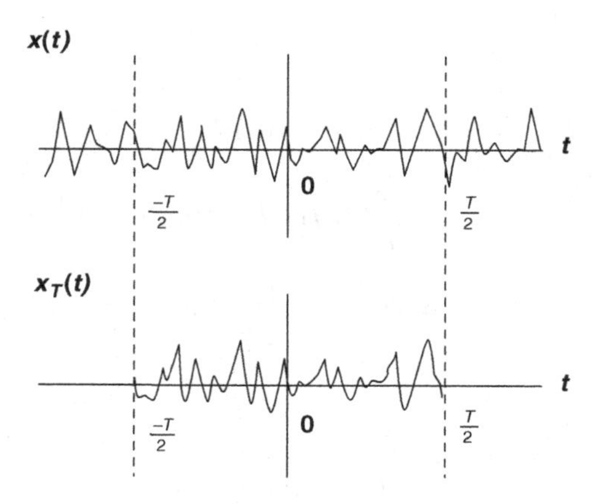

Figure 4.8 Random signals.

Consider a member a function $x(t)$ of a stationary random process. As the signal may be assumed to have commenced at $t = -\infty$ and must be presumed to continue until $t = \infty$, we cannot define its Fourier transform $A(if)$. There is no difficulty, however, in determining the Fourier transform $A_T(if)$ of a signal $x_T(t)$ which we define to be identical with $x(t)$ over the interval $-\frac{T}{2} < t < \frac{T}{2}$ and to be zero at all other times, as shown in Figure 4.8.

We can therefore use Equation (4.32) to express the mean-square value of $x_T(t)$ in terms of $A_T(if)$, as follows:

$$\begin{aligned} < x_T^2(t) > &= \frac{1}{T}\int_{-T/2}^{T/2} x_T^2(t)dt \\ &= \frac{1}{T}\int_{-\infty}^{\infty} x_T^2(t)dt \\ &= \frac{2}{T}\int_{0}^{\infty} |A_T(if)|^2\, df \end{aligned} \tag{4.33}$$

If we now let $T \to \infty$, we obtain an expression for the mean-square value of $x(t)$, i.e.

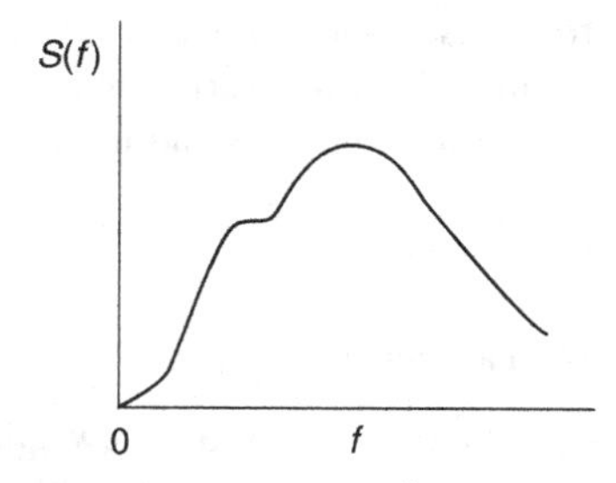

Figure 4.9 Power spectra.

$$< x^2(t) > = \int_0^\infty \lim_{T\to\infty} \left[\frac{2}{T} |A_T(if)|^2\right] df$$
$$= \int_0^\infty S(f)\, df \tag{4.34}$$

Where we have written

$$S(f) = \lim_{T\to\infty} \left[\frac{2}{T} |A_T(if)|^2\right] \tag{4.35}$$

The quantity $S(f)$ is called the *spectral density* of the function $x(t)$. It is formally defined by Equation (4.35) but its properties are best appreciated by considering Equation (4.34). From the latter it can be seen that the form of $S(f)$ indicates the manner of the distribution of the harmonic content of the signal over the frequency range from zero frequency to infinite frequency (as in Figure 4.9): the amount of $< x^2(t) >$ associated with a narrow band of frequency Δf is simply $S(f)\Delta f$. (This can be considered as the mean-square value of the signal passed by a narrow band filter of bandwidth Δf). Its dimension will depend on those of $x(t)$: if for example, $x(t)$ is acceleration then $S(f)$ might well be expressed as g^2/cps.

Spectral density is more precisely – though not apt for our purposes – described as 'power spectral density', the nomenclature derived from its use in electrical problems, where a randomly varying current $x(t)$ through unit resistance gives mean power consumption $< x^2(t) >$.

The spectral density $S(f)$ determined for a particular signal $x(t)$ is also applicable to a whole range of similar functions because of its derivation by way of the modulus $|A_T(if)|$, which being independent of phase – is shared by many Fourier transforms $A_T(if)$. Different member functions $x(t)$ of a stationary ergodic random process $\{x(t)\}$ may be expected to have a common $S(f)$, although it is not the case that a knowledge of the spectral density alone is sufficient to define a random process (Robson, 1963). However, the harmonic content of all member functions of a stationary ergodic random process $\{x(t)\}$ can therefore be described by a single spectrum in which $S(f)$ is plotted against the frequency f as in Figure 4.9.

4.4 Numerical Integration Scheme for Frequency Content

4.4.1 Introducing Discrete Fourier Transform (DFT)

The Fourier integral transform, as we have seen, leads to the so-called 'power spectrum', by which the input (such as ground motion) and resulting responses are represented in the frequency domain. The time-history represented by a given power spectrum is non-deterministic and non-periodic despite its characterization of frequency content. In its basic form it relates to a 'stationary process', i.e. one that continues with constant statistical characteristics. This is attractive as a model for earthquake ground motion because the trace recorded on a specific instrument is the local result of a transmission process so complex as to be effectively non-deterministic. A similar instrument nearby would record a signal of similar overall appearance. But the characteristic spikiness would prove on closer inspection not to correspond in detail.

The frequency content as revealed in the power spectra should in principle be the same, although practical problems arise concerning the adequacy of data for robust evaluation.

Further treatment of Fourier integral transform and the associated numerical technique such as the Fast Fourier Transform (FFT) is beyond the scope of the book. The reader is referred to Newland (1993).

However, useful insight into the concept of frequency content of the record can be obtained from Discrete Fourier Transform (DFT) as follows.

Having selected a nominal duration of the event, T_p defines the lowest *frequency* that can be captured in the analysis. i.e. $\Delta f = \frac{1}{T_p}$

The load period is divided into N equal time increments: $\Delta t = T_p/N$ and the load is defined for the discrete times $t_n = n\Delta t$; n ranging from $1 \ldots\ldots. N$ and the frequencies $f_{(m)} = m\Delta f$ with m ranging from $1 \ldots\ldots. M$.

With reference to Equations (4.27) and (4.28) we may express the discretized form as

$$x_T(t_{(n)}) = 2\Delta f \sum_{m=1}^{m=M} A_T(f_{(m)})\, e^{i2\pi f_{(m)} t_{(n)}} \quad \text{or}$$

$$x_T(t_{(n)}) = \frac{2}{T_p} \sum_{m=1}^{m=M} A_T(f_{(m)})\, e^{i2\pi f_{(m)} t_{(n)}} \tag{4.36}$$

$$A_T(f_{(m)}) = \Delta t \sum_{n=1}^{n=N} x_T(t_{(n)})\, e^{-i2\pi f_{(m)} t_{(n)}} \tag{4.37}$$

Equations (4.36) and (4.37) are the DFT pair and its similarity to integrals transforms of Equations (4.27) and (4.28) may be noted.[2]

Thus Equation (4.35), in computational terms, is reduced to:

$$S(f) = \frac{2}{T_p}(R_{e(f)}^2 + I_{m(f)}^2) \tag{4.38}$$

where

$R_{e(f)}$ = real part of Equation (4.37),
$I_{m(f)}$ = imaginary part of Equation (4.37).

We are thus left with the task of collecting the real and imaginary parts and squaring them. This is the scheme that has been used in the following worked example.

4.5 A Worked Example (Erzincan, 1992)

The acceleration time history recorded on a seismogram during the Earthquake in Erzincan in Turkey is shown in Figure 4.10. The particulars are:

[2] Note that while $A_T\,(f_{(m)})$ is complex, in the series in Equation (4.36) all terms ($m = 1$ to M) can be arranged in conjugate pairs such that the imaginary parts *cancel* and the sum is *real*.

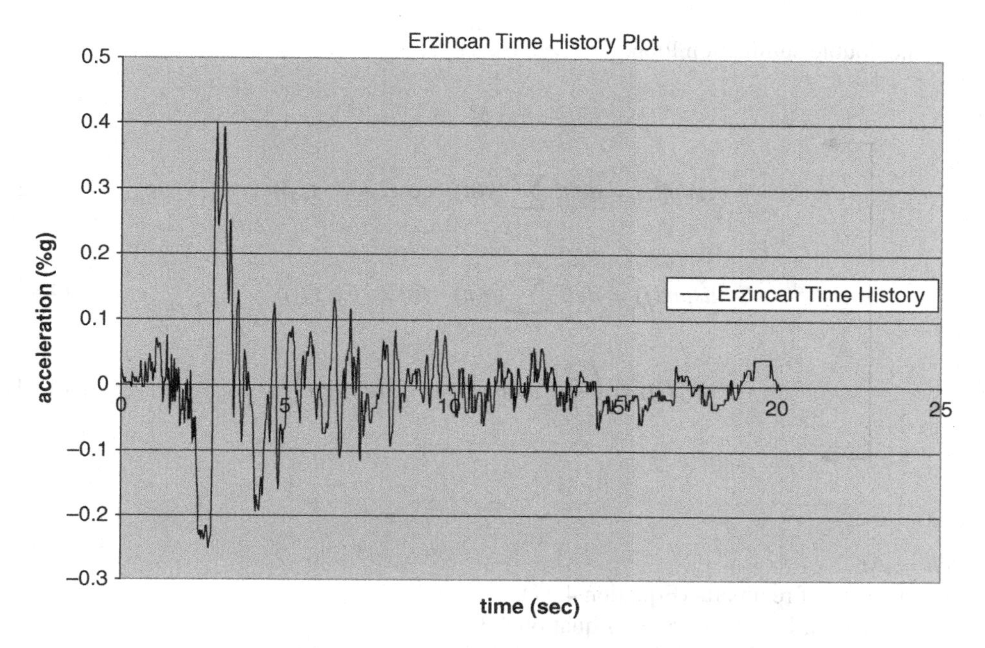

Figure 4.10 Acceleration time history record (Erzincan, 1992).

Duration:	T_P	20.47 sec.
Time interval of recording:	Δt	0.01 sec
Lowest frequency captured $(1/T_p)$:	Δf_p	0.04485

We need to determine the frequency content of the record.
As a preliminary step we need to remember

$$e^{ix} = \cos x + i \sin x \tag{4.39}$$

$$e^{-ix} = \cos x - i \sin x \tag{4.40}$$

Step 1
Compute 'M' frequency values:

$$f_{(m)} = m\Delta f \tag{4.41}$$

Step 2
Collect the real and imaginary parts of Equation (4.37).

Set up the double summation loop:

$$j = M$$

$$arsum(j) = delt^* \sum_{n=1}^{n=N} (y(n) \cdot \cos(2\pi f_{(j)} \cdot t_{(n)})) \tag{4.42}$$

$$aisum(j) = delt^* \sum_{n=1}^{n=N} (y(n) \cdot \sin(2\pi f_{(j)} \cdot t_{(n)})) \tag{4.43}$$

$$S(f(j)) = \left(\frac{2}{T_p}(arsum(j)^2 + aisum(j)^2)\right) \tag{4.44}$$

$$j = 1$$

where

$delt = \Delta t$
arsum sum of real parts (Equation 4.37),
aisum sum of imaginary parts (Equation 4.37),
$t_{(n)}$, $y(n)$ – acceleration time history record (from seismogram).

The plot of the frequency content [$S(f)$] versus '(f)' obtained for the Erzincan time history record (Figure 4.10) is shown in Figure 4.11.

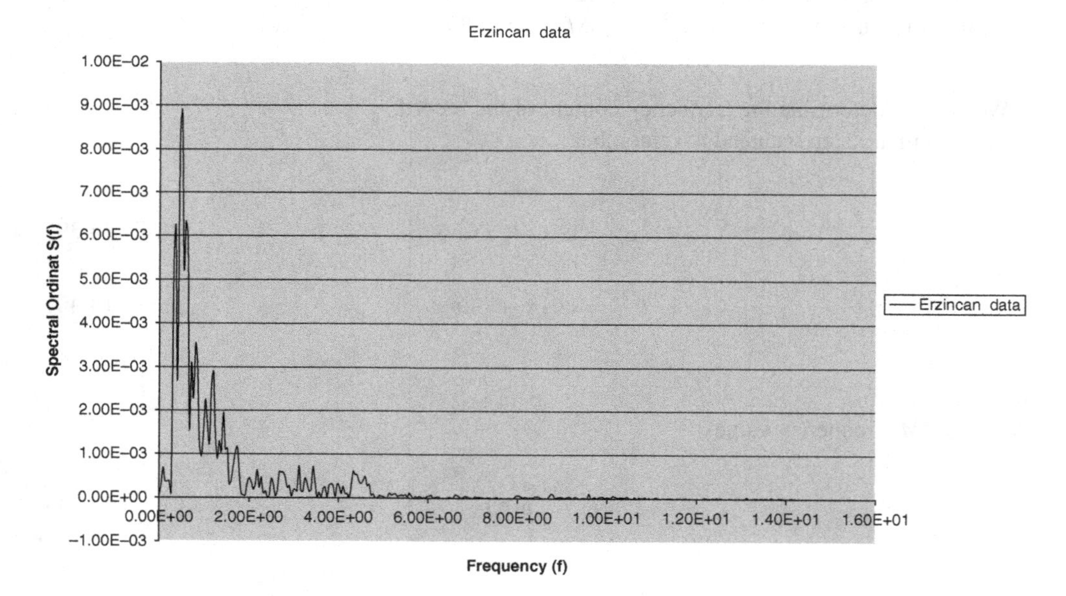

Figure 4.11 Frequency content for time history plot (Figure 4.10).

It will be noted from Equation (4.38) that the units of spectral function are the square of the physical quantity **times** time. In fact, the area under the curve represents the variance (mean square) of the physical process. Because the response bandwidth of the frequency response function is also relative (in this case the natural frequency of the structure), a plot of $f, (S(f))$ on a logarithmic scale of frequency (f) gives the clearest picture of the significance of the frequency content. It will be seen that this plot format retains the interpretation of the area under the curve as the variance, because $df = fd(\ln(f))$.

Computational Check

With mean approximately equal to zero ($\cong$ 0.1233E-02)
The area under curve in Figure 4.11 = 0.546E-02
The variance of the record Figure 4.10 = 0.508E-02
This is a good check on the computational process.

As we can see most of the energy is concentrated between 0.5 and 1.0 Hz. Therefore buildings with a natural period around that frequency range might be subjected to severe damage. It is a useful indication for the designers of the worst frequency range. The spikiness we see in the time history record (Figure 4.10) is the result of a complex earthquake transmission process that has taken place.

Frequency domain representation also offers a key to generation of synthetic input records (time histories of ground motion) for the time domain solution of non-linear structural behaviour, including dissipative processes and progressive loss of stiffness if desired.

The discussions in this chapter have focused mainly on the frequency content as it is of significant importance to earthquake studies. It is now standard computational procedure to generate a stationary time series with a given frequency content as defined by the power spectrum (Newland, 1993). To produce time histories representative of equally probable materializations of ground motion at Erzincan, appropriate lengths of the stationary series can be scaled by a shaping function to represent the build-up and decay of the event. This can be repeated to build up a robust estimate of possible structural response.

Random vibration theory has wide applicability in today's modern-day engineering world. There are many good texts on the subject by Newland (1993), Roberts and Spanos (1990), Robson (1963), Yang (1986), Manolis and Koliopoulos (2001) to name a few.

4.6 References

Ang, A.H-S. and Tang, W.H. (1975) *Probability Concepts in Engineering Planning and Design*. John Wiley & Sons, Inc., New York.

Crandall, S.H. (eds) (1958) *Random Vibration*. M.I.T. Press and John Wiley & Sons, Inc., New York.

Hald, A. (1952), *Statistical Tables and Formulas*. John Wiley & Sons, Inc., New York, USA.

Hoel, P.G., Port, S.C. and Stone, C.J. (1971) *Introduction to Probability Theory*. Houghton Mifflin, Boston.

Manolis, G.D. and Koliopoulos, P.K. (2001) *Stochastic Structural Dynamics in Earthquake Engineering*. WIT Press, Southampton, UK.

Newland, D.E. (1993) *An Introduction to Random Vibration and Spectral Analysis*. Longman Group Ltd.

Roberts, J.B. and Spanos, P.D. (1990) *Random vibration and statistical linearization*. John Wiley & Sons, Ltd, Chichester.

Robson, J.D. (1963) *An Introduction to Random Vibration*, Edinburgh University Press, Edinburgh.

Yang, C.Y. (1986) *Random Vibration of Structures*. John Wiley & Sons, Inc., New York.

5

Ground Motion Characteristics

5.1 Characteristics of Ground Motion

The motion produced during an earthquake is complex. In addition to being a complex subject, it is also a very dynamic subject with new information being made available after each earthquake. The databases of earthquake records are being updated continuously, and predicted empirical relationships becoming obsolete possibly faster than in any other field. A complete picture of the earthquake process has yet to emerge. There are many areas of '*known unknowns*', and the knowledge base is updated with each additional piece of new information. There are surely many areas of '*unknown unknowns*'. For example, we know the severity of the Loma Prieta earthquake (1989) exceeded all predictions with widespread damage. However, to put the reader at ease, in this chapter we will deal only with areas of '*known unknowns*'.

A ground particle at any instant of time generally follows a three-dimensional path. Its velocity, acceleration and displacement changes with time and the earthquake record encompasses a broad range of frequencies. It is to be noted that several factors influence the ground motion that reaches the site of interest. The major influence comes from the source, travel path and local site conditions. These effects are discussed very briefly below.

5.1.1 Ground Motion Particulars

General

Fault location and configuration affect the ground motion. A fault is a zone of the Earth's crust where two sides have been displaced. The fault is caused by sliding or slippage during the resultant ground motion. The Imperial Valley earthquake of 1940, for example, produced fault movements of 5.8 m horizontally and 1.2 m vertically (Bolt, 1988, 2004). In general, most earthquakes occur at faults. Faults exist along plate boundaries and also away from the plate boundaries.

Earthquakes occurring along the different types of plate boundaries across the world are called *inter-plate* earthquakes. They account for the vast majority of global earthquakes (see Chapter 1, Figure 1.5, Epicentral locations around the world). Earthquakes may also

Fundamentals of Seismic Loading on Structures Tapan Sen

occur in the interior of plates and these are called *intra-plate* earthquakes. It should be mentioned that there are many 'inactive faults' scattered around the world where slippages have ceased long ago and future earthquakes are very unlikely to happen. The primary investigations are for 'active faults' where movements are expected to occur.

Faults vary in length, from less than a metre up to few hundred kilometres and they may be up to 200 km deep. Few faults can be seen on the ground surface. During an earthquake the fault ruptures (breaks or bursts suddenly). The mechanism of rupture does not last long – several seconds to a couple of minutes at most. However the seismic body waves generated propagate long after the rupture has stopped. In most cases earthquakes cause damage in the near field (50–60 km) but there are exceptions to this such as the Mexico City earthquake of 1985 where catastrophic damage occurred approximately 400 km from the epicentre. The Mexico City earthquake is unique in many respects and is discussed separately in Chapter 10.

Fault mechanisms and related empirical models have been studied in detail by seismologists. Many believe that the most damaging earthquakes occur on faults and for that reason much of their investigative effort has focused on these. Expressions to correlate fault lengths with the size (magnitude) of earthquakes have been suggested.

Empirical relationships are reported in the literature on correlating magnitude with respect to rupture length, rupture displacement and rupture area. Slemmons (1977) suggested relationships between magnitude and rupture length and again with displacement. Bonilla *et al.* (1984) suggested a similar relationship with respect to rupture length. Wyss's (1979) relationship correlates magnitude and rupture area instead of rupture length. Nuttli (1983) reported on fault rupture characteristics of mid-plate earthquakes.

Faulting Origins

It is now widely believed that virtually all earthquakes are caused by sudden displacements on faults, which are breaks in the rock across where some previous displacement has occurred and are hence relatively weak. The faulting origins of earthquakes were realized as far back as 1891 after the Mino-Awari (Japan) earthquake where evidence pointed towards these origins. However, faulting origins of the earthquake were only accepted gradually, partly because of limited observational data due to the fact that faults only occasionally break the surface.

Surveys following the 1906 San Francisco earthquake, with its surface faulting, confirmed the theory of faulting origins. Observations in these surveys led Reid to formulate the 'elastic rebound theory' (1911, see Chapter 1). When the strain accumulation reaches a threshold level of the rock material, abrupt frictional sliding occurs, releasing the accumulated strain energy. As mentioned briefly in Chapter 1, much of the strain energy is dissipated in the form of heat generated and fracturing of the rocks but a portion is converted into seismic waves that propagate outward. A new cycle of strain accumulation gets underway, leading to many such cycles during the active lifetime of the fault.

In many significant and damaging earthquakes no visible evidence of fault trace reaching the surface has been observed due to the deep-seated nature of their foci. In Chile, for example, no fault ruptures have broken through to the surface in recent geological times, partly because of the depth of the subduction zones (Dowrick, 1987). In areas where earthquake foci are located at shallower depths and surface soils are stiff enough for fracture to proceed right through, fault lines are visible on the surface (e.g. the San Andreas fault, California region). In some cases faults may reach the surface but are difficult to recognize due, for example, to thick vegetation covering the fault. Alternatively a low degree of fault activity may create less evidence and for identification of the fault it may be necessary to wait until the next major movement.

Fault Rupturing Process

Following on from the elastic rebound theory, a broad picture of the rupturing process emerges. Static friction which resists motion is overcome, sliding motions are initiated at a point (hypocentre) and a slip front expands outward over the fault. Figure 5.1 illustrates this.

The rupturing process, as envisaged above, is far from smooth in reality and is a heterogeneous process. The rupture front has to overcome friction and comes across resistance in the form of protrusions on fault surfaces. These protrusions known as *asperities (Latin asperitas, asper (rough))* are welded contacts between two sides of a fault (see Figure 5.2 for a schematic representation). These welds must be overcome to allow sliding. This localized resistance is overcome to allow sliding and once the welds give away, sliding proceeds at a reduced friction level.

5.1.2 After Shocks and Before Shocks

If you have ever wondered about the cause of 'after shocks' which occur after the main earthquake, the patches in Figure 5.2 show the location of 'after shocks', representing the release of stress through some mechanism within the crust.

Similarly, after the rupture process has begun, there may be an initial burst of energy which might precede the main shock by several hours or perhaps even days. This is termed a 'before shock'. The Chilean earthquake (1985) is an example of an earthquake preceded by 'before shocks'.

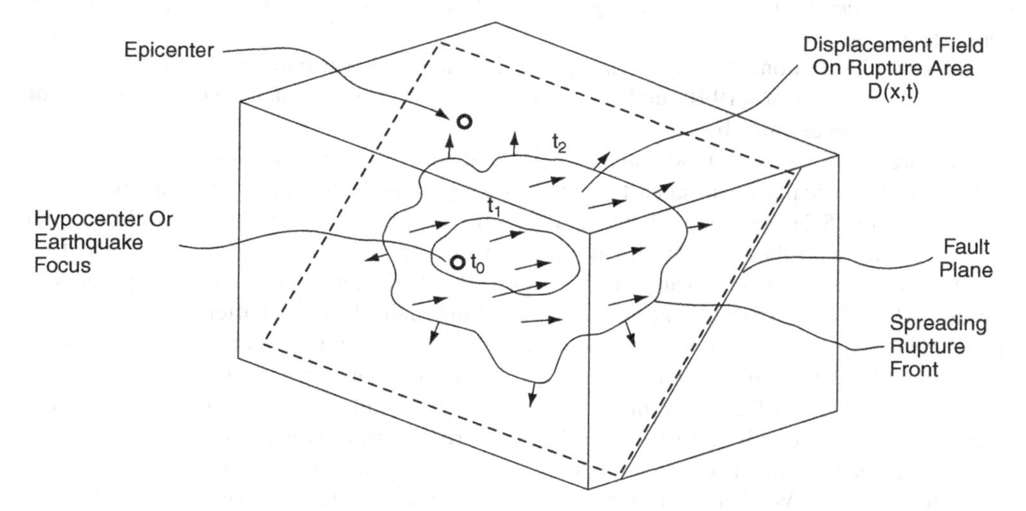

Figure 5.1 A schematic diagram of rupture on a fault spreading from the hypocentre (or earthquake focus) over the fault plane. All regions that are sliding continually radiate *P* and *S*-wave energy. The displacement field $D(x, t)$, varies over the surface of the fault. Note that the direction of the rupture propagation does not generally parallel the slip direction. Reproduced from Lay and Wallace (1995) by permission of Academic Press.

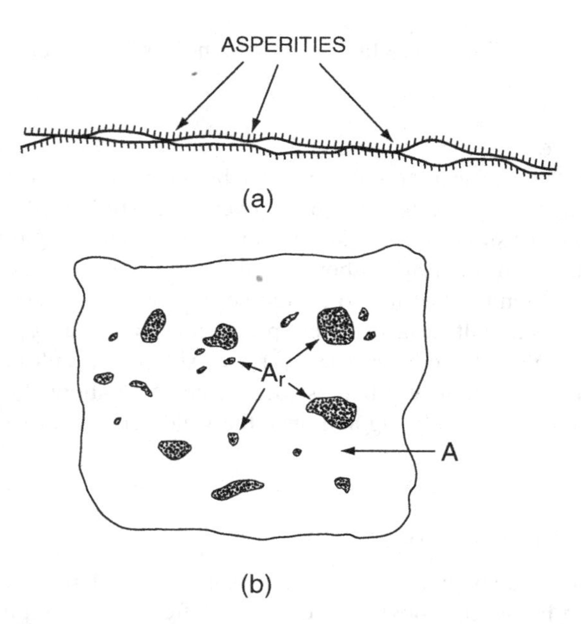

Figure 5.2 A schematic of the Asperity model, in section and plan view of contacting surfaces, the stippled areas (patches) in the plan view represent the areas of asperity contact, which together comprise real contact area A_r. Reproduced from Scholz (1990) by permission of Cambridge University Press.

The stored elastic strain energy in the source region is partly released as heat in overcoming friction and partly as seismic waves which begin the process of fault slippage. The rupturing process continues until the strain energy is insufficient to overcome friction and eventually comes to an end.

The slip distributions recorded during the Loma Prieta earthquake (1989) have been reported by Wald *et al.* (1991) and are shown in Figure 5.3. The heterogeneous nature of the rupturing process may be noted.

In Figure 5.3, a region of moderate slip is located between two asperities.

Asperity has usually been taken to mean a region of high moment release as those protrusions (Figure 5.2) are broken down. This is a region where the stress drop would be high as well. Along with the concept of asperities the concept of *barrier* has been suggested by many researchers and refers to areas of high strength which slow down or stop the process of rupture. Aki (1984) proposed that the existence of the main shock and after shocks indicates that some strong stressed patches behave as barriers whilst other regions behave as asperities.

Another observation put forward is that of a barrier with low strength in which the rupture dies out; the concept of *a 'relaxation barrier'* (Lay and Wallace, 1995) has been proposed. The concepts of strength and relaxation barriers are generally consistent with the asperity model if adjacent segments of the fault are considered (Lay and Wallace, 1995).

In a further study, Wald *et al.* (1996) presented a rupture model of the Northridge earthquake (1994) determined from joint inversion (a mathematical technique, refer to section 1.10) of near source strong ground motion recordings, *P* and *SH* teleseismic body waves, Global Positioning System (GPS) displacement vectors, and permanent uplift measured along levelling lines. Figure 5.4 shows the slipping portion of the fault and the amount of slip during 1-sec

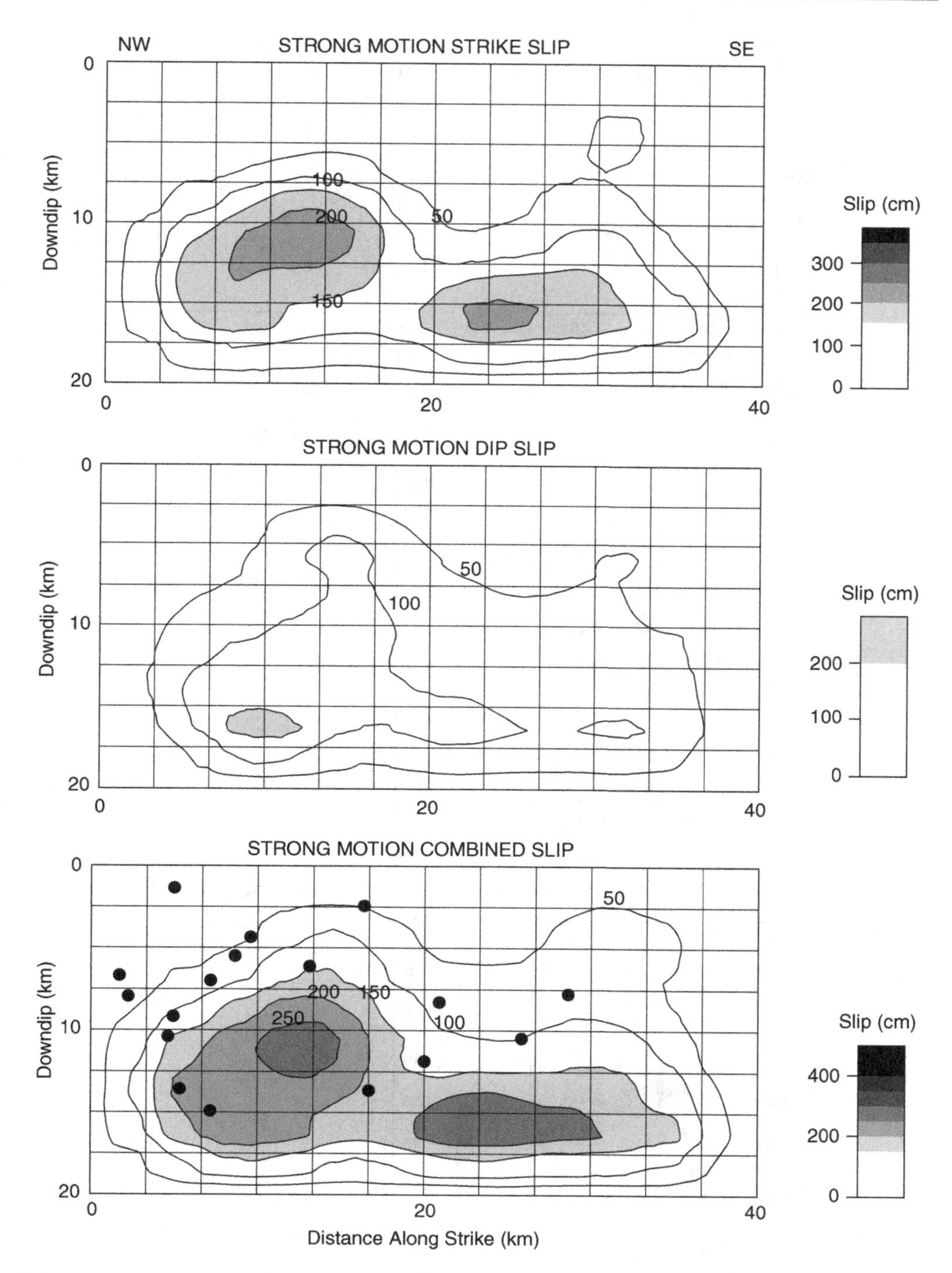

Figure 5.3 Slip distribution on the fault associated with the 1989 Loma Prieta earthquake (North-west end on the fault on the left). There are two prominent regions of slip, known as asperities. Reproduced from Wald *et al.* (1991), (c) the Seismological Society of America.

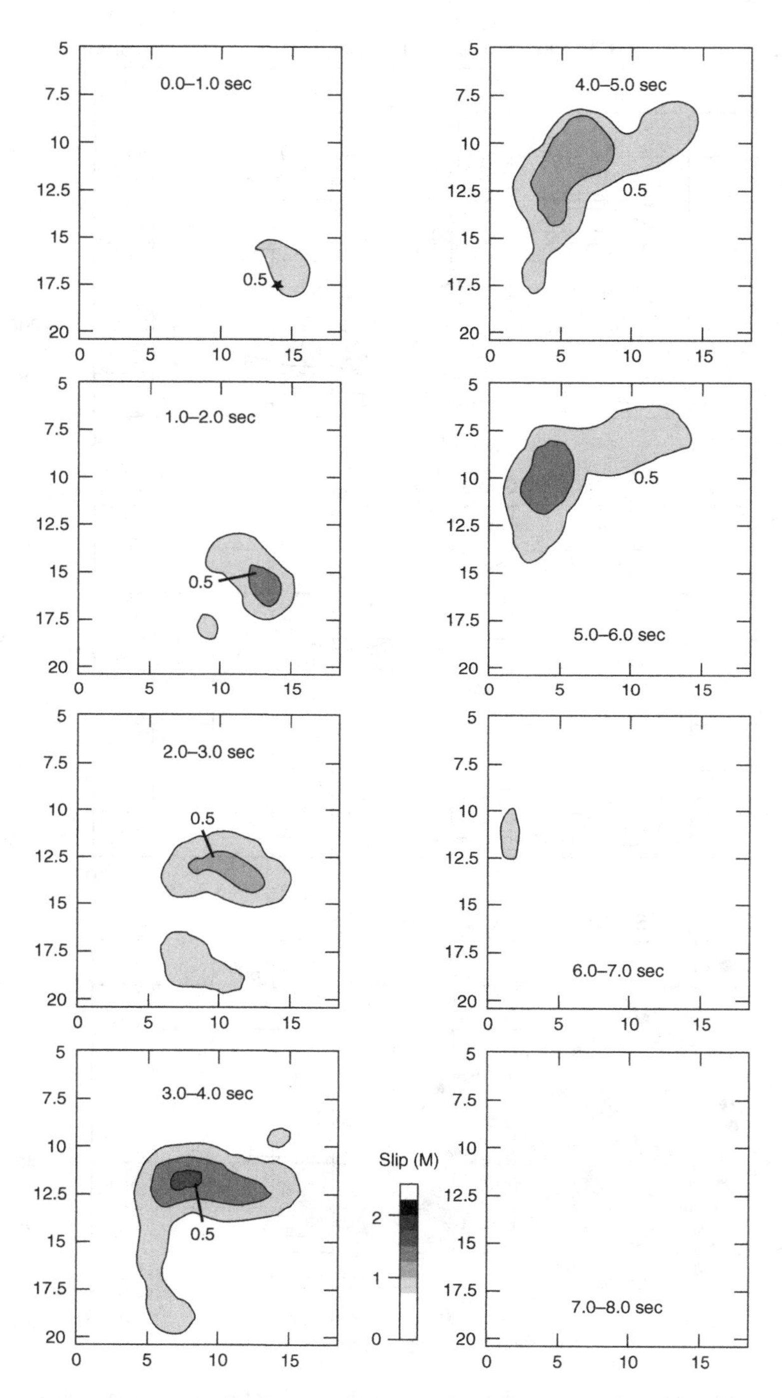

Figure 5.4 Time progression of the combined rupture model at intervals of 1 sec as labelled. The contour interval is 0.5 m. The rupture appears to have begun at the centre. Reproduced from Wald *et al.* (1996), (c) the Seismological Society of America.

'slices' in time. Hence the images also depict slip velocity. The total rupture duration is about 7 seconds. Rupture began at the epicentre and propagated North-westerly. The figure indicates that only a portion of the rupture surface is slipping at one time. Such models hold a great deal of promise in exploring the complex physics of earthquake faulting.

The dynamics of frictional sliding are complex. Laboratory experiments show that sliding displacements under applied shear are not smooth. Earthquake rupture may be viewed as a process consisting of two steps: (1) formation of a crack and (2) growth of a crack. It is known that there will be stress concentration at the tip of the crack. If the stress at the crack tip exceeds a critical value then the crack will grow in an unstable manner and a sudden slip will follow. In general, there is no slip on the fault surface unless the critical value of the shear stress (τ) is reached and then the sudden slip occurs followed by a drop in stress. This causes a period of no-slip when with the accumulation of strain, the stress builds up to the critical value and then the sudden slip occurs again.

Figure 5.5 shows the variation of stress pattern, at a point (say a) along the fault which is in the process of rupturing. Before the rupture front reaches the neighbourhood, the stress at point a is less than the strength of the fault. As the rupture progresses the stress at point a rises till it reaches the critical value and the slip at a begins. As the point slips the stress comes down. When the slippage stops the stress at a adjusts itself to a final value (Figure 5.5).

5.1.3 Earthquake Source Model

The subject of source models is an important area of investigation for seismologists, the results of which are fundamental to our understanding of the nature of ground motion. The subject of source spectra is briefly introduced and for a fuller understanding specialized text books should be consulted. Seismologists have used 'box car' time function representation to model faults, including actual slip at a point on the fault.

Consider a point source on a finite length of fault L. Let the slip on the fault be represented by a ramp function so that the displacement $u(t)$ at time t at a point x on the fault is represented by:

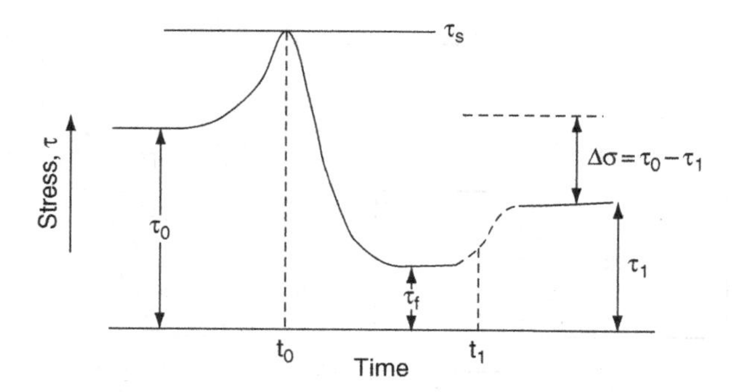

Figure 5.5 Stress at a point (say a) on the fault surface. As the rupture front approaches the point, stress increases to a value of τ_s, after which failure occurs at the point. The point slips by distance d and the stress is reduced to some value τ_f. The difference between the initial stress and final stress, shown as $\Delta\sigma$, is the stress drop. Reproduced from Yamashita (1976), (c) the Seismological Society of Japan and the Volcanological Society of Japan.

$$u(t) = u_f Z(t - \frac{x}{v_r})$$

where Z is a ramp function (see Figure 5.6a) which increases linearly from $t=0$ to unity at $t=\tau_r$, τ_r is the rise time, u_f is the final displacement and v_r is the rupture velocity; the velocity at which the rupture propagates.

The finite fault does not all break at the same time. Instead waves arrive from the initial point of rupture and later from points further along. We will observe the rupture from each point at different times. The far end of the fault ruptures a time L/v_r later. To an observer standing at some distance away, the apparent duration time τ_d will depend upon the orientation of the fault relative to the observer. If the fault is in line with the observer and the rupture propagates towards the observer, τ_d will be shortest. In general τ_d is a function of the orientation of the fault relative to the observer and the direction and velocity of the rupture. It is referred to as the *Haskell* fault model in the literature. The far field displacement pulse $u(t)$ will be a 'box car'.

The rupture velocity v_r has been variously estimated to be 0.7–0.8 times the shear wave velocity of the rupturing material.

The *Haskell* fault model, which is valid for a simple model of a line source, is shown in Figure 5.6b. The shape of the far-field displacement is given by the '*convolution*' of two box car functions, of width τ_r and τ_d where τ_r – is the rise time (duration of the box car pulse) and τ_d – the rupture duration time.

A schematic representation of the radiated pulses in the direction of rupture and in the opposite direction is shown in Figure 5.7b.

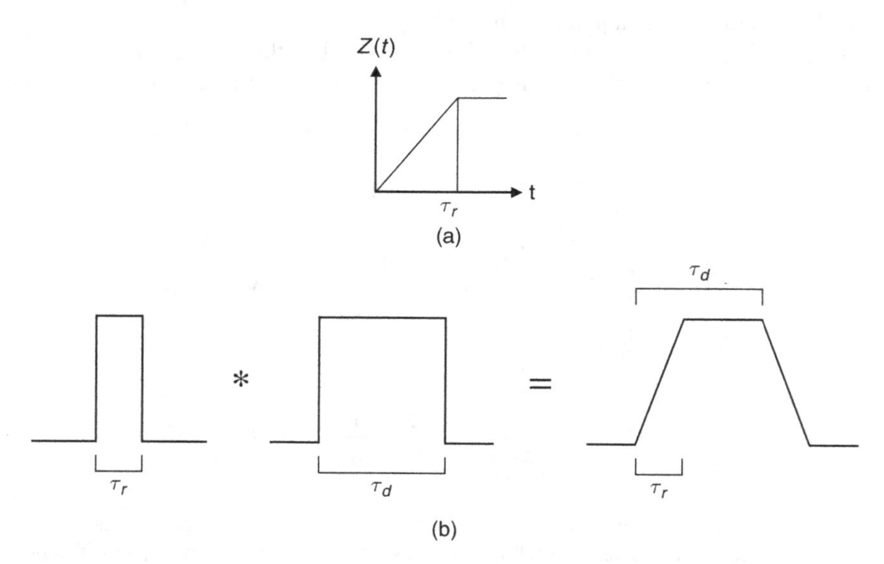

Figure 5.6 The Haskell fault model consists of the convolution of two box car functions with widths given by the rise time and rupture duration time. Reproduced from Shearer (1999) by permission of Cambridge University Press.

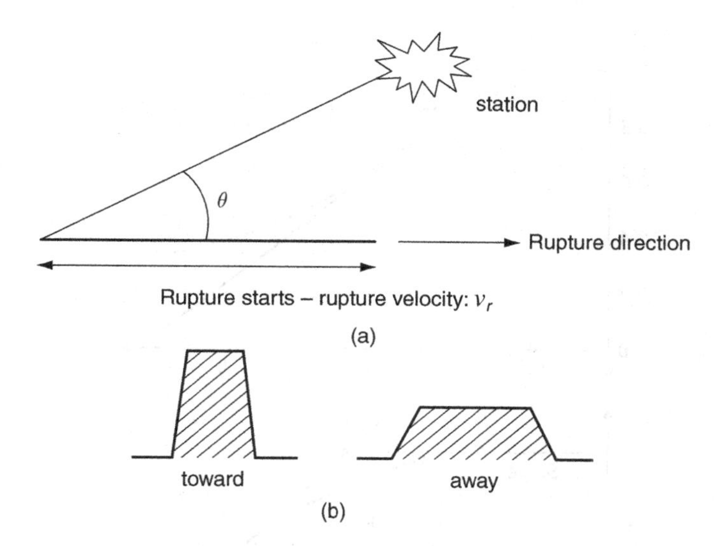

Figure 5.7 Schematic representation of the displacement pulses radiated in the direction of rupture propagation will be higher in amplitude but shorter in duration than pulse radiated in the opposite direction. Reproduced from Shearer (1999) by permission of Cambridge University Press.

Directivity

Referring to Figure 5.7, it can be shown that the area of the trapezoid is proportional to the scalar moment M_0 for the entire event (Shearer, 1999). It follows that the width of the trapezoid will vary with the angle the direction of rupture propagation makes with the azimuth (Figure 5.7a). This is also referred to as the 'directivity' effect.

Note: M_0 is obtained from the area under the far field $u(t)$ (displacement), only after correcting for geometrical spreading, the radiation pattern and the material properties at the source, the properties will be different in the two directions as shown (Shearer, 1999).

Source Spectra

It is convenient to revert back to box car representation of displacement to formulate expressions for source spectra. In the previous paragraph, we used the Haskell fault model in which the time function was the convolution of two box car time functions due to the finite length of the fault and the finite rise time of the faulting at any point.

The transform of a box car of height $1/T$ and length T is:

$$F(\omega) = \int_{-T/2}^{T/2} \frac{1}{T} e^{i\omega t} dt \tag{5.1}$$

$$= \frac{1}{Ti\omega}(e^{i\omega T/2} - e^{-i\omega T/2}) \tag{5.2}$$

$$= \frac{\sin(\omega T/2)}{\omega T/2} \tag{5.3}$$

(Stein and Wysession, 2003)

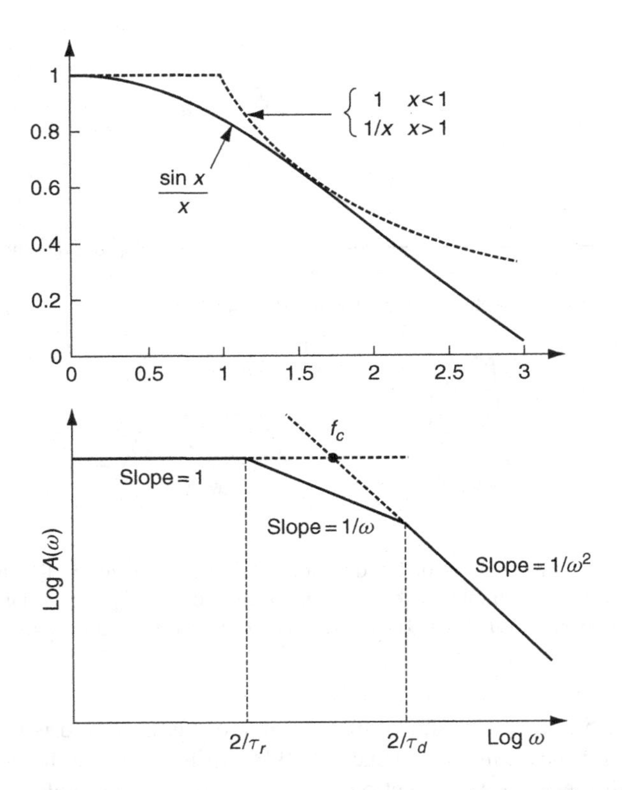

Figure 5.8 Top: the approximation to the $[(\sin x)/x]$ function used in modelling the source spectrum; bottom: theoretical source spectrum of an earthquake, modelled as three regions with slopes of 1, ω^{-1} and ω^{-2}, divided by angular frequencies corresponding to the rupture duration time and rise times, τ_r and τ_d. Another common approximation uses a single corner frequency, f_c, at the intersection of the first and third spectrum segments. The flat segment extending to zero frequency gives M_0. Reproduced from Stein and Wysession (2003) by permission of Blackwell Publishing.

This function, $(\sin x)/x$, is commonly referred to as sin cx and appears in many applications in which only part of a signal is selected. It is interesting to note that a box car function in the time domain produces a sin c function in the frequency domain (Shearer, 1999). The Haskell fault model has been used by seismologists for investigating source spectra, and the frequency domain plot of the amplitude spectrum as presented by Stein and Wysession (2003) is shown in Figure 5.8.

Corner Frequencies

As may be seen, the plot in Figure 5.8 is divided into three regions by the frequencies $2/\tau_r$ and $2/\tau_d$ which are called '*corner frequencies*'; the spectrum is flat (slope = 1) for frequencies less than the first corner; the slope becomes $1/\omega$ between the corners and is followed by a $1/\omega^2$ fall-off at higher frequencies. This is sometimes referred to as the ω^{-2} source model. Thus the spectrum is characterized by three parameters: moment, rise time, and rupture time.

It is worth mentioning that other source spectral models may be used. For example, it is possible to introduce a third corner frequency as a ω^{-3} segment at high frequencies representing the effects of fault width and yielding (Stein and Wysession, 2003).

By studying the spectra of real earthquakes, it is possible in principle to recover M_0, τ_r and τ_d for this model. Sometimes we are able to identify a single corner frequency, f_c (dashed line in Figure 5.8, *bottom*).

On a cautionary note, any interpretation of an observed spectrum directly in terms of source properties, should take note of *attenuation* and *near-source* effects as these can distort the spectrum, particularly at high frequencies.

5.1.4 Empirical Relations of Source Parameters

A different approach, which has proved extremely useful for engineering purposes, is that of empirical relations. These have been developed from regression analyses on a compilation of data of source parameters from historical earthquakes worldwide. Wells and Coppersmith (1994) reported such relationships between moment magnitude, surface rupture length, subsurface rupture length, and down dip rupture width, rupture area and maximum and average displacement per event. Relationships between surface rupture lengths, average displacement and moment magnitude, are shown in Figure 5.9 a, b. As mentioned, these relationships provide useful information to engineers about past earthquakes and guidelines for potential future earthquakes. For example, it may be inferred that an earthquake on a 50 km fault would have an average slip of 1 m and a moment magnitude of approximately 7.

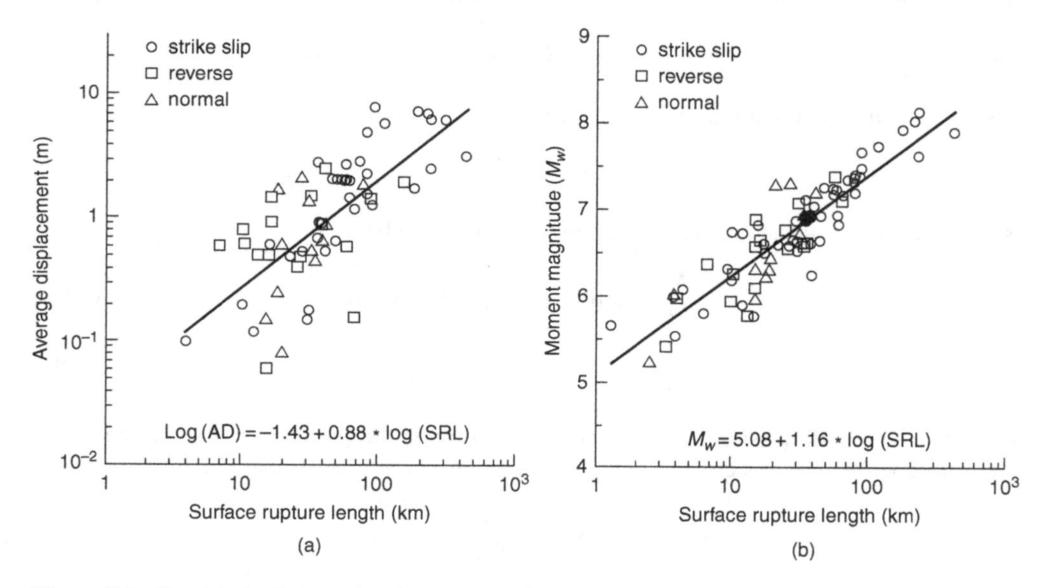

Figure 5.9 Empirical relations showing average slip, fault length, and moment magnitude for a compilation of earthquakes. Reproduced from Wells and Coppersmith (1994), (c) the Seismological Society of America.

Stress Drop

The concept of stress drop is shown schematically in Figure 5.5. Basically, the stress drop that takes place during slip in an earthquake is tied to the relationship that exists between the fault's dimensions, its seismic moment and the amount of energy released. Since energy released is associated with stress release it is referred to as *stress drop*.[1]

In simple terms the stress drop is defined as the difference between the average stress on the fault before an earthquake and that after an earthquake and may be expressed as:

$$\Delta\sigma = \frac{1}{A}\int_S [\sigma(t_2) - \sigma(t_1)]dS \tag{5.4}$$

where the integral is over $S =$ the surface of the fault and
$A =$ the area of the fault.

Consider a fault of length L and width $b << L$ and average displacement $\overline{D}$ in the direction of L. The change in shear strain along the fault may be roughly approximated as

$$\varepsilon_{xx} = \frac{\partial u_x}{\partial x} = \frac{\overline{D}}{L};$$

so the stress drop averaged over the fault is approximately

$$\Delta\sigma = \mu\overline{D}/L; \tag{5.5}$$

where μ is the shear modulus.

To estimate $\Delta\sigma$ from seismic data we need to know the fault dimensions and $\overline{D}$ (average displacement). $\overline{D}$ may be estimated from the seismic moment

$$\overline{D} = \frac{M_0}{\mu A}$$

substituting this into Equation (5.5) we obtain

$$\Delta\sigma = \frac{M_0}{AL};$$

assuming $A = aL^2$ where a is an aspect ratio parameter, we have

$$\Delta\sigma = \frac{M_0}{aL^3}; \text{ more generally we may write...} \tag{5.6a}$$

$$\Delta\sigma = \frac{cM_0}{aL^3}; \tag{5.6b}$$

where 'c' is a non-dimensional constant that depends upon the geometry of the rupture.

For the special case of a circular fault of radius R, we have (e.g. Brune, 1970)

$$\Delta\sigma = \frac{7M_0}{16R^3} \tag{5.6c}$$

[1] In relation to discussions on stress drop, stress refers to shear stress across the fault plane.

These equations may be used to estimate the stress drop from an observed seismic moment and inferred fault dimensions. Inferring source dimensions from corner frequency or source time functions requires that we assume the rupture velocity and fault geometry. Moreover, as we see that $\Delta\sigma$ depends inversely upon the cube of the fault dimensions (Equation (5.6b, c)), so uncertainty in fault dimensions causes a large uncertainty in $\Delta\sigma$.

We can postulate that without independent knowledge of fault dimensions, estimating stress drop has not proved easy. Observed stress drops are generally between 10 and 100 bars[2]. The stress drop is essentially constant over five orders of magnitude in moment, although there appears to be a difference when the tectonic makeup of the region differs. Stress drops for *inter-plate* events average about 30 bars and *intra-plate* events average 100 bars. Differences have also been noted among earthquakes occurring at different types of plate boundaries (Lay and Wallace, 1995).

At present, identifying these fault features with sufficient accuracy is not possible. It follows that present day seismic hazard analyses do not include these features.

Travel Path

Travel path effects are reflected in the attenuation of the propagating waves. Seismologists have developed attenuation models from recorded earthquakes and the models vary from region to region. The amplitude decreases with distance and the level of attenuation depends upon a variety of factors as the waves travel through layers of varying soil strata.

Local Site Conditions

Local site conditions can greatly affect the ground motions. Let us consider as an example the Mexico City (1985) earthquake. The epicentre was approximately 400 km away. Buildings some ten to twenty storeys high in a specific area of the city's downtown district were severely damaged and the death toll was over 10,000. The soil layer in this part of the city consisted of sediments from an ancient lakebed. The natural period of the soil layer was approximately two seconds, which coincided with the predominant period of the ground motion. The natural periods of the buildings were also around two seconds and they were damaged due to the resonant condition. At another location, only a few kilometres away, buildings on a volcanic rock site suffered little or no damage. The natural period of the rock site was away from the predominant period of the ground motion. Events in Mexico City demonstrated that over a distance of only a few kilometres there could be great differences in ground motion as soil conditions changed from alluvium to rock.

Local site conditions are defined in terms of the materials that lie directly beneath the site. The preferred definition is in terms of shear-wave velocity and the depth of sediment beneath the site. These two parameters can be directly related to the dynamic response models of the soil layer. In dynamic models the ground motion acts at the bedrock level and the motion propagates from below. This type of simulation can now routinely be carried out by various finite element (FE) packages such as ABAQUS. An example of such analysis is discussed in Chapter 6.

---

[2] 1 Mpa = 10 bars.

5.2 Ground Motion Parameters

The first recorded ground motion was during the Long Beach earthquake, California, in 1933. Since then earthquake records have steadily increased, with seismograms spread across the globe. A collection of several earthquake records obtained from various parts of the world is shown in Figure 5.10. It is clear from this collection of records that no two earthquakes are the

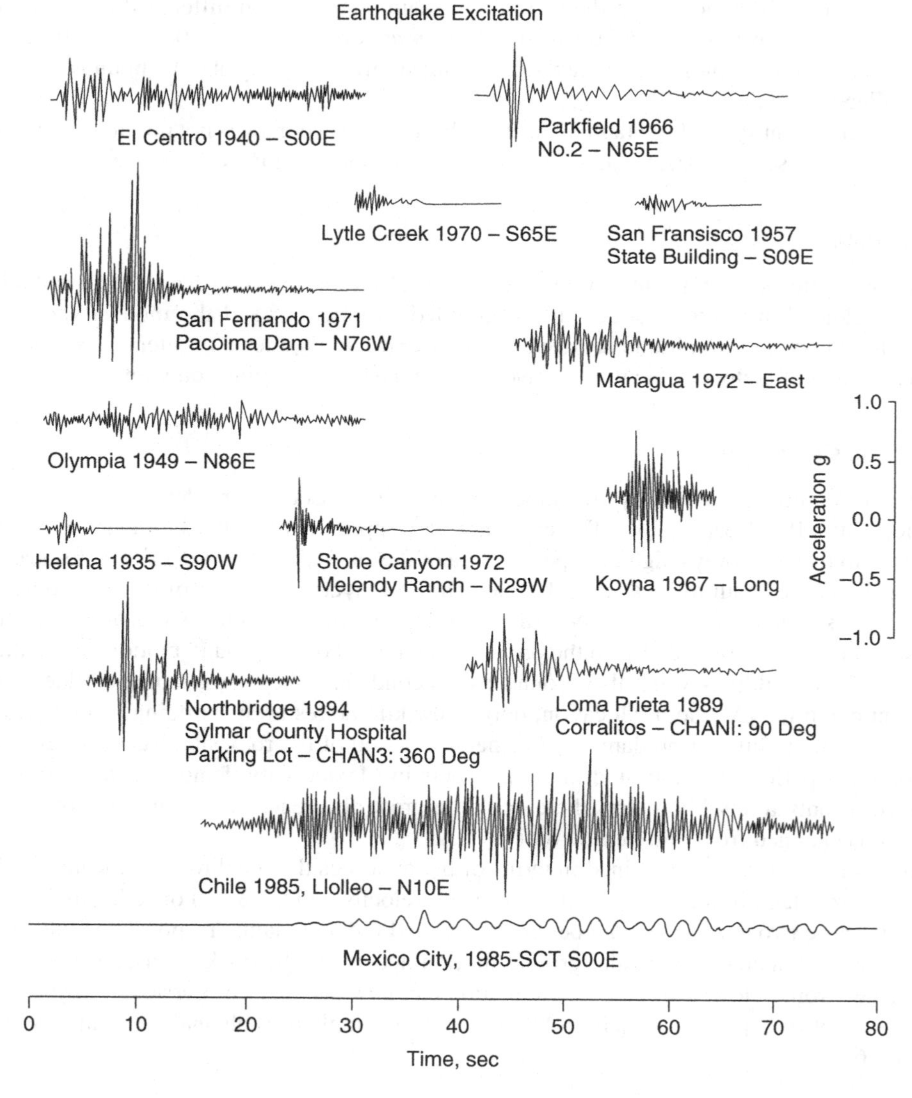

Figure 5.10 General ground motion recorded during several earthquakes. Reproduced from Hudson (1979) by permission of the Earthquake Engineering Research Institute.

same and the random variation of acceleration with time is what makes assessment of seismic loading a challenging field.

The main horizontal component from each record is plotted to the same time and acceleration scale. As may be seen, there is wide variation of (a) amplitude and (b) duration. Also to be noted is that the variation of acceleration with time is highly irregular. The ground motion would also consist of a number of periods (or frequencies).

At first glance, it does seem that the period of the ground motion, for example for the Mexico City (1985) earthquake, is different to that of the other records. What is not clear at this stage is the frequency content of the various records. Another observation to be made is that analytical, closed form solutions to the equations of motion are not possible. The seismograms record the accelerations at closely spaced, discrete intervals of time (refer to the acceleration record of the Erzincan earthquake in Chapter 4).

Specification or identification of ground motion parameters is an important step and of practical use to structural engineers. It is clear from the observations made in the preceding paragraph, that the specification of a single parameter that described the important characteristics of ground motion would not be possible (Jennings, 1985). The important parameters identified are:

1. time history record
2. frequency content
3. peak ground acceleration
4. duration of ground motion.

Time History Record

The data obtained during the El Centro (1940) earthquake, although not the first recorded, has since been used extensively for analysis by scientists and engineers.

The time history of motion recorded during the El Centro (1940) earthquake is shown in Figure 5.11. The first plot is the variation of ground acceleration with time, obtained from seismogram recordings. The peak ground acceleration is 0.32 g. The second plot is the variation with time of ground velocity and the third plot is that of ground displacement. The velocity plot is obtained by integrating the acceleration record with time and the displacement plot by integrating the velocity record obtained with time.

Frequency Content

Apart from time history of acceleration (velocity and displacement time histories through numerical integration) ground motion may be represented in another way – in the frequency domain rather than the time domain. As we have seen in Chapter 4, standard Fourier analysis enables the analyst to transform the time history into amplitude and phase spectra, which directly depicts the frequency dependent characteristics of the seismic record. The frequency content of an earthquake record is an intrinsic property of the record, just as the natural frequency of a structure is an intrinsic property of the structure. From our engineering experience we know certain analyses are carried out more easily in one domain than in the

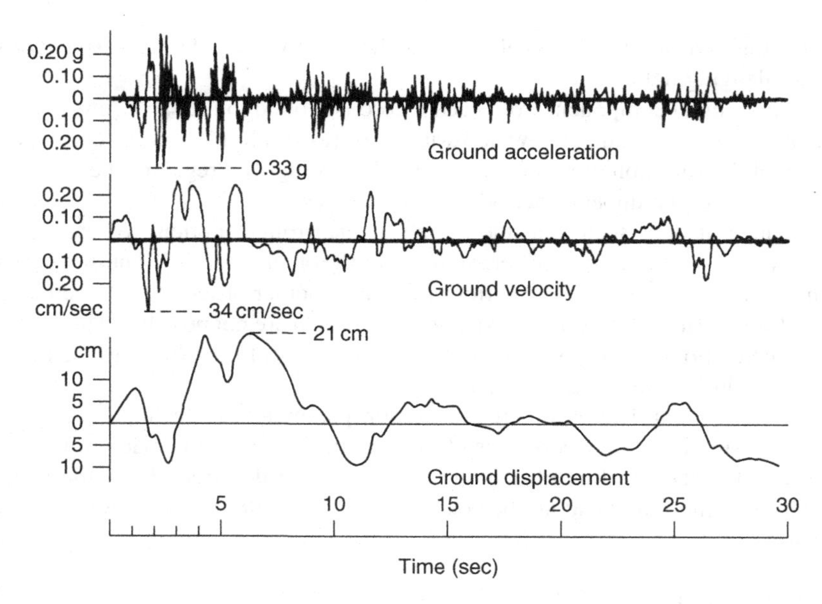

Figure 5.11 North-south component of horizontal ground acceleration recorded at the Imperial Valley Irrigation District substation, El Centro, California, during the Imperial Valley earthquake of 18 May 1940. The ground velocity and ground displacement were computed by integrating the ground acceleration. Reproduced from Dowrick (1987) by permission of John Wiley & Sons, Ltd.

other. Fourier analysis techniques are extensively used by seismologists to study stress drops (Hanks, 1982) and other source effects.

Peak Ground Acceleration (PGA)

Peak ground acceleration is the most commonly used parameter for describing ground motion. Usually the horizontal component is specified and it is the highest value in the record, irrespective of the sign. This has been a preferred parameter as it is easy for engineers to relate this parameter with inertia forces.

The ground motion produced is three dimensional in nature with two horizontal components and a vertical component. In current design practice, it is common to assume that the maximum ground acceleration in the two horizontal directions is equal and the vertical acceleration is two-thirds or more of the horizontal component.

In a separate study, Dietrich (1973) showed, applying finite element techniques to various source models, that the peak ground acceleration is proportional to stress drop.

In the 1970s various investigators (Newmark and Rosenbleuth, 1971; Trifunac and Brady, 1975b) estimated peak velocity and displacement in terms of distance and magnitude. Seed, Murarka, Lysmer, and Idriss (1976) have presented values of peak ground accelerations recorded on rock and stiff soil sites during past earthquakes. These values are useful and have been extensively used by investigators to develop empirical relations for estimating horizontal ground motions (attenuation relationships).

Further amplitude parameters have been suggested by various investigators and are available in the literature, including Kramer (1996).

Peak ground acceleration is also affected by another source effect known as directivity (see Section 5.1).

Duration of Ground Motion

The duration of shaking is a useful parameter for determining the strength of shaking.

The damage caused during the earthquake, as may be expected, is strongly related to the duration of the earthquake. There would be a proportionate increase in the number of cyclic stresses with the increase in duration. Available seismic records confirm this. For example, the 1966 Parkfield and 1972 Stone Canyon, California earthquakes have both produced high peak ground acceleration, in the order of 0.5 g. But since the duration was less (Figure 5.10), the damage caused was not heavy. On the other hand, the El Centro (1940) earthquake produced greater damage, with peak ground acceleration of 0.33 g, as the duration was much longer (Figure 5.10).

There are strong indications that if an earthquake persists for a long time, this will result in liquefaction (Seed and Idriss, 1971). Further detailed studies on earthquakes and their effects (Trifunac and Brady, 1975a; Trifunac and Westermo, 1977) indicate that the three most important interlinking and often-complementary parameters that cause heavy damage are:

1. amplitude,
2. duration,
3. dominant period at which the shaking takes place.

If the amplitudes are small this will not cause high stresses. A short duration earthquake does not cause enough stress reversals for any significant damage. If the dominant period coincides with the natural period of the structure, this gives rise to resonant condition and severe damage consequently. During the Mexico City earthquake such conditions developed; the dominant period of shaking coincided with the natural period of some of the buildings. Interestingly 10–20 storey buildings were severely damaged while other buildings nearby suffered little or no damage.

Seismologists have proposed several quantitative definitions of duration of strong motion. The two most widely accepted definitions are known as the 'significant duration' and the 'bracketed duration'.

Arias (1970) showed that the value of the integral $\int a^2(t)dt$ is a measure of the strength of ground shaking (a(t): variation of acceleration with time 't').

The significant duration is defined as the time required to build up from 5 % to 95 % of the integral of acceleration record $\int a^2(t)\,dt$ (Dobry *et al.*, 1978). The significant duration, it may be surmised, is a function of the strength of ground shaking.

Bracketed duration is defined as the time elapsed between the first and last occurrence of the record above a certain value 0.03 g (Ambraseys and Sarma, 1967) or 0.05 g (Bolt, 1969).

Dobry *et al.* (1978) have provided an empirical correlation of magnitude with 'significant duration' T_s, of strong ground motion:

$$\log(T_s) = 0.423.\mathrm{M} - 1.83$$

The correlation was developed for the Western United States with magnitude ranging from M = 4.5–7.6. The duration on soil sites could be twice that on rock sites.

5.2.1 The Nature and Attenuation of Ground Motion

Needless to say, the mechanics of wave motion through solid media are very complex, especially when layers with different soil properties are present. As stress waves spread out and travel from the focus through the soil media, energy is dissipated in two ways: (1) due to material damping and (2) due to radiation damping. (See Chapter 10, Section 10.2.1 for a brief note on material and radiation damping).

The magnitude of an earthquake is a measure of the energy release. Hence, the ground motion characteristics at a local site (after accounting for the energy dissipation) would be related in some way to the magnitude. Analysis of the simplified models of the Earth's crust has yielded the following results:

(a) body waves decay at the rate of $1/r$
(b) surface waves decay at the rate of $1/\sqrt{r}$
'r' = radial distance from the focus

Surface waves decay at a much slower rate and can travel a considerable distance where tremors may be felt. In the Mexico City (1985) earthquake the surface waves travelled approximately 400 km and damaged many buildings.

5.2.2 PGA and Modified Mercalli Intensity (MMI)

Researchers have investigated the correlation of peak ground acceleration with the magnitude of the earthquake and its variation with distance, reflected in the various attenuation relationships (discussed in the following section). The magnitude of earthquakes may be estimated only for those where instrumental recordings are available. As mentioned earlier, the first seismograms were installed in 1933. Prior to 1933, the size of earthquakes could only be estimated from intensity descriptions. The relationships are not precise, but should be viewed as useful since the only information available from previous historical records is that of estimates of intensity.

Murphy and O'Brien (1977) carried out a detailed study and derived the following relationship from world wide data involving 900 observations:

$$\log \sigma_H = 0.25 I_{mm} + 0.25 \tag{5.7}$$

where σ_H is the peak horizontal ground acceleration in cm/sec^2 and I_{mm} is the Modified Mercalli Intensity.

A variety of statistical analyses were carried out in this study. Dependence of the correlation (Equation (5.7)) on other variables such as local earthquake magnitude M, epicentre distance R and the geographical region in which the earthquakes occurred were investigated. The correlation equation with the best fit for the data is as follows:

$$\log \sigma_H = 0.14 I_{MM} + 0.24M - 0.68 \log R + \beta_k \tag{5.8}$$

where R is in kilometres and

$$\beta_{WesternUnitedStates} = 0.60$$
$$\beta_{Japan} = 0.69$$
$$\beta_{SouthernEurope} = 0.88$$

Southern European data shows much higher values of peak horizontal acceleration than the corresponding values for the Western United States and Japan. The authors reckon that the differences may be due to due to a consistent measurement bias associated with the assignment of intensities in Southern Europe or due to variations in the regional tectonic makeup. Further investigation is required.

A number of other relationships that have been proposed are shown in Figure 5.12.

A comparison of Murphy and O'Brien's relationship with that of Trifunac and Brady (1975b) is shown in Figure 5.13.

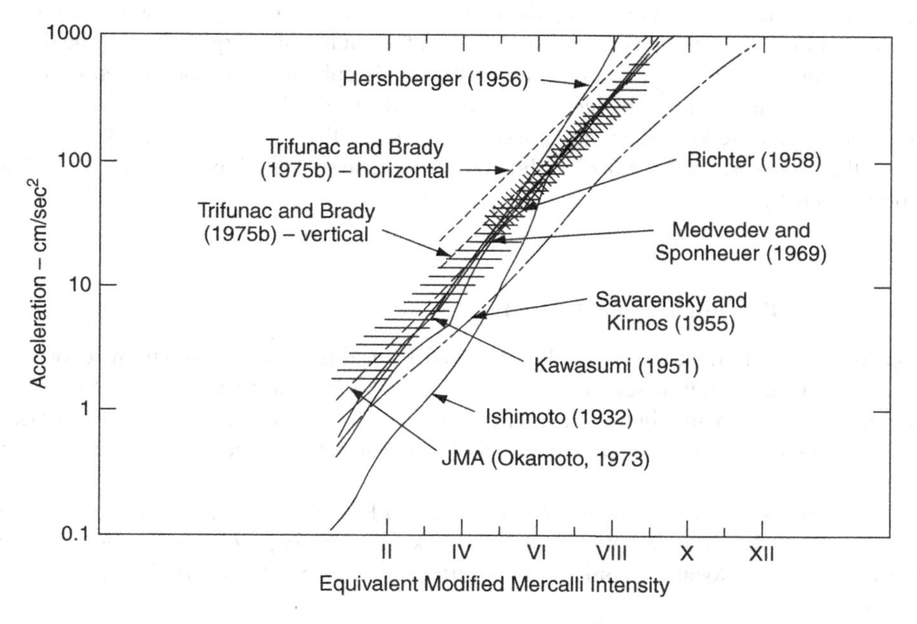

Figure 5.12 Proposed relationships between PHA and MMI. Reproduced from Trifunac and Brady (1975b), (c) the Seismological Society of America.

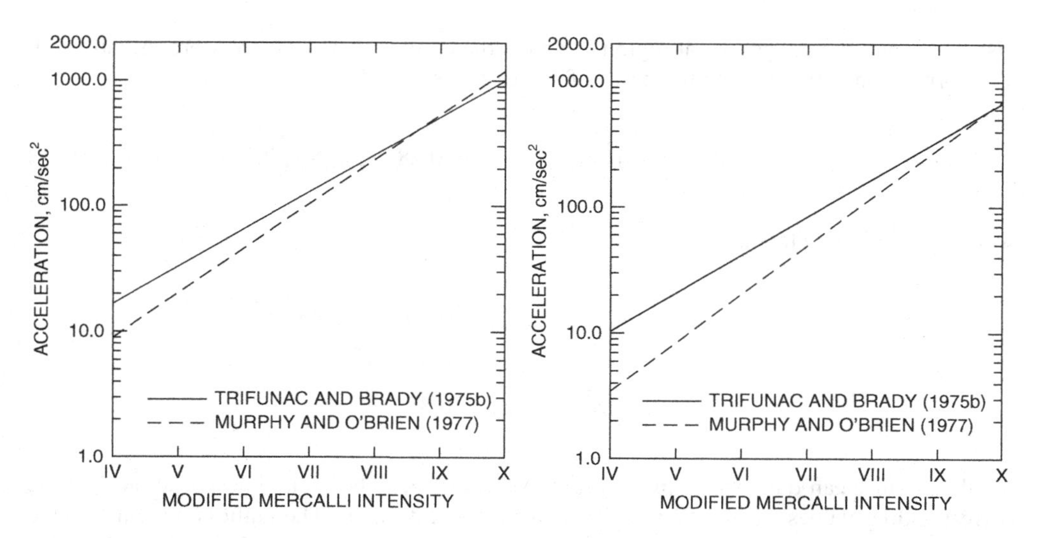

Figure 5.13 Comparison of intensity/peak acceleration relationship for the horizontal (left) and vertical (right) components of ground motion obtained with the results obtained by Trifunac and Brady (1975b). Reproduced from Murphy and O'Brien (1977), (c) the Seismological Society of America.

5.2.3 Engineering Models of Attenuation Relationships

Prediction of ground motions from earthquakes is a key element in any seismic hazard analysis. Ground motion attenuates with distance from the source of energy release. The distance referred to could be from the fault or the earthquake's epicentre. As mentioned earlier, the first strong motion seismogram record was obtained during the Long Beach (1933) earthquake. Seismogram records have increased steadily since then, not only in USA, but in other parts of the world as well. Major earthquake zones across the globe are well linked with seismograms today.

Loma Prieta (1989) Earthquake Record

The ground motion attenuation may be observed from this set of seismogram records. The peak values of acceleration recorded at different locations during the Loma Prieta (1989) earthquake are shown in Figure 5.14. The epicentre is shown with a concentric dark circle, and accelerations (PGA) shown are highest nearer the epicentre and decrease with distance.

The key point is that these values are for the Loma Prieta earthquake with its regional soil conditions. With the same magnitude of earthquake, the ground motions obtained at similar distances at another location could well be different due to a totally different set of local conditions.

Theoretical ground motion modelling and numerical simulation based on detailed understanding of the earthquake energy release mechanism and wave propagation have not reached a sufficient degree of maturity for use in engineering design offices.

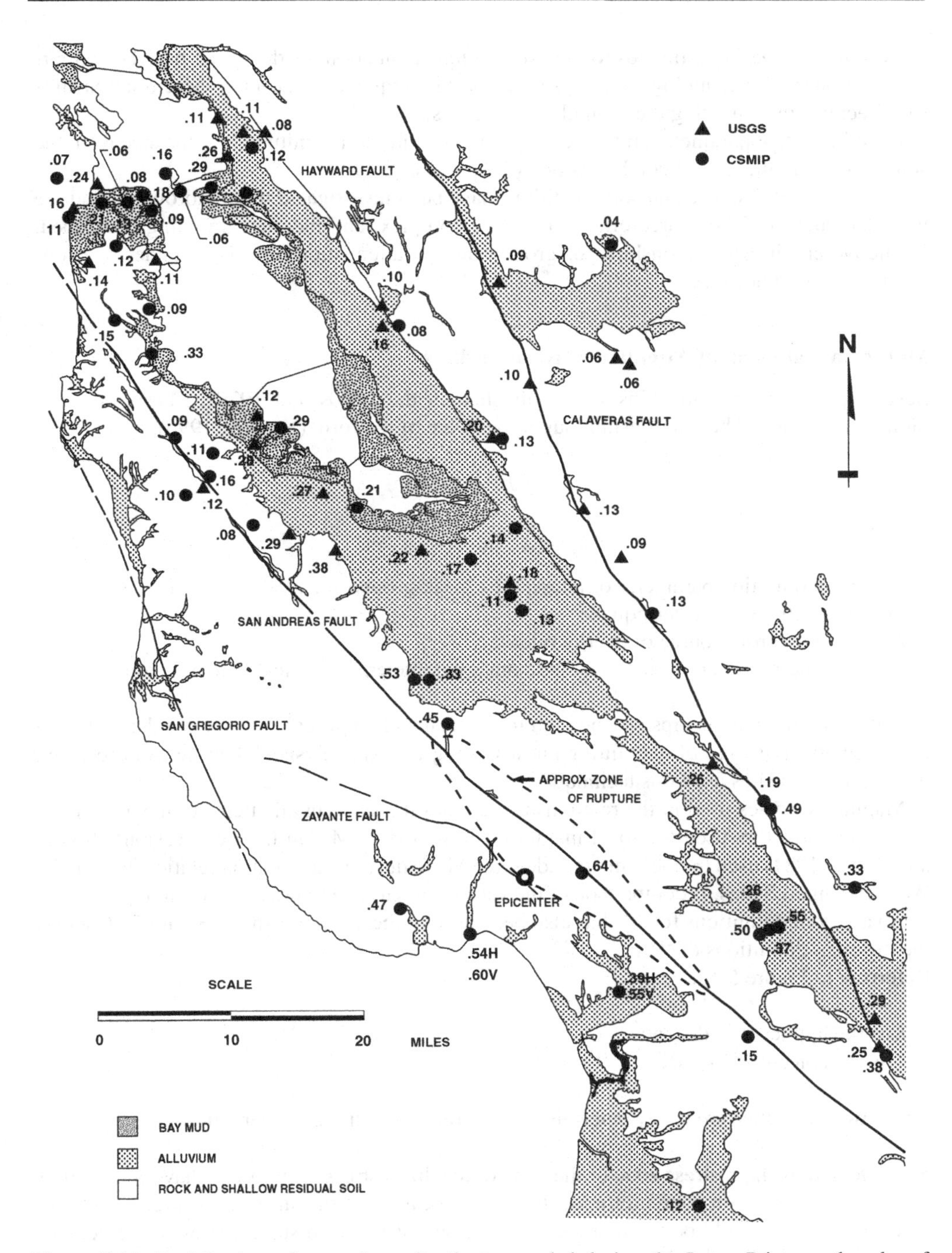

Figure 5.14 Peak horizontal ground accelerations recorded during the Loma Prieta earthquake of 17 October 1989. Reproduced from R.B. Seed (1990) by permission of the Earthquake Engineering Research Center.

Current engineering estimates for ground motion attenuation are therefore, based on empirical relationships. Seismologists have studied past earthquakes to develop these relationships for direct use in estimating the ground motion at a site.

Most important parameters influencing attenuation are (a) magnitude, (b) distance from the source or the fault, (c) soil conditions and (d) source effects.

The effects of distance and soil conditions have been investigated more thoroughly and are well documented. The source effects are more complex, and theories are being developed. Predictive empirical relationships for ground motion attenuation have been under development for several decades.

Mathematical Form of Attenuation Relationships

Several predictive relationships were published in the 1960s and 1970s. These were the pioneering efforts. The relationships suggested were of the form (Kramer, 1996):

$$p = f(M, R, C_1)$$

where

p = ground motion parameter of interest (ex. peak ground acc., vel. or disp.)
M = magnitude of the earthquake
R = distance from source of energy release
C_1 = a general term or terms to include other parameters if applicable to a region.

The attenuation relationships rely on both theoretical and empirical data and are derived from a mathematical process of data fitting known as regression analysis. All subsequent research work in this area has used this technique.

Magnitude scales used in the relationships are not the same in all attenuation expressions. Earlier investigators used the local magnitude scale (M or M_L) in their expressions. Joyner and Boore (1981) used moment magnitude scale M_w to develop attenuation relationships in the Western United States. A comparison of the various magnitude scales is shown in Figure 5.15.

A variety of definitions for 'R' has also been used in these relationships. Figure 5.16 shows the different definitions of 'R'.

Referring to Figure 5.15,

j_1 is the hypocentral distance,
j_2 is the epicentre distance.

These two quantities are easily determined from recording station observations.

- j_3 refers to the high stress zone or the zone of localized stress drop. As we have seen earlier in the discussion on source effects, this is the location of the strongest source of ground motion, and j_3 would be the most correct measure of source distance. However, extensive analyses are required to locate these zones. The other important point is that from the view of predictive relationships, it is impossible to predict where these zones will be in the fault in the case of a future earthquake.

- j_4 is the closest distance to the zone of rupture. It was introduced by Schnabel and Seed (1973).
- j_5 is the distance to the surface projection of the rupture. Joyner and Boore (1981) included this term in their attenuation expression.
- j_4 and j_5 have both figured extensively in predictive relationships (Kramer, 1996).

Early Attenuation Relationships

An early attempt at an attenuation relationship, relating peak acceleration amplitude, magnitude and distance to the epicentre was by Milne and Davenport (1969).

$$A = \frac{0.69e^{1.64M}}{1.1e^{1.10M} + \Delta^2} \tag{5.9}$$

where A is the peak acceleration amplitude in percentage of gravity and Δ, the distance to the epicentre, in kilometres and M is the magnitude. It is shown that the empirical equation fits Cloud's (1963) data adequately.

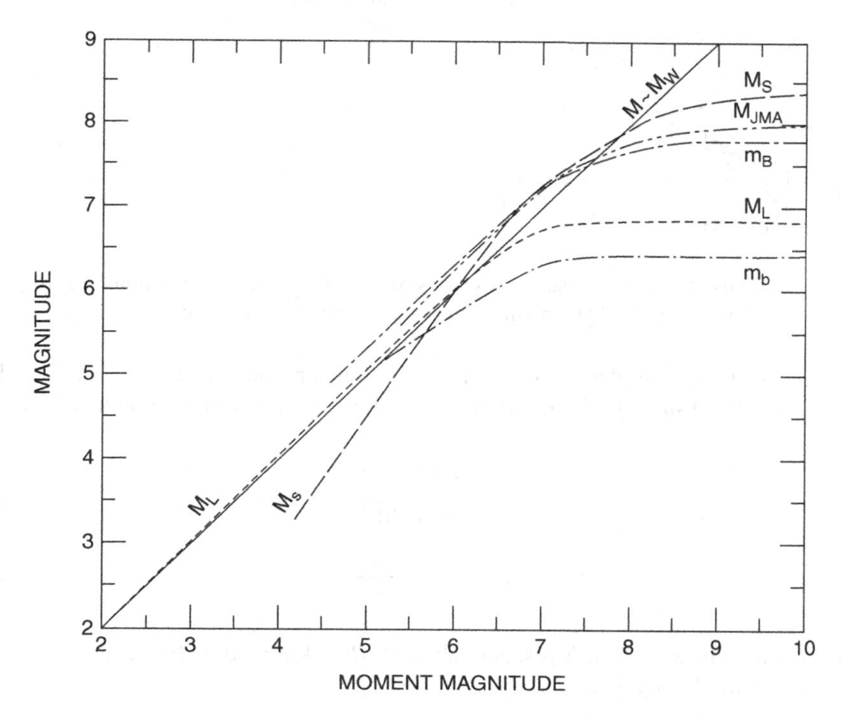

Figure 5.15 Comparison of magnitude scales: M_w (moment magnitude), M_L (Richter local magnitude), M_S (surface wave magnitude, m_b (short-period body wave magnitude), m_B (long-period body wave magnitude, M_{JMA} (Japanese Meteorological Agency magnitude). Reproduced from Heaton *et al.* (1986) by permission of Springer.

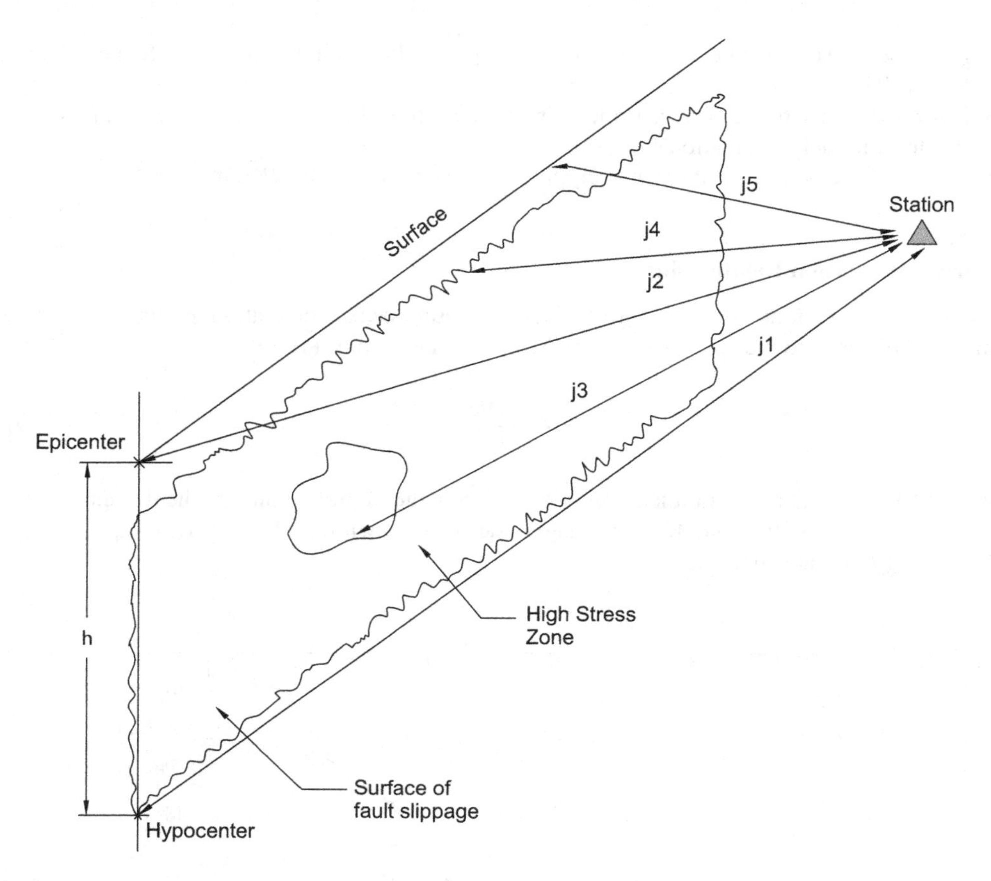

Figure 5.16 Various measures of distance used in strong motion predictive relationships. Reproduced from Shakal and Bernreuter (1981) by permission of the Lawrence Livermore National Laboratory.

Following are two examples of attenuation relationships published in the 1970s. Esteva and Villaverde (1973) suggested the following relationships for peak ground acceleration and velocity:

$$a = \frac{5600e^{0.8M}}{(R+40)^2} \tag{5.10a}$$

$$v = \frac{32e^{M}}{(R+25)^{1.7}} \tag{5.10b}$$

where a and v are cm/s^2 and cm/s respectively and 'R' is kilometres. In the above expression 'R' is the hypocentral distance.

An Estimate for Displacement

Newmark and Rosenbleuth (1971) provided the following empirical relationship for ground displacement:

$$5 \leq \frac{ad}{v^2} \leq 15 \tag{5.11}$$

McGuire (1974) suggested the following the relationship:

$$y = b_1' 10^{b_2 M}(R + 25)^{-b_3} \tag{5.12a}$$

or

$$\log y = b_1 + b_2 M - b_3 \log(R + 25) \tag{5.12b}$$

The coefficients are shown in Table 5.1

With each earthquake event additional information becomes available and seismologists have been able to improve the predictive relationships to reflect the ground motion process as closely as possible.

More Recent Attenuation Relationships

Campbell (2003a) describes the following form for attenuation equations:

$$Y = c_1 e^{c_2 M} R^{-c_3} e^{-c_4 R} e^{c_5 F} e^{c_6 S} \varepsilon \tag{5.13}$$

The logarithmic form would be:

$$\ln Y = c_1 + c_2 M - c_3 \ln R - c_4 R + c_5 F + c_6 S + \varepsilon \tag{5.14}$$

where the distance term R is given by one of the alternative expressions:

$$R = r + c_7 \exp(c_8 M) \tag{5.15a}$$

$$\text{or} \quad R = \sqrt{r^2 + [c_7 + \exp(c_8 M)]^2} \tag{5.15b}$$

In the above expressions, Y is the strong motion parameter of interest, M is the magnitude, F is the faulting mechanism of the earthquake, S is a description of the local site conditions beneath the site, ε is a random error term with a mean of zero and standard deviation of $\sigma_{\ln Y}$ (the standard error of estimate of $\ln Y$), and r is a measure of the shortest distance from the site to the source of the earthquake. In the more complicated form of Equations 5.13 to 5.15b,

Table 5.1 Coefficients for Mcguire's Attenuation Relationship (1974).

	b_1'	b_1	b_2	b_3	Coeff. of var. of y
$a(cm/\text{sec}^2)$	472.3	2.649	0.278	1.301	0.548
$v(cm/\text{sec})$	5.640	0.714	0.401	1.202	0.696
$d(cm)$	0.393	−0.460	0.434	0.885	0.883

the coefficients c_3, c_6 and c_7 are defined in terms of M and R. Many of these coefficients have also been found to be dependent on the tectonic environment of the regions in which the earthquakes occurred.

It is worth noting discussions on the origins of attenuation relationships, as it would greatly aid our understanding of ground motion modelling. The roots of the relationships can be traced back to theoretical findings (Lay and Wallace, 1995; Campbell, 2003a). An erudite insight into the relationships has been provided by Campbell (2003a) as follows:

1. The expression $Y \propto c_2 M$ and $\ln Y \propto c_2 M$ are consistent with Richter's definition of earthquake magnitude (Richter, 1935).
2. The expressions $Y \propto R^{-c_3}$ and $\ln Y \propto -c_3 \ln R$ are consistent with geometric attenuation of the seismic wave front as it propagates away from the source of energy release. Incidentally, $c_3 = 1$, is the theoretical value for spherical spreading of the wave front from a point source in a homogenous whole space. The value of c_3 is varied to correlate with the regional differences in geometric attenuation.
3. The expressions $Y \propto e^{-c_4 R}$ and $\ln Y \propto -c_4 R$ are consistent with the attenuation that would result from material damping and scattering as the waves propagate through the crust.
4. The relationship between Y and the remaining parameters has been established over the years from empirical and theoretical ground motion modelling.

Fault Mechanism

Only some empirical relationships include faulting. Faulting mechanisms in the relationships at present include strike slip, reverse slip and normal slip. It is believed that reverse and thrust faulting cause higher ground motion than strike slip or normal faulting. It has been empirically demonstrated by Campbell (1981) and many subsequent studies have shown this to be universal.

Present codes of practice define site classes. The International Building Code (IBC) 2003 provides site class definitions in Table 1615.1.1 based on shear wave velocity. Where the soil shear wave velocity, v_s, is not known, site class shall be determined, as permitted in Table 1615.1.1 (IBC 2003), from standard penetration resistance, $\overline{N}$, or from soil undrained shear strength, s_u, calculated in accordance with section 1615.1.5 (IBC 2003).

Attenuation Relationships Commonly Used

Earthquakes

A large number of attenuation relationships have been reported in the literature during the last four decades. A selection of such relationships in common use today will be discussed. The important thing to note is the context in which the attenuation relationships are used in the engineering evaluation and assessment of seismic loading.

All relationships have limitations which, as might be expected, result foremost from the paucity of data. Theoretical assumptions and seismological parameters chosen by the authors to define the source path and site effects also impose limitations on the model.
Table 5.2 summarizes discussions of relationships reported in literature.

Table 5.2 Summary of discussions of relationships reported in literature.

Region	Relationship
Active Crustal Regions	
Western North America	Boore *et al.* (1997)
	Campbell and Bozorgnia (2003)
Eastern North America	Atkinson, G.M. and Boore, D.M. (1995)
	Campbell (2003b)
Europe	Ambraseys *et al.* (1996a) (Horiz. Acc.)
	Ambraseys *et al.* (1996b) (Vert. Acc.)
	Ambraseys *et al.* (2005a) (Horiz. Acc.)
	Ambraseys *et al.* (2005b) (Vert. Acc.)
Japan	Molas and Yamazaki (1996)
Japan	Tamura, K., Matsumoto, S. and Nakao, Y. (2003)
Intra Plate	Dahle *et al.* (1990)
	Dahle *et al.* (1991)
(North America, Europe, China, Australia) Worldwide Active Crustal Regions	Campbell and Bozorgnia (1994)

Horizontal Acceleration Spectra

The main thrust of investigation by seismologists has been towards horizontal response spectra as this causes the most damage.

Western North America (WNA)

The two relationships presented are commonly used to develop engineering estimates of strong ground motion in WNA.

The relationship presented by Boore, Joyner and Fumal (1997) has the following functional form:

$$\ln Y = b_1 + b_2(M - 6) + b_3(M - 6)^2 + b_5 \ln r + b_v \ln {}^{V_s}\!/_{V_A} \tag{5.16}$$

where

$$r = \sqrt{r_{jb} + h^2} \tag{5.17}$$

and

$b_1 = b_{1ss}$ ...... for strike slip faults
$b_1 = b_{1RS}$ ...... for reverse slip earthquakes
$b_1 = b_{1ALL}$ ...... if mechanism not specified
r_{jb} is the closest horizontal distance from the station to a point on the earth's surface that lies directly above the rupture.

In the above equation Y is the ground motion parameter (peak horizontal acceleration or pseudo acceleration response in 'g'). The variables are moment magnitude (M), distance (r_{jb} in km) and average shear-wave velocity to 30 m (V_s in m/sec). Coefficients to be determined are $b_{1ss}, b_{1RS}, b_{1ALL}, b_2, b_3, b_5, h, b_v$, and V_A.

Note: The authors have used h (Equation 5.17) as one of the coefficients to be determined from the regression analyses.

The mean plus one sigma value of the of the natural logarithmic of the ground-motion value from Equation 5.16 is $\ln Y + \sigma_{\ln Y}$ where $\sigma_{\ln Y}$ is the square root of the overall variance of the regression.

The coefficients for the peak horizontal ground acceleration derived by the authors are shown in Table 5.3.

Note: The equations are to be used for $M5.5 - 7.5$ and d no greater than 80 km

The authors have also provided in tabular form (Boore *et al.*, 1997) smoothed coefficients for use in Equation 5.16 to estimate pseudo-acceleration response spectra (g) for the random horizontal component at 5 % damping.

The shear wave velocity term accounts for site response through the now accepted 30-m velocity V_{s30}. According to the authors the use of average shear wave velocity to a depth of 30 m as a variable to characterize site conditions is a choice dictated by the scarcity of velocity data for greater depths. The authors' preferred parameter in this case would be the average shear wave velocity to a depth of one-quarter wavelength for the period of interest (Joyner and Fumal, 1984). By the quarter wavelength rule 30 m is the approximate depth for a period of 0.19 sec for the typical rock site (average velocity 620 m/sec) and for a period of 0.39 sec for the typical soil site (average velocity 310 m/sec). Some typical values for V_{s30} are provided in the International Building Code (IBC), 2003.

Campbell and Bozorgnia (2003)

The authors have presented attenuation relationships in earlier publications (Campbell and Bozorgnia, 1994; Campbell, 1997). This study is an update on the ground motion relations for peak ground acceleration (PGA) at 5 % damped pseudo-acceleration spectra. The database used consisted of strong motion data recorded between 1957 and 1995 and includes 960 uncorrected accelerograms from 49 earthquakes and 443 processed records from 36 earthquakes. All the events were from tectonically active, shallow crustal regions located throughout the world.

The relationship is of the form:

$$\begin{aligned}\ln Y = c_1 + f_1(M_w) + c_4 \ln \sqrt{f_2(M_w, r_{seis}, S)} + f_3(F) + f_4(S) \\ + f_5(HW, F, M_w, r_{eis)} + \varepsilon\end{aligned} \tag{5.18a}$$

Table 5.3 Coefficients for Peak Horizontal Acceleration at 5 % damping.

b_{1ss}	b_{1RV}	b_{1ALL}	b_2	b_3	b_5	b_V	V_A	h	$\sigma_{\ln Y}$
−0.313	−0.117	−0.242	0.527	0.000	−0.77	−0.371	1396	5.57	0.520

Source: Reproduced from Boore *et al.* (1997), (c) the Seismological Society of America (SSA).

where the magnitude scaling effect is represented by

$$f_1(M_w) = c_2 M_w + c_3(8.5 - M_w)^2 \qquad (5.18b)$$

the distance scaling effect is represented by

$$f_2(M_w, r_{seis}, S = r_{seis}^2 + g(S)(\exp[c_8 M_w + c_9(8.5 - M_w)^2])^2 \ldots \qquad (5.18c)$$

the near-source effect of local site conditions is represented by

$$g(S) = c_5 + c_6(S_{VFS} + S_{SR}) + c_7 S_{FR} \ldots\ldots \qquad (5.18d)$$

the effect of faulting mechanism is given by

$$f_3(F) = c_{10} F_{RV} + c_{11} F_{TH} \qquad (5.18e)$$

the far-source effect is represented by

$$f_4(S) = c_{12} S_{VFS} + c_{13} S_{SR} + c_{14} S_{FR} \qquad (5.18f)$$

and the effect of hanging wall 'HW' (the block overhanging the fault plane) is represented by

$$f_5(HW, F, M_w, r_{seis}) = HW f_3(F) f_{HW}(M_w) f_{HW}(r_{seis}) \ldots \qquad (5.18g)$$

where

$$\begin{aligned} HW &= S_{VFS} + S_{SR} + S_{FR} \to 0 \quad r_{jb} \geq 5 \text{ km} \qquad (5.18h) \\ &= (5 - r_{jb})/5 \quad r_{jb} \leq 5\text{km otherwise } \delta > 70^0 \\ f_{HW}(M_w) &= 0 \qquad \text{for} \quad M_w < 5.5 \qquad (5.18i) \\ &= M_w - 5.5 \quad \text{for} \quad 5.5 \leq M_w \leq 6.5 \\ &= 1 \qquad \text{for} \quad M_w > 6.5 \end{aligned}$$

and

$$\begin{aligned} f_{HW}(r_{seis}) &= c_{15}(r_{seis}/8) \text{ for} \quad r_{seis} < 8 \text{ km} \qquad (5.18j) \\ &= c_{15} \qquad \text{for} \quad r_{seis} \geq 8 \text{ km} \end{aligned}$$

In these Equations (5.18 a–j), Y is either the vertical component Y_V, or the average horizontal component, Y_H, of PGA or 5 % damped PSA in $g(g = 980cm/\sec^2)$; M_W is moment magnitude; r_{seis} is the closest distance to seismogenic rupture in kilometres; r_{jb} is the closest distance to the surface projection of fault rupture in kilometres; δ is fault dip in degrees; $S_{VFS} = 1$ for very firm soil; $S_{SR} = 1$ for soft rock, $S_{FR} = 1$ for firm rock, and $S_{VFS} = S_{SR} = S_{FR} = 0$ for firm soil; $F_{RV} = 1$ for reverse faulting, $F_{TH} = 1$ for thrust faulting, and $F_{RV} = F_{TH} = 0$ for strike-slip and normal faulting; and ε is random error term with zero mean and standard deviation equal to $\sigma_{\ln Y}$.

The standard deviation, $\sigma_{\ln Y}$, is defined either as a function of magnitude,

$$\begin{aligned} \sigma_{\ln Y} &= c_{16} - 0.07M_w \quad \text{for} \quad M_w < 7.4 \\ &= c_{16} - 0.518 \quad \text{for} \quad M_w \geq 7.4 \end{aligned} \tag{5.18k}$$

or as a function of PGA,

$$\begin{aligned} \sigma_{\ln Y} &= c_{17} + 0.351 && \text{for} \quad \text{PGA} \leq 0.07g \\ &= c_{17} - 0.132\ln(PGA) && \text{for} \quad 0.07g < PGA < 0.25g \\ &= c_{17} + 0.183 && \text{for} \quad PGA \geq 0.25g \end{aligned} \tag{5.18l}$$

where PGA is either uncorrected PGA or corrected PGA. The authors have provided coefficients for predicting uncorrected and corrected PGA and/or PSA (pseudo-spectral acceleration). The hanging wall effect is defined as a 5 km margin around the surface projection of the rupture surface, which has been defined as r_{jb} by Boore *et al.* (1997).

Europe

Ambraseys *et al.*'s (1996a and b) relationships cover the shallow active crust of Europe. The dataset used is shown to be representative of European strong motion in terms of attenuation of peak ground acceleration. The equations are recommended for use in the range of magnitudes from M_s 4.0 to 7.5 and for source distances of up to 200 km.

The recommended relationship is of the form:

$$\log(y) = C_1' + C_2M_s + C_4\log(r) + C_AS_A + C_SS_S + \sigma P \tag{5.19}$$

where y is the parameter being predicted, in this case peak horizontal ground acceleration in g. The other terms are as follows:

M_s is the surface wave magnitude
$r = \sqrt{d^2 + h_0^2}$
d is the shortest distance from the station to the surface projection of the fault rupture
σ is the standard deviation of $\log y$
P takes a value of 0 for mean and 1 for 84-percentile values of log (y)

h_0, C_1', C_2, C_4, C_A, C_S are constants to be determined. S_A takes the value of 1 if the soil is classified as stiff (A) and 0 otherwise. S_S is defined similarly for soft (S) soil sites. S_A and S_S are set to 0 for rock sites. The value of h_0 reported is 3.5. Regression analysis on 416 records yielded the following equation:

$$\log(a) = -1.48 + 0.266M_s - 0.922\log(r) + 0.117S_A + 0.124S_S + 0.25P \tag{5.20a}$$

In their previous study (Ambraseys and Bommer, 1991) demonstrated that the site independent predictive equations for the European area are very similar to those derived for Western

North America and a similar comparison with the site-dependent equation produced by Boore *et al.* (1993) for the larger component of peak horizontal acceleration (PHA):

$$\log(a) = -1.334 + 0.216M_s - 0.777\log(r) + 0.158G_B + 0.254G_C + 0.21P \qquad (5.20b)$$

with $h_0 = 5.48$ and where G_B and G_C are variables which are equivalent to the variables S_A and S_S in Equation (6.20a). Soil classification Class D in Boore *et al.*'s study is equivalent to very soft soil (L) in Ambraseys *et al.*'s (1996) study. However the dataset for soft soil being poorly represented was ignored.

Ambraseys *et al.* (1996a) reported an interesting comparison of prediction of peak ground acceleration with predictive relationships derived for Western North America (Boore, Joyner and Fumal, 1993) and Italian earthquakes (Sabetta and Pugliese (1987). The comparison is shown in Figures 5.17 and 5.18. The authors have drawn attention to the dichotomous term in site geology in the Italian study.

The figures also show predictions by Sabetta and Pugliese (1987).

The comparisons underline the similarity of tectonic makeup of Western North America and active regions of Europe.

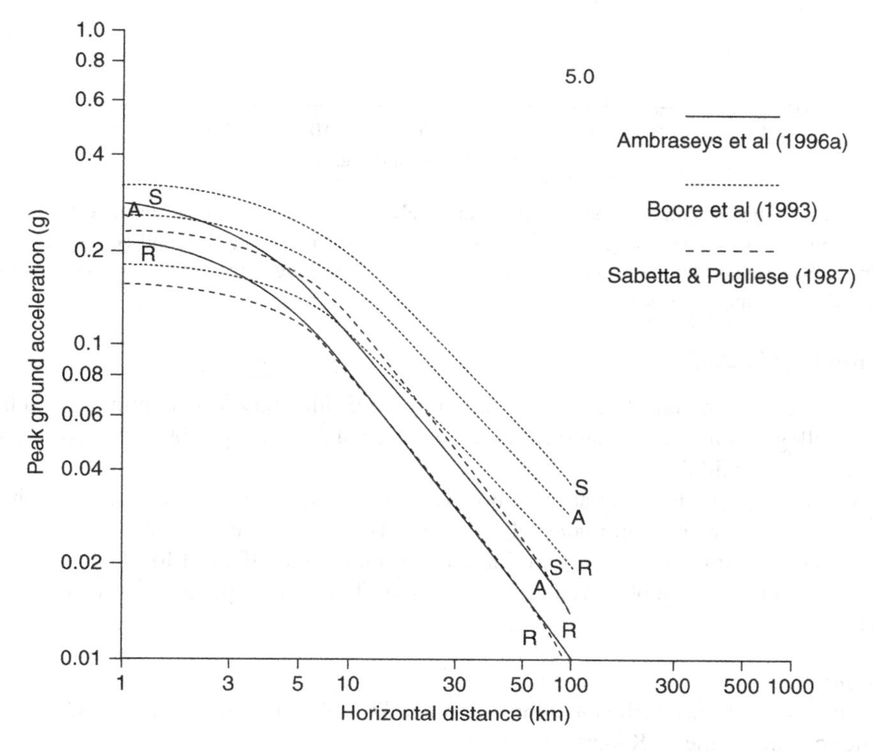

Figure 5.17 Predicted values of peak horizontal acceleration as a function of distance for an earthquake magnitude M_s 5, according to Equations (6.20a and b) for rock, stiff and soft soil sites and according to Sabetta and Pugliese (1987) for rock soil sites. Reproduced from Ambraseys *et al.* (1996a) by permission of John Wiley & Sons, Ltd.

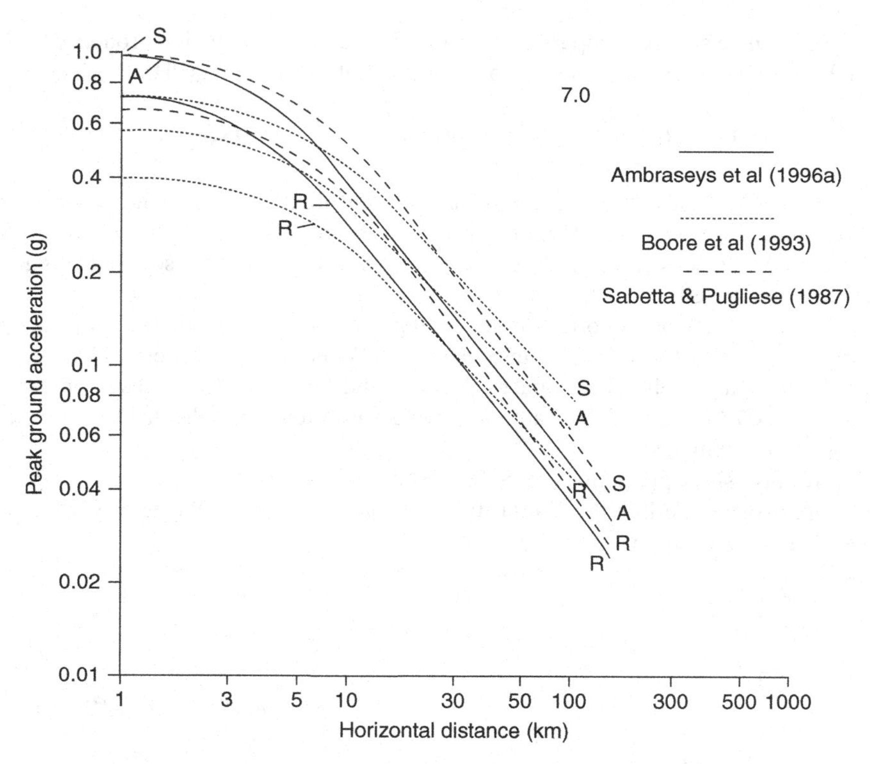

Figure 5.18 Predicted values of peak horizontal acceleration as a function of distance for an earthquake magnitude M_s 7, according to Equations (6.20a and b) for rock, stiff and soft soil sites and according to Sabetta and Pugliese (1987) for rock soil sites. Reproduced from Ambraseys *et al.* (1996a) by permission of John Wiley & Sons, Ltd.

Ambraseys *et al.*, 2005a

This paper presents the latest set of attenuation relationships based on continuing studies at Imperial College, London. The database used consisted of 595 strong-motion records recorded in Europe and the Middle East.

The relationships are for estimation of horizontal strong ground motions caused by shallow crustal earthquakes with magnitudes $M_w \geq 5$ and distances to the surface projection of the fault less than 100 km. Coefficients are included to model the effect of local site effects and faulting mechanism on the observed ground motions. The various parameters included are as follows:

Magnitude

The magnitude scale used is the moment magnitude M_w defined by $M_w = 2/3 \log M_0 - 6$ where M_0 is the seismic moment (Kanamori, 1977).

Source-to-Site Distance

The distance to the surface projection of the fault has been used as the distance measure. This has the advantage that it does not require an estimate of the depth of the earthquake, which can be associated with more major errors.

Faulting Mechanism

Fault classification basis is that due to Frohlich and Apperson (1992). Normal and thrust faults (also referred to as reverse faults in the literature) are identified and all other faults are defined as odd faults.

Local Site Conditions

The categories of soil class included are:

- soft soil (S) $V_s \leq 360\,\mathrm{m/sec}$
- stiff soil (A) $V_s \leq 360 \leq 750\,\mathrm{m/sec}$
- rock (R) $V_s \geq 750\,\mathrm{m/sec}$

The recommended form of relationship (horizontal peak ground acceleration $\mathrm{m/sec^2}$) is of the form:

$$\log y = a_1 + a_2M_w + (a_3 + a_4M_w)\log\sqrt{d^2 + a_5^2} + a_6S_S + a_7S_A + a_8F_N + a_9F_T + a_{10}F_O \quad (5.21)$$

where $S_S = 1$ for soft soil and 0 otherwise, $S_A = 1$ for stiff soil sites and 0 otherwise, S_A and S_S are set to 0 for rock sites. $F_N = 1$ for normal faulting earthquakes and 0 otherwise, $F_T = 1$ for thrust faulting earthquakes and 0 and otherwise, $F_0 = 1$ for odd faulting earthquakes and 0 otherwise. The coefficients for PGA are shown in Table 5.4.

Comparisons with Previous Equations

The predictions of ground response derived from this study (Ambraseys *et al.*, 2005a) have been compared to previous studies.

In Figure 5.19 the estimated ground response is compared to those derived by Ambraseys et al (1996a). The estimated responses are similar from the two studies for moderate and large earthquakes at all distances.

Figure 5.20 shows the comparison of Ambraseys *et al.* (1996a) to Campbell and Bozorgnia (2003) and is derived using the same magnitude scale and distance metric as used by Ambraseys *et al.* (2005a).

Japan

Tamura et al. (2003)

The authors have analysed 216 pairs of two horizontal components of strong motion accelerations from 40 earthquakes in Japan to develop a consistent set of practical attenuation

Table 5.4 Derived coefficients for the estimation of horizontal peak ground acceleration for 5 % damping (Ambraseys *et al.*, 2005a).

a_1	a_2	a_3	a_4	a_5	a_6	a_7	a_8	a_9	a_{10}
2.522	−0.142	−3.184	0.314	7.60	0.137	0.050	−0.084	0.062	−0.044

Source: Reproduced from Ambraseys *et al.* (2005a) by permission of Springer.

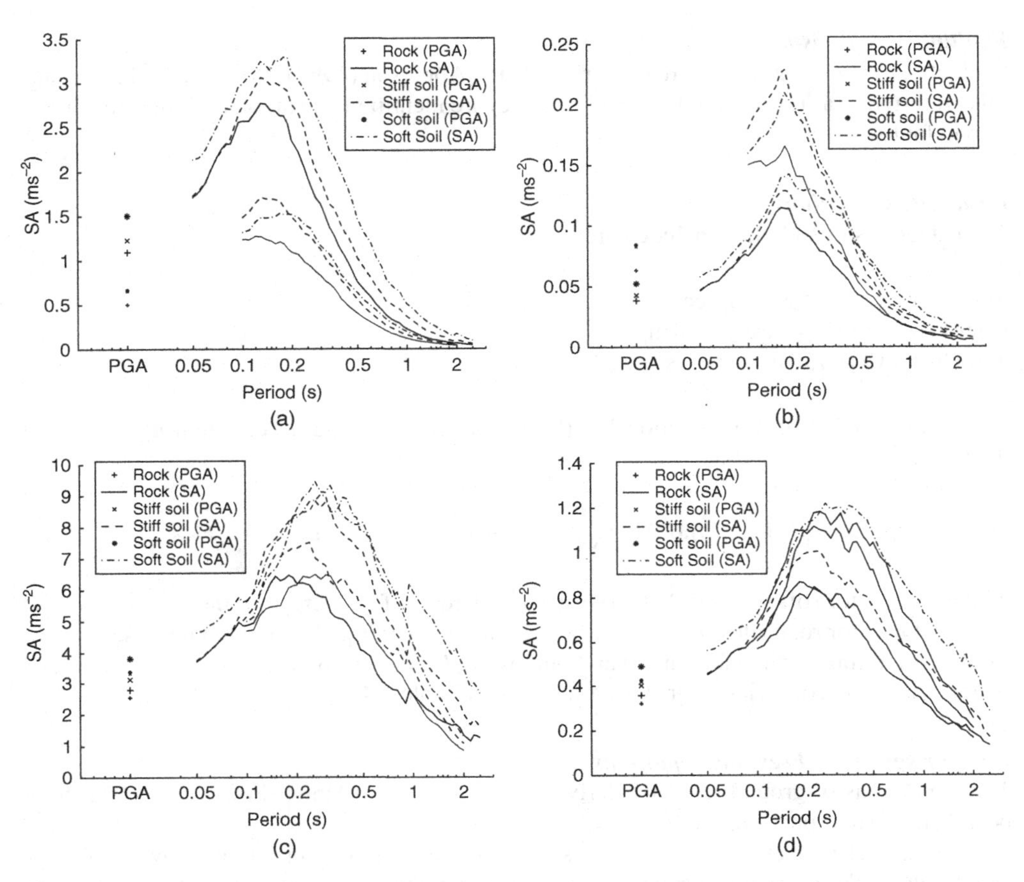

Figure 5.19 Comparison of the estimated median response spectra given by the equation in this study for strike slip faulting (thick lines) and those presented by Ambraseys *et al.* (1996a) (thin lines), which are independent of faulting mechanism: (a) $M_w = 5.0$ ($M_s = 4.3$), $d_f = 10$ km, (b), $M_w = 5.0$ ($M_s = 4.3$), $d_f = 100$ km, (c) $M_w = 7.0$ ($M_s = 6.9$), $d_f = 10$ km, (d) $M_w = 5.0$ ($M_s = 4.3$), $d_f = 100$ km. Reproduced from Ambraseys *et al.* (2005a) with kind permission of Springer Science and Business Media and Ambraseys.

relations for peak ground acceleration and 5 % damped spectral acceleration in terms of earthquake magnitude and distance. Selection of strong motion records was restricted to those obtained from the Japanese Public Works Research Institute to maintain consistency in terms of installation of accelerographs and data processing. Earthquakes analysed are from the 1967 Kinkazan-oki earthquake through to the 1995 Hyogo-ken (Kobe) earthquakes. The earthquakes were selected based on the following criteria: (1) earthquake magnitude ≥ 5.0, (2) focal depth ≤ 60 km and iii) three or more records were obtained for each earthquake. The authors have used Japan Meteorological Agency (JMA) magnitude M_j as a magnitude scale (see comparison of magnitude scales – Figure 5.15). The shortest distance between the recording station and the fault plane is used as the distance measure in attenuation relation when the earthquake source parameters were available. In cases where source parameters are not

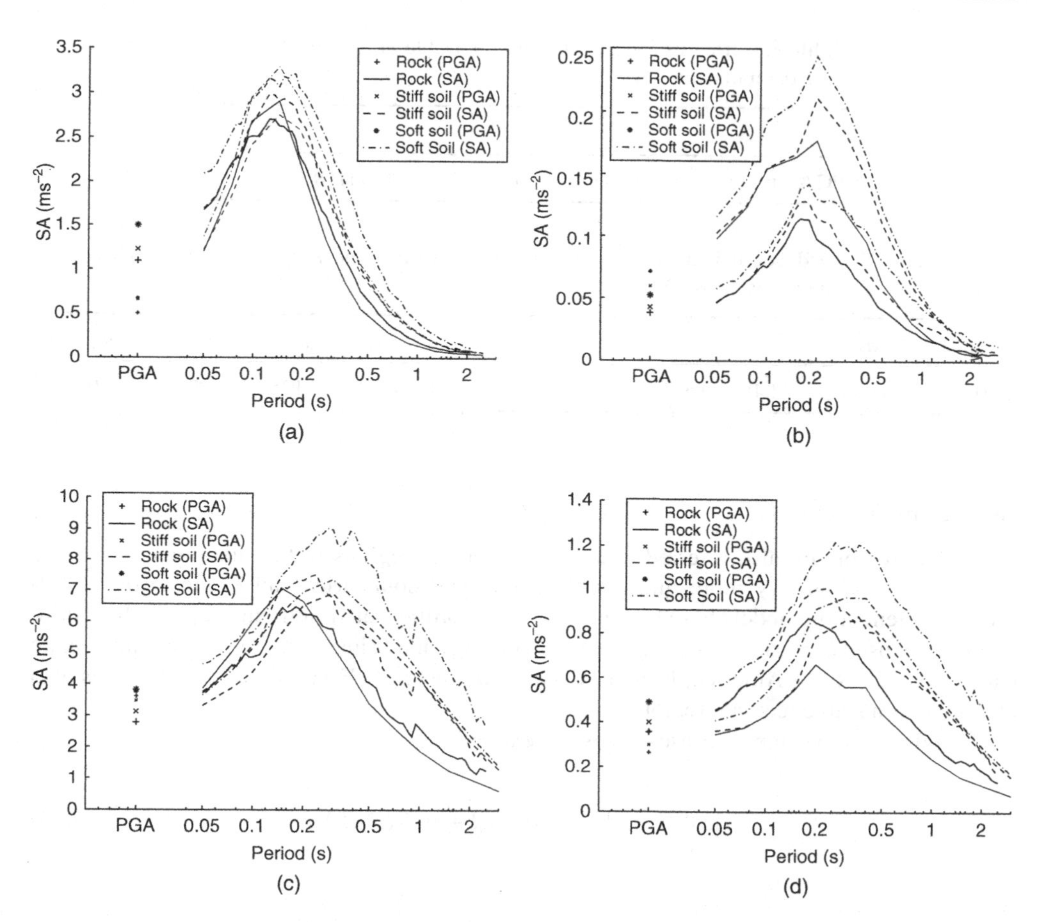

Figure 5.20 Comparison of the estimated median response spectra given by the equation in this study (thick lines) and those presented by Campbell and Bozorgnia (2003) (thin lines), for strike-slip faulting: (a) $M_w = 5.0$, $d_f = 10\,\text{km}$, ($d_s = 10.4\,\text{km}$), (b), $M_w = 5.0$, $d_f = 100\,\text{km}$, ($d_s = 100\,\text{km}$),, (c) $M_w = 7.0$, $d_f = 10\,\text{km}$, ($d_s = 10.4\,\text{km}$), (d) $M_w = 7.0$, $d_f = 100\,\text{km}$, ($d_s = 100\,\text{km}$). Reproduced from Ambraseys *et al.* (2005a) with kind permission of Springer Science and Business Media and Ambraseys.

available, epicentral distances were used as the distance measure. Soil conditions of recording sites were classified into three groups, according to fundamental natural period of ground. The equation for peak ground acceleration is given by:

$$\log Y = \log a + b.M - c\log(X + X_0) \tag{5.22}$$

where Y is either peak ground acceleration or 5 % damped response spectral acceleration at a specified natural period T(cm/sec^2), M is JMA magnitude, X is distance (km), X_0 is a constant accounting for ground motion saturation at short distance, and a, b, and c are regression coefficients to be determined. In the paper X_0 is assumed to be 30 km after Kawashima *et al* (1986).

Table 5.5 Regression coefficients for Dahle *et al.*'s relationship (5 % damping).

	c_1	c_2	c_4	$\sigma_{\ln A}$
PGA (m/sec^2)	1.471	0.849	−0.00418	0.83

Table 5.6 Derived coefficients for the estimation of vertical peak ground acceleration for 5 % damping (Ambraseys *et al.*, 2005a).

a_1	a_2	a_3	a_4	a_5	a_6	a_7	a_8	a_9	a_{10}
0.835	0.083	−2.489	0.206	5.6	0.078	0.046	−0.126	0.005	−0.082

Intra-Plate Regions

Intra-plate region are also referred to as stable tectonic regions in the literature. Intra-plate regions are geologically more uniform than the plate boundaries and Dahle *et al.* (1990) presents attenuation models based on earthquake recordings at various sites across the world. The dataset used consisted of strong motion recordings at 87 sites from 56 different intra-plate earthquakes in North America, Europe, China and Australia. Eastern North America is also classified as a stable tectonic region.

The relationship can be represented by the expression;

$$\ln A = c_1 + c_2 M_s + c_4 R + \ln G(R, R_0) \tag{5.23}$$

where

A is the horizontal component of peak ground acceleration,
R is the hypo central distance,
M_s is the surface wave magnitude.

Coefficients c_1, c_2, c_4 and spreading function $G(R, R_0)$ are to be determined. The spreading function is of the form;

$$G(R, R_0) = R^{-1} \quad \text{for} \quad R \leq R_0$$

$$G(R, R_0) = R^{-1}\left(\frac{R_0}{R}\right)^{5/6} \quad \text{for} \quad R \geq R_0$$

The recommended value for R_0 is 100 km.

Dahle's regression coefficients for PSA at 40 Hz at 5 % damping (Dahle *et al.*, 1990) are shown in Table 5.5.

Worldwide

Campbell and Bozorgnia (1994) used worldwide data recorded between 1957 and 1993 to develop the following attenuation relationship for peak ground acceleration (PGA):

$$\begin{aligned}\ln(pga) = &-3.512 + 0.904M_w - 1.328\ln\sqrt{R_s^2 + [0.149\exp(0.647M_w)]^2} \\ &+ [1.125 - 0.112\ln R_s - 0.0957M_w]F + [0.440 - 0.171\ln R_s]S_{sr} \\ &+ [0.405 - 222\ln R_s]S_{hr} + \varepsilon\end{aligned} \tag{5.24}$$

where

pga: mean of the two horizontal components of peak ground acceleration (g)
M_w: moment magnitude
R_s: closest distance to seismogenic rupture on the fault (km)

$$\begin{aligned} F &= 0 \quad \text{(strike slip and normal faults)} \\ &= 1 \quad \text{(reverse, reverse oblique and thrust faults)} \\ S_{sr} : &= 1 \quad \text{(soft rock)} \\ &= 0 \quad \text{(hard rock)} \\ S_{hr} &= 1 \quad \text{(hard rock)} \\ &= 0 \quad \text{(soft rock and alluvium)} \end{aligned}$$

ε is the random error term with zero mean and a standard deviation equal to $\ln(pga)$.

$$\begin{aligned} \sigma_{\ln(pga)} &= 0.55 \qquad pga < 0.068 \\ &= 0.173 - 0.140\ln(pga) \quad 0.068 \le pga \le 0.21 \\ &= 0.39 \qquad pga > 0.21 \end{aligned}$$

with a standard error estimate of 0.021.

Vertical Acceleration Spectra

Vertical acceleration during an earthquake must also be considered in the analyses. Guidance on the construction of a design vertical response spectrum may be found in the work of Hays (1980). Attenuation relationships with regard to response spectra (vertical) have been reported by several researchers (Ambraseys, 1996b, 2005b; Campbell, 1997).

Ambraseys et al., 2005b

This paper presents the latest set of attenuation relationships based on the studies at Imperial College, London. The database used consisted of 595 strong-motion records recorded in Europe and the Middle East.

The relationships are for estimation of vertical strong ground motions caused by shallow crustal earthquakes with magnitudes $M_w \ge 5$ and distances to the surface projection of the fault lass than 100 km. Coefficients are included to model the impact of local site effects and faulting mechanism on the observed ground motions. The various parameters included are as discussed earlier for the horizontal peak ground acceleration (Ambraseys *et al.*, 2005a).

The recommended form of relationship (vertical peak ground acceleration $\mathrm{m/sec^2}$) is of the form:

$$\log y = a_1 + a_2 M_w + (a_3 + a_4 M_w) \log \sqrt{d^2 + a_5^2} + a_6 S_S + a_7 S_A + a_8 F_N + a_9 F_T + a_{10} F_O \quad (5.25)$$

where $S_S = 1$ for soft soil and 0 otherwise, $S_A = 1$ for stiff soil sites and 0 otherwise, S_A and S_S are set to 0 for rock sites. $F_N = 1$ for normal faulting earthquakes and 0 otherwise, $F_T = 1$ for thrust faulting earthquakes and 0 and otherwise, $F_0 = 1$ for odd faulting earthquakes and 0 otherwise. The coefficients for PGA are shown in Table 5.6.

5.2.4 Next Generation Attenuation (NGA) Models for Shallow Crustal Earthquakes in Western United States (2008)

This section provides the updated references for the NGA models developed.

Background

The Next Generation of Ground Motion Attenuation Models project is a multidisciplinary research programme conducted over the past several years in the USA The objective of the project is to develop new ground motion prediction relationships.

Five sets of ground motion models were developed by teams working independently but interacting with other groups as well. There were several important components of the programme:

1. Develop separate ground motion models by five teams (with each team developing their model independently of the others).
2. Develop an updated and expanded ground motion database for the research teams to develop the ground motion models (Pacific Earthquake Engineering Research Center (PEER) database).
3. Organize and conduct a number of supporting research databases to provide an improved scientific basis for evaluating the functional forms of and constraints on the models.
4. Organize several project interaction meetings between the teams.

A good overview of the NGA Project components, process and products is presented by Power *et al.* (2008).

Consideration of Effects on Ground Motion (Power *et al.*, 2008)

- moderate to large magnitude scaling at close distances;
- distance scaling at both close and far distances;
- rupture directivity;

- footwall vs hanging wall for dipping faults;
- style of faulting (strike-slip, reverse, normal);
- depth to faulting (buried vs. surface rupture);
- static stress drop (or rupture area);
- site amplification relative to a reference 'rock' condition;
- three-dimensional (3-D) sedimentary basin amplification/depth to basement rock.

Based on their past experience and judgement (all had developed attenuation models earlier), the teams decided which of these effects should be included in their model.

The main product of the NGA project was the development of improved ground motion attenuation relations. These were the five sets of ground motion attenuation equations, and were developed by Abrahamson and Silva (2008), Boore and Atkinson (2008), Campbell and Bozorgnia (2008), Chiou and Youngs (2008) and Idriss (2008). Each of the models was developed independently by the previously mentioned teams but there were interactions between these teams.

A notable feature of the project was that the research teams used a common database of recorded ground motions and supporting information. Thus an extensive update and expansion of the PEER (Pacific Earthquake Engineering Research) Center had to be undertaken. Also noteworthy, is that all the formulations of ground motion were based on the average shear wave velocity in the upper 30 m of the sediment. V_{S30}. This helped to create a common basis for characterizing site amplification of ground motion due to sediment stiffness. Dependence of amplification on the level of ground shaking was also considered by the research teams working on the development of the models. The research teams used basic seismological theory, simple seismological models and supporting analyses to make decisions about their models.

Others who contributed to the project were Spudich and Chiou (2008), Baker and Jayaram (2008), and Huang *et al.* (2008) while a good description of the motion database used by the project group is provided by Chiou *et al.* (2008).

Briefly summarizing, the following may be noted:

1. Five sets of ground motion models developed for shallow crustal earthquakes in the Western United States and similar active tectonic regions;
2. The models were developed for wider ranges of magnitude, distance, site conditions and response spectral period of vibration than existing models;
3. Systematic method for evaluation of predictor parameters for predicting earthquake effects;
4. The predictor parameters variously incorporated included:-

   (a) earthquake magnitude;
   (b) style of faulting;
   (c) depth to top of fault rupture;
   (d) source to site distance;
   (e) site location on hanging wall or foot wall of dipping faults;
   (f) near-surface soil stiffness and
   (g) sedimentary basin depth/depth to rock.

5. A series of supporting research projects conducted to provide an improved scientific basis defining and constraining the functional form used in the ground motion models.

Availability of Ground-Motion Models and Reports

Reports documenting NGA models are available electronically from the PEER website (http://peer.berkeley.edu/products/rep_nga_models.html 01/08/2008) and in hard copy from PEER. Computer-coded NGA models in various forms, convenient for use, are available with each report on the website.

5.3 References

Abrahamson, N. and Silva, W. (2008) 'Summary of the Abrahamson and Silva NGA ground-motion relations'. *Earthquake Spectra* **24**(1), Special Issue, 3–22.

Aki, K. (1984) 'Prediction of strong motion using physical models of earthquake faulting'. *Proceedings of the 8th World Conference on Earthquake Engineering*, San Francisco, 433–40.

Ambraseys, N.N. and Bommer, J.J. (1991) 'The attenuation of ground accelerations in Europe'. *Earthquake Engineering and Structural Dynamics* **20**(12): 1179–1202.

Ambraseys, N.N. and Sarma, S.K. (1967) 'The response of earth dams to strong earthquakes'. *Geotechnique* **17**(2): 181–283.

Ambraseys, N.N., Simpson K.A. and Bommer, J.J. (1996a) 'Prediction of horizontal response spectra in Europe'. *Earthq. Eng. Struct. Dyn.* **25**: 371–400.

Ambraseys, N.N. and Simpson K.A. (1996b) 'Prediction of vertical response spectra in Europe'. *Earthq. Eng. Struct. Dyn.* **25**, 401–12.

Ambraseys, N.N., Douglas, J. and Smit, P.M. (2005a) 'Equations for estimation of strong ground motions from shallow crustal earthquakes using data from Europe and the Middle East: horizontal peak ground acceleration and spectral acceleration'. *Bull. Earthquake Eng.* **3**(1): 11–53.

Ambraseys, N.N., Douglas, J., Sarma, S.K. and Smit, P.M. (2005b) 'Equations for estimation of strong ground motions from shallow crustal earthquakes using data from Europe and the Middle East: vertical peak ground acceleration and spectral acceleration'. *Bull. Earthquake Eng.* **3**(1): 55–73.

Arias, A., (1970) 'A measure of earthquake intensity', in R. Hanson (ed.), *Seismic Design of Nuclear Power* Plants, *Massachusetts Institute of Technology Press*. Cambridge, Massachusetts.

Atkinson, G.M. and Boore, D.M. (1995) 'Ground motion relations for eastern North America'. *Bull. Seism. Soc. Am.* **85**(1): 17–30.

Baker, J.W. and Jayaram, N. (2008) 'Correlation of spectral acceleration values from NGA ground motion values'. *Earthquake Spectra* **24**(1), Special Issue, 299–318.

Bolt, B.A. (1969) 'Duration of strong Motion'. *Proceedings of the 4th World Conference on Earthquake Engineering*, Santiago, 1304–1315.

Bolt, B.A. (1988) *Earthquakes*. W.H. Freeman, New York.

Bolt, B.A. (2004) *Earthquakes*. W.H. Freeman, New York.

Bonilla, M.G., Mark, R.K. and Lienkaemper, J.J. (1984) 'Statistical relations among earthquake magnitude, surface rupture length, and surface fault displacement'. *Bull. Seism. Soc. Am.* **76**(6); 2379–411.

Boore, D.M. and Atkinson, G.M. (2008) 'Ground-motion prediction equations for the average horizontal component of PGA, and 5 %-Damped PSA at spectral periods between 0.01s and 10.0s', *Earthquake Spectra* **24**(1), Special Issue, 99–138.

Boore, D.M., Joyner, W.B. and Fumal, T.E. (1993) *Estimation of Response Spectra and Peak Acceleration from Western North American Earthquakes: An Interim Report*. Open-File Report 93–509, United States Geological Survey.

Boore, D.M., Joyner, W.B. and Fumal, T.E. (1997) 'Equations for estimating horizontal response spectra and peak acceleration from western North American Earthquakes: a summary of recent work'. *Seism. Res. Lett.* **68**, 128–153.

Brune, J.N. (1970), 'Tectonic stress and spectra of seismic shear waves from earthquakes', *J. Geophys. Res.*, **75**, No. 26, 4997–5009.

Campbell, K.W. (1981) 'Near-Source Attenuation of peak horizontal acceleration'. *Bull. Seism. Soc. Am.* **71**(6): 2039–70.

Campbell, K.W. (1997) 'Empirical near source attenuation relationships for horizontal and vertical components of peak ground acceleration, peak ground velocity and pseudo-absolute acceleration response spectra'. *Seism. Res. Lett.* **68**: 154–179.

Campbell, K.W. (2003a) 'Engineering models of strong ground motion', in Chen, W-F., Scawthorn, C.S. (eds), *Earthquake Engineering Handbook*. CRC Press LLC, Boca Raton, Florida.

Campbell, K.W. (2003b) 'Prediction of strong ground motion using the hybrid empirical method and its use in the development of ground motion (attenuation) relations in eastern North America'. *Bull. Seism. Soc. Am.* **93**(3): 1012–33.

Campbell, K.W. and Bozorgnia, Y. (1994) 'Near-source attenuation of peak horizontal acceleration from worldwide accelerograms recorded from 1957 to 1993'. *Proceedings of the 5th US National Conference on Earthquake Engineering*, Chicago, Illinois, 283–92.

Campbell, K.W. and Bozorgnia, Y. (2003) 'Updated near-source ground motion (attenuation) relations for the horizontal and vertical components of peak ground acceleration and acceleration response spectra'. *Bull. Seism. Soc. Am.* **93**(1): 314–31.

Campbell, K.W. and Bozorgnia, Y. (2008) 'NGA ground motion model for the geometric mean horizontal component of PGA, PGV, PGD and 5 % damped linear elastic response spectra for periods ranging from 0.01 to 10s', *Earthquake Spectra* **24**(1), Special Issue, 139–72.

Chiou, B.S.J. and Youngs, R.R. (2008) 'An NGA model for the average horizontal component of peak ground motion and response spectra', *Earthquake Spectra* **24**(1), Special Issue, 173–216.

Chiou, B.S.J., Darragh, R., Gregor, N. and Silva, W. (2008) 'NGA project strong-motion database', *Earthquake Spectra* **24**(1), Special Issue, 23–44.

Chopra, A.K. (2001) *Dynamics of Structures: Theory and Applications to Earthquake Engineering*. Prentice Hall, New Jersey.

Cloud, W.K. (1963). 'Maximum accelerations during earthquakes', *Proceedings of the Chilean Conference on Seismology and Earthquake Engineering*.

Dahle, A., Bungum, H. and Kvamme, L.B. (1990) 'Attenuation models inferred from intraplate earthquake recordings'. *Earthq. Eng. Struct. Dyn.* **19**, 1125–41.

Dahle, A, Bungum, H. and Kvamme, L.B. (1991) 'Empirically derived PSV spectral attenuation models for intraplate conditions'. *European Earthquake Eng.* **3**: 42–52.

Dietrich, J.H. (1973) 'A deterministic near-field source model'. *Proceedings of the 5th US National Conference on Earthquake Engineering*, Chicago, Illinois, 2385–96.

Dobry, R., Idriss, I.M. and Ng, E. (1978) 'Duration characteristics of horizontal components of strong motion earthquake records'. *Bull. Seism. Soc. Am.* **68**(5): 1478–1520.

Dowrick, D.J. (1987) *Earthquake Resistant Design: For Engineers and Architects*. John Wiley & Sons, Ltd, Chichester, UK.

Esteva, L. and Villaverde, R. (1973) 'Seismic risk, design spectra and structural reliability'. *Proceedings of the 5th World Conference on Earthquake Engineering*, Rome, 2586–96.

Frohlich, C. and Apperson, K.D. (1992) 'Earthquake focal mechanisms, moment tensors and the consistency of seismic activity near plate boundaries'. *Tectonics* **11**(2): 279–96.

Hanks, T.C. (1982) 'f_{max}'. *Bull. Seism. Soc. Am.* 71(6), Part A, 1867–79.

Hays, W.W. (1980) *Procedures for Estimating Earthquake Ground Motion*. Geological Survey Professional Paper 1114, US Government Printing Office, Washington, DC.

Heaton, T.H., Tajima, F. and Mori, A.W. (1986) 'Estimating ground motions using recorded accelerograms'. *Surveys in Geophysics* **8**, 25–83.

Huang, Y-N., Whittaker, A.S. and Luco, N. (2008) 'Maximum spectral demands in the near-fault region', *Earthquake Spectra* **24**(1), Special Issue, 319–41.

Hudson, D.E. (1979) *Reading and Interpreting Strong Motion Accelerograms*. Earthquake Engineering Research Institute, Berkeley, California.

Idriss, I.M. (2008) 'An NGA empirical model for estimating the horizontal spectral values generated by shallow crustal earthquakes', *Earthquake Spectra* **24**(1), Special Issue, 217–42.

International Code Council. (2003) *International Building Code (IBC) 2003*. Washington, DC. ICC.

Jennings, P.C. (1985) 'Ground motion parameters that influence structural damage'. In R.E. Scholl and J.L. King (eds), *Strong Ground Motion Simulation and Engineering Applications*, EERI Publication 85-02, Earthquake Engineering Research Institute, Berkeley, California.

Joyner, W.B. and Boore, D.M. (1981) 'Peak horizontal acceleration and velocity from strong motion records from the 1979 Imperial Valley, California earthquake'. *Bull. Seism. Soc. Am.* **71**(6): 2011–38.

Joyner, W.B. and Boore, D.M. (1988) 'Measurement, characterisation and prediction of strong ground motion'. *Proceedings of the Specialty Conference Earthquake Eng. and Soil Dynamics II – Recent Advances in Ground-Motion Evaluation*, Geotechnical Special Publication 20, ASCE, New York, 43–102.

Joyner, W.B. and Fumal, T.E. (1984). 'Use of measured shear-wave velocity for predicting geologic site effects on strong ground motion'. *Proceedings of the 8th World Conference on Earthquake Engineering*, San Francisco, 777–83.

Kanamori, H. (1977) 'The energy release in great earthquakes'. *J. Geophys. Res.* **82**(20): 2981–87.

Kawashima, K., Aizawa, K., and Takahashi, K. (1986) 'Attenuation of peak ground acceleration, velocity and displacement based on multiple regression analysis of Japanese strong motion records', *Earthq. Eng. Struct. Dyn.* **14**(2): 199–215.

Kramer, S.L. (1996) *Geotechnical Earthquake Engineering*. Prentice Hall, New Jersey.

Lay, T. and Wallace, T.C. (1995) *Modern Global Seismology*. Vol. 58 (International Geophysics Series). Academic Press, San Diego; London.

McGuire, R.K. (1974) *Seismic Structural Response Risk Analysis, Incorporating Peak Response Regressions on Earthquake Magnitude and Distance*. Research Report R74-51, Dept of Civil Engineering, Massachusetts Institute of Technology (MIT), Cambridge, Massachusetts.

Milne, W.G. and Davenport, A.G. (1969) 'Distribution of earthquake risk in Canada'. *Bull. Seism. Soc. Am.* **59**(2): 754–79.

Molas, G.L. and Yamazaki, F. (1996) 'Attenuation of response spectra in Japan using new JMA records'. *Bull. Earthquake Resistant Structure* **29**: 115–28.

Murphy, J.R. and O'Brien, L.J. (1977) 'The correlation of peak ground acceleration amplitude with seismic intensity and other physical parameters'. *Bull. Seism. Soc. Am.* **67**(3): 877–915.

Newmark, N.M. and Rosenbleuth, E. (1971) *Fundamentals of Earthquake Engineering*. Prentice Hall, New Jersey.

Nuttli, O.W. (1983) 'Average seismic source-parameter relations for mid-plate earthquakes'. *Bull. Seism. Soc. Am.* **73**(2): 519–535.

Pacific Earthquake Engineering Research (PEER) Center website. http://peer.berkeley.edu/products/rep_nga_models.html 01/08/2008

Power, M., Chiou, B., Abrahamson, N., Bozorgnia, Y., Shantz, T. and Roblee, C. (2008) An Overview of the NGA Project, *Earthquake Spectra* **24**(1) Special Issue, 3–22.

Reiter, L. (1991) *Earthquake Hazard Analysis: Issues and Insights*. Columbia University Press, New York.

Richter, C.F. (1935) 'An instrumental earthquake magnitude scale'. *Bull. Seism. Soc. Am.* **25**(1): 1–32.

Sabetta, F. and Pugliese, A. (1987) 'Attenuation of peak horizontal acceleration and velocity from Italian strong-motion records'. *Bull. Seism. Soc. Am.* **77**(5): 1491–1513.

Schnabel, P.B. and Seed, H.B., (1973), 'Accelerations in rock for earthquakes in the Western United States', *Bull. Seism. Soc of Am.* **62**: 1649–1664.

Scholz, C.H. (1990) *The Mechanics of Earthquakes and Faulting*. Cambridge University Press, Cambridge.

Seed, H.B. and Idriss, I.M. (1971) 'Simplified procedure for evaluating soil liquefaction potential'. *J. Soil Mechanics and Foundations Division*, ASCE, **97**(9): 1171–82.

Seed, H.B., Murarka, R., Lysmer, J. and Idriss, I.M. (1976), 'Relationships of maximum acceleration, maximum velocity, distance from source and local site conditions for moderately strong earthquakes', *Bull. Seism. Soc of Am.*, **66**(4): 1323–42.

Seed, R.B. *et al.* (1990) *Preliminary Report on the Principal Geotechnical Aspects of the October 17, 1989 Loma Prieta Earthquake*. EERC Report 90-05. Earthquake Engineering Research Center, University of California, Berkeley.

Shakal, A.F. and Bernreuter D.L. (1981) *Empirical Analysis of Near Source Ground Motion*. Nuclear Regulatory Committee Report, NUREG/CR-1095, Washington, DC.

Shearer, P.M. (1999) *Introduction to Seismology*. Cambridge University Press, Cambridge.

Slemmons, D.B. (1977) *State-of-the-art for Assessing Earthquake Hazards in the United States; Report 6, Faults and Earthquake Magnitude*. Misc. Paper S-173-1, US Army Corps of Engineers, Waterways Experiment Station, Vicksburg, Mississippi.

Spudich, P. and Chiou, B.S.J. (2008) 'Directivity in NGA earthquake ground motions: analysis using isochrone theory', *Earthquake Spectra* **24**(1), Special Issue, 279–98.

Spudich, P., Joyner, W.B., Lindh, A.G., Boore, D.M., Margaris, B.M. and Fletcher, J.B. (1999) 'SEA99: a revised ground motion prediction relation for use in extensional tectonic regimes'. *Bull. Seism. Soc. Am.* **89**(5): 1156–70.

Stein, S. and Wysession, M. (2003) *An Introduction to Seismology, Earthquakes and Earth Structure*. Blackwell Publishing, USA.

Tamura, K., Matsumoto, S. and Nakao, Y. (2003) 'Attenuation relations of peak ground acceleration and acceleration response spectra applications'. *Res. Rept of Public Works Res. Inst.* **199**: 79–95.

Trifunac, M.D. and Brady, A.G. (1975a) 'A study on the duration of strong earthquake ground motion'. *Bull. Seism. Soc. Am.* **65**(3): 581–626.

Trifunac, M.D. and Brady, A.G. (1975b) 'On the correlation of seismic intensity scales with the peaks of recorded ground motion'. *Bull. Seism. Soc. Am.* **65**(1): 139–62.

Trifunac, M.D. and Brady, A.G. (1976) 'Correlations of peak acceleration, velocity, and displacement with earthquake magnitude, and site conditions'. *Int. J. Earthquake Eng. and Structural Dynamics* **4**: 455–471.

Trifunac, M.D. and Westermo, B.D. (1977) *Dependence of the Duration of Strong Earthquake Ground Motion on Magnitude, Epicentral Distance, Geologic Condition at the Recording Station, and Frequency of Motion.* University of Southern California, Dept of Civil Engineering, Report CE76-02, 64pp.

Wald, D.J., Helmberger, D.V. and Heaton, T.H. (1991) 'Rupture model of the 1989 Loma Prieta from the inversion, of strong-motion and broadband teleseismic data'. *Bull Seism. Seism. Soc. Am.* **81**(5): 1540–1672.

Wald, D.J., Heaton, T.H. and Hudnut, K. (1996) 'The slip history of the 1994 Northridge, California, earthquake determined from strong-motion teleseismic, GPS and leveling data'. *Bull. Seism. Soc. Am.* **86**(1B), S49–70.

Wells, D.L. and Coppersmith, K.J. (1994) 'New empirical relations among, magnitude, rupture length, rupture width, rupture area, and surface displacement'. *Bull Seism. Seism. Soc. Am.* **84**(4): 974–1002.

Wyss, M. (1979) 'Estimating maximum expectable magnitude of earthquakes from fault dimensions'. *Geology* **7**(7): 336–40.

Yamashita, T. (1976) 'On the dynamical process of fault motion in the presence of friction and inhomogeneous initial stress: Part 1, Rupture propagation'. *J. Phys. Earth* **24**: 417–24.

6

Introduction to Response Spectra

6.1 General Concepts

6.1.1 A Designer's Perspective

The problem faced by the designer is to provide for a safe and economical structure subject *not* to a known time history or loading, but to a future earthquake about which very little can be predicted with any certainty. Each earthquake is different and the wide range of possible earthquake time history records that are available makes it impractical to base the design on a single record. The designer must extract the pertinent information about past earthquakes in general and apply such information to his/her design.

The response spectrum method, developed over the course of the last century, has become a key process in earthquake engineering and design which yields a conservative design for any site. The method was originally proposed by Biot (1943) and popularized as a design tool by Biot and Housner. The method essentially describes the frequency content of a given time history of excitation.

6.1.2 A Practical Way Forward

Consider the single degree of freedom (SDOF) system shown in Figure 6.1 (see also Chapter 2).

Duhamel's Integral for responses to arbitrary loading, as discussed in Chapter 2 takes the form

$$y(t) = \frac{1}{m\omega_d} \int_0^t F(\tau) \cdot e^{-\xi\omega(t-\tau)} sin\, \omega_d(t - \tau) d\tau \tag{6.1}$$

Fundamentals of Seismic Loading on Structures Tapan Sen

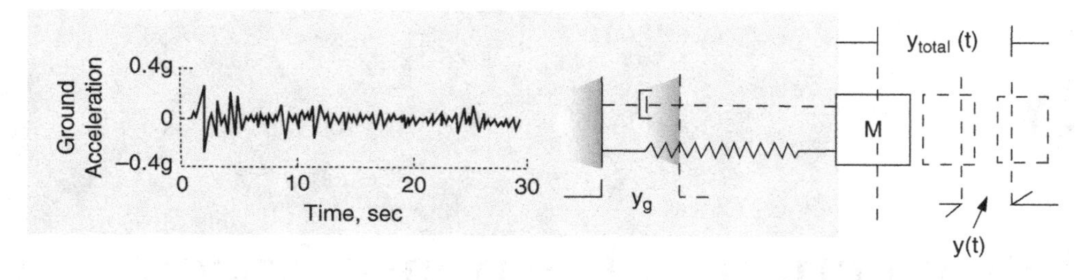

Figure 6.1 Single degree of freedom system.

Seismic loading is a good example of arbitrary loading. In this case the loading term is replaced by $-m\ddot{y}_g(t)$ (Section 2.3.3) then for seismic excitation the response is given by:

$$y(t) = -\frac{1}{\omega_d}\int_0^t \ddot{y}_g(\tau)\cdot e^{-\xi\omega(t-\tau)} Sin\,\omega_d(t-\tau)d\tau \tag{6.2}$$

ω, ω_d, ξ and τ are as previously explained (see Chapter 2).

The quantities of interest are relative displacement, y, relative velocity $\dot{y}(t)$ and peak acceleration. Note, peak acceleration is represented by $\ddot{y}_{total}(t) = \ddot{y}(t) + \ddot{y}_g(t)$

where $y(t)$ is a function of ξ, ω_d and $\ddot{y}_g(t)$ is the ground acceleration.

Differentiating Equation (6.2) will yield an expression for the velocity response.

$$\dot{y}(t) = -\int_0^t \ddot{y}_g(\tau)\cdot e^{-\xi\omega(t-\tau)} Cos\,\omega_d(t-\tau)d\tau + \text{a term involving } \xi \tag{6.3}$$

The equation for the acceleration $\ddot{y}_{total}(t)$ may be obtained by differentiating again Equation (6.3) and adding the ground acceleration $\ddot{y}_g(t)$.

It can be seen that the expressions for $y(t)$ contains ω_d, and $\dot{y}(t)$ contains ω_d and a term involving ξ and $\ddot{y}_{total}(t)$ would contain ω_d and terms involving ξ and ξ^2.

Another point to note, discussed in greater detail later, is that the relative-velocity spectrum and acceleration spectrum are plots of $\dot{y}(t)_{\max}$ and $\ddot{y}_{total}(t)_{\max}$, the peak values of $\dot{y}(t)$ and $\ddot{y}_{total}(t)$, respectively, as functions of period, T_n.

In order to be more readily usable and obtain simpler expressions without sacrificing too much of accuracy, certain approximations have been introduced.

We have seen earlier, $\omega_d = \omega\sqrt{1-\xi^2}$

First, as the damping ratio ξ is small in the region of structural response of interest (2 %–15 %), ω_d has been replaced by ω. Next, the non-integral terms of the right hand side of equations for $\dot{y}(t)$ and $\ddot{y}_{total}(t)$ which are multiplied by ξ and ξ^2 have been neglected and introducing

y_a for $\ddot{y}_g$ for conciseness. The simplified version of three new quantities, S_d, S_v and S_a, are represented by

$$S_d(\omega, \xi) = \left| -\frac{1}{\omega} \int_0^t y_a(\tau) \cdot e^{-\xi\omega(t-\tau)} \cdot sin\, \omega(t-\tau) d\tau \right|_{max} \tag{6.4}$$

$$S_v(\omega, \xi) = \left| - \int_0^t y_a(\tau) \cdot e^{-\xi\omega(t-\tau)} \cdot cos\, \omega(t-\tau) d\tau \right|_{max} \tag{6.5}$$

$$S_a(\omega, \xi) = \left| \omega \int_0^t y_a(\tau) \cdot e^{-\xi\omega(t-\tau)} \cdot sin\, \omega(t-\tau) d\tau \right|_{max} \tag{6.6}$$

These approximations lead to

$$S_v = \omega \cdot S_d \tag{6.7}$$

$$S_a = \omega^2 \cdot S_d \tag{6.8}$$

Maximum values of S_v and S_a plotted against period T_n are known in literature as 'pseudo' velocity and acceleration spectra. The term 'pseudo' has been introduced because of the mathematical approximations introduced. The plot of $S_v = \omega \cdot S_d$ is not exactly the same as the plot of relative velocity $\dot{y}(t)$ and similarly the plot of $S_a = \omega^2 \cdot S_d$ is not exactly the same as the plot of $\ddot{y}_{total}(t)$. It is worthwhile to note the significance of the different response spectra. The deformation spectrum provides information about the peak deformation of the system. The 'pseudo' velocity spectrum is related to the strain energy stored in the system and the 'pseudo' acceleration spectrum may be used to compute the equivalent static force and base shear. The pseudo-response spectra have proved to be of great practical significance, being widely used in engineering for design of structures in regions prone to seismic hazard.

Construction of Pseudo-response Spectra

The steps for construction of the pseudo-response spectra may now be outlined.

Step 1 Select an earthquake record.

Step 2 Check the digitization. The digitization should report acceleration at equal time steps.

Step 3 Select damping coefficient ξ.

Step 4 Select period T_n.

Step 5 Compute $y(t)$ the deformation response due to input motion.
A numerical procedure, for example the Duhamel Integral, may be adopted for computing $y(t)$.

Step 6 Record **peak** value of $y(t)$ – identified as S_d
Note: the sign of the ordinate is not important in earthquake engineering.
Step 7 Compute $S_v = \omega S_d$ and $S_a = \omega^2 S_d$
Step 8 Repeat steps 4–7 for another value of T_n

The information required may be organized in a tabular form as shown in Figure 6.2a. Only three points for steps 4–8 are shown (see Figures 6.2b–e), but these need to be repeated for more points.

Steps 4–8 constitute only *one set* of computation for a particular value of damping coefficient ξ. All the above steps are repeated for another value of ξ.

The procedure illustrated is the technique proposed by pioneers in the field Benioff (1934), and Biot (1943).

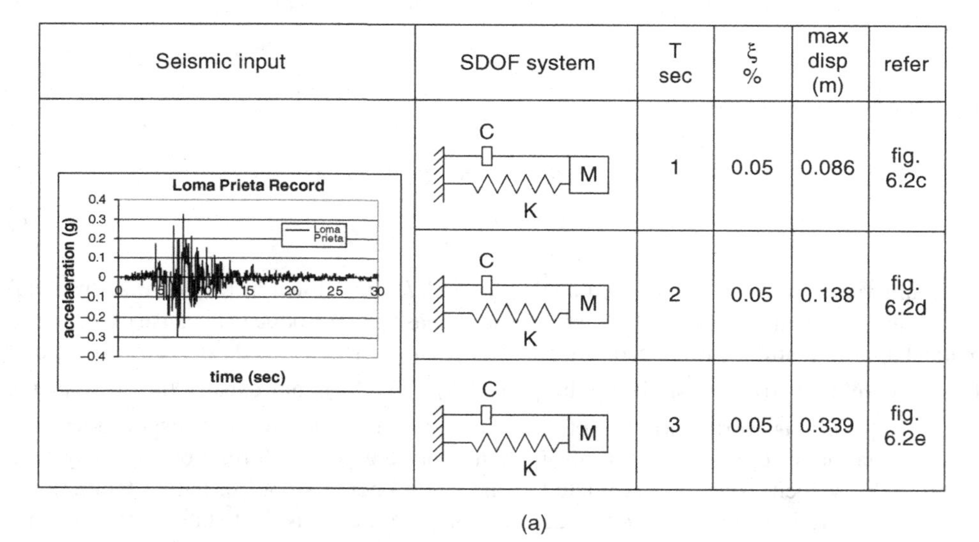

Seismic input	SDOF system	T sec	ξ %	max disp (m)	refer
Loma Prieta Record	C K M	1	0.05	0.086	fig. 6.2c
	C K M	2	0.05	0.138	fig. 6.2d
	C K M	3	0.05	0.339	fig. 6.2e

(a)

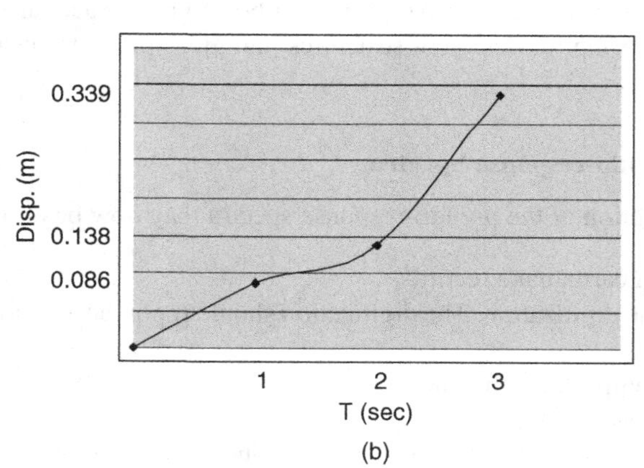

(b)

Figure 6.2 Computational steps for the construction of response spectra.

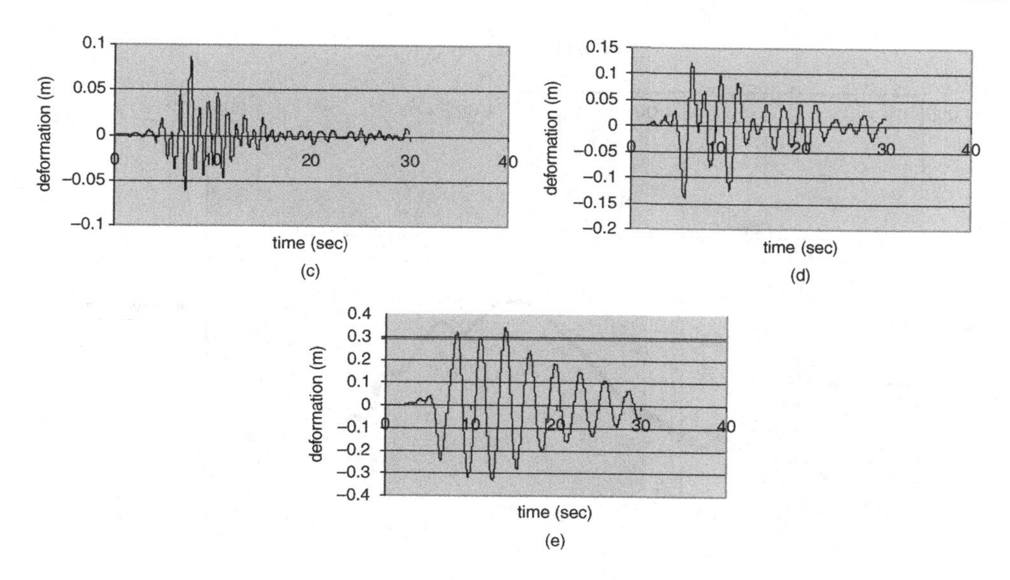

Figure 6.2 *(continued)*.

6.1.3 Constructing Tripartite Plots

It may be observed that the 'pseudo' spectra are simply obtained from $S_v = \omega \cdot S_d$ and $S_a = \omega^2 \cdot S_d$ Thus if one of the spectra is known, the other two can be obtained from algebraic manipulations. The composite result can be plotted on a special graph paper. The procedure is illustrated with an earthquake record. Consider the earthquake record of the Fruili after shock (Northern Italy, 1976).

The earthquake time history record was accessed from the website for European Strong Motion Data (ISESD) (www.isesd.cv.ic.ac.uk 20/07/08, maintained by Imperial College, London). The website also provided the spectral data. The numerical data for period, S_d, S_v, S_a were obtained from this website data. The construction of four-way representation of this data is illustrated below. The special graph paper is a four-way logarithmic paper.

Step 1

The plot of Frequency vs. S_v is shown in Figure 6.3.

The next step consists of determining the logarithmic scales for S_d and S_a in Figure 6.3.

In order to illustrate the next step and preserve the clarity, only a few points of the Freq. vs. PSV (pseudo spectral velocity) plot are shown in Figure 6.4.

Step 2

The scales for displaying S_d and S_a are introduced next and shown in Figure 6.5. The vertical and horizontal scales are standard logarithmic scales. The scale for S_d is sloping at $-45°$ and that of S_a at $+45°$ respectively. These are also logarithmic scales but not identical to the logarithmic vertical scale. The values S_d and S_a may be read off the same plot after introducing

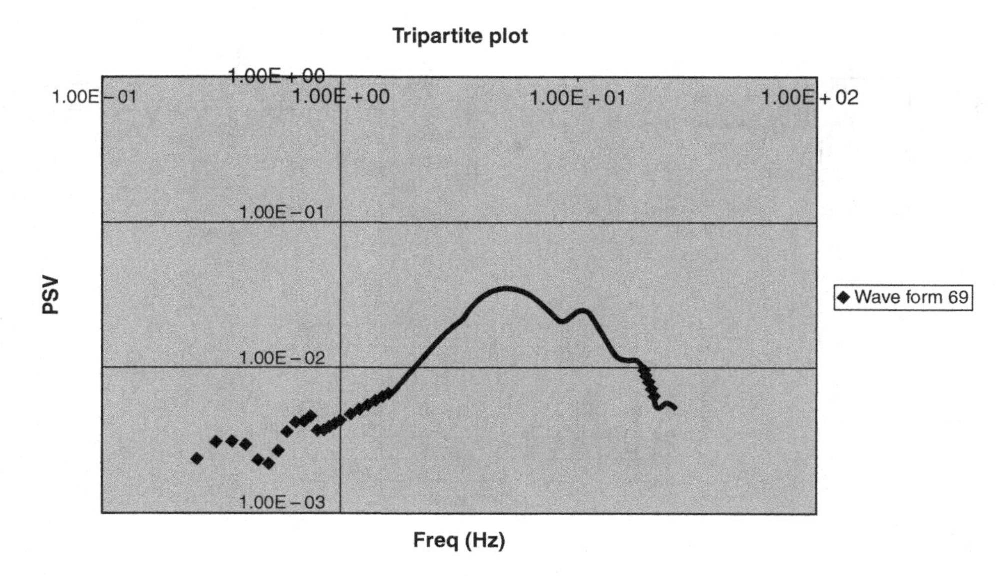

Figure 6.3 Response spectra (Freq. vs. PSV – 5 % damping).

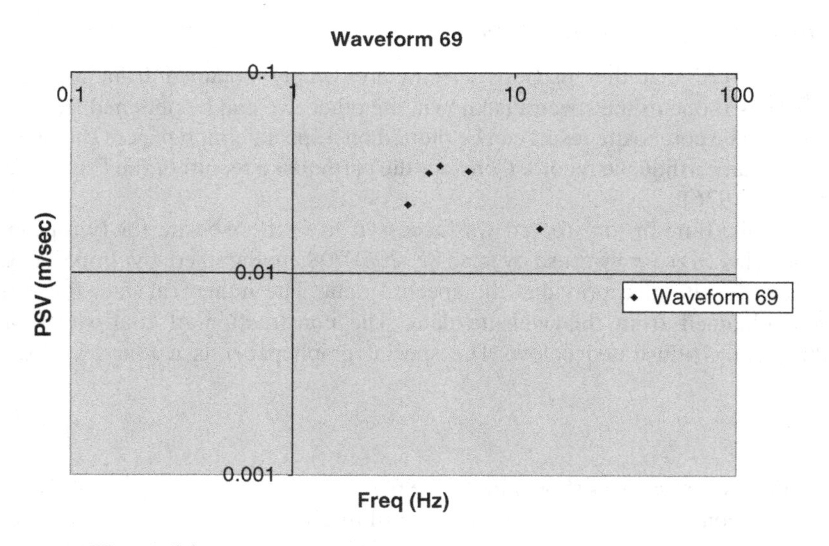

Figure 6.4 Response spectra – Response spectra (Freq. vs. PSV).

the logarithmic scales for S_d and S_a on Figure 6.3. The integrated presentation is possible because the three spectral quantities are inter-related:

$$\frac{S_a}{\omega} = S_v = \omega S_d$$

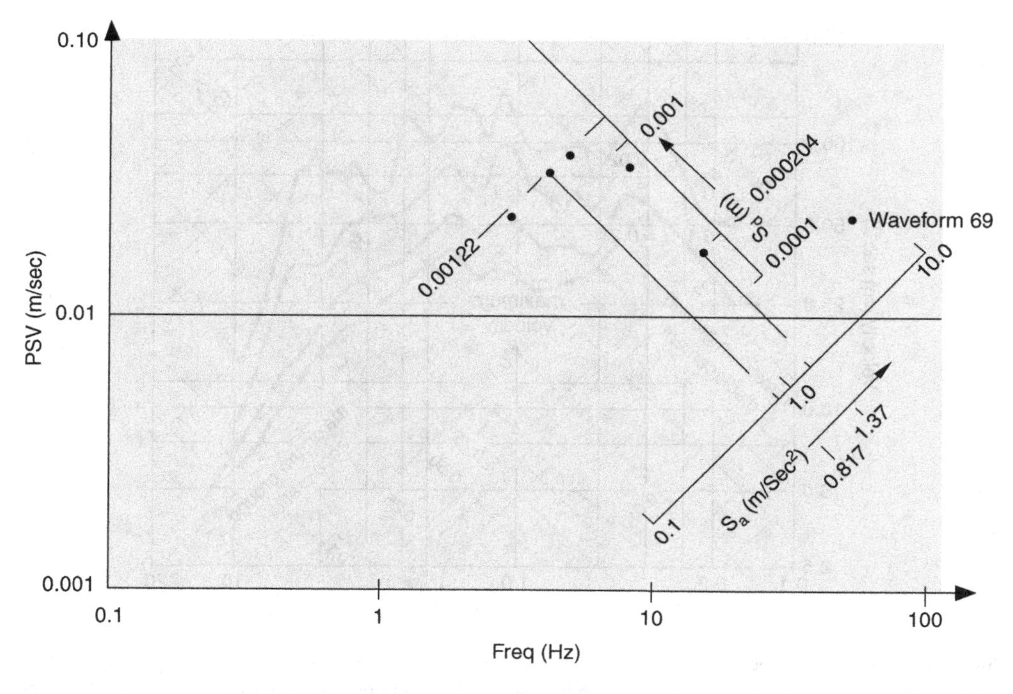

Figure 6.5 Determining logarithmic scales for S_d and S_a.

The S_d and S_a values for two points (identified in the large circle) are shown in Figure 6.5. The two points chosen for S_d have values: 0.000204 and 0.00122. From these two points the logarithmic scale for S_d may be determined (Figure 6.5). Similarly, the same two points for S_a have values: 0.817 and 1.370. From these two points the logarithmic scales may be determined.

The scales obtained from Figure 6.5 may now be introduced on Figure 6.3 to obtain the four-way logarithmic plot.

The response spectra for a single mass elastic system for the El Centro (1940) earthquake is shown in Figure 6.6.

Each earthquake is different and hence will have its own characteristic spectra. Many digitized records are available throughout the world and the spectra obtained from the majority of these are of the same general shape as that produced for the El Centro earthquake.

Note, the maximum displacement S_d is plotted on the lines sloping downward at $-45°$ against a horizontal logarithmic frequency scale. On the same plot ωS_d can be read on a vertical logarithmic scale. The value of $\omega^2 S_d$ can be read on the lines sloping upward at $+45°$.

Further Comments on Pseudo Spectra

If we know the natural period of the structure and the level of damping of an SDOF system, its peak spectral acceleration S_a can be read off the acceleration spectrum plot. In the case of an un-damped system, the peak response will occur at the point of maximum displacement of the spring.

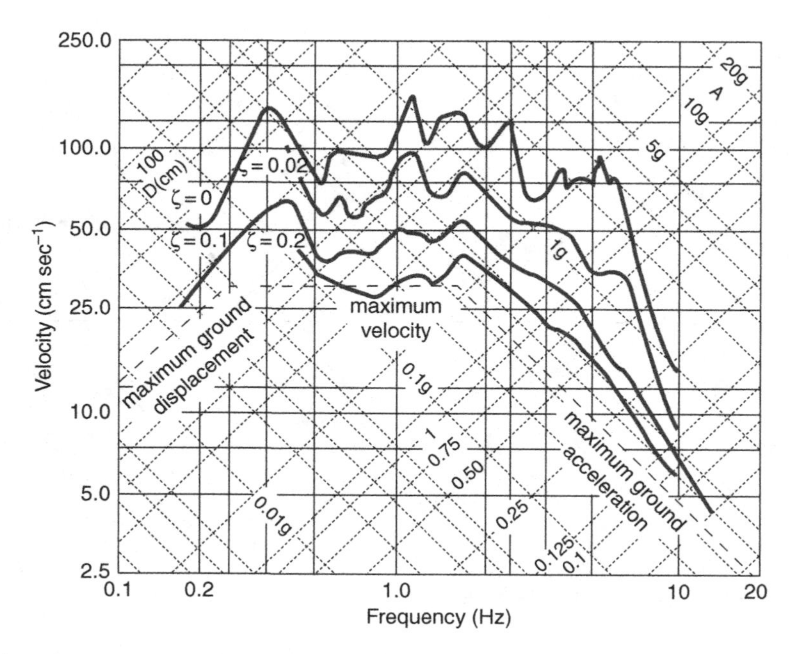

Figure 6.6 Response spectra for a single mass elastic system, El Centro (1940) earthquake, N-S component. Reproduced from Levy and Wilkinson (1976) by permission of the Portland Cement Association, Skokie, Illinois.

The maximum force in the spring is exact and is given by:

$$F_{\max} = kS_d = m\omega^2 S_d = mS_a \tag{6.9}$$

The peak spectral displacement S_d is given by

$$S_d = \frac{F}{k} = \frac{mS_a}{k} = \frac{mS_a}{m\omega^2} = \frac{S_a T^2}{4\pi^2} \tag{6.10}$$

For a practical range of structures damping is relatively small, hence the forces are given by the above relationships.

Note that the maximum acceleration occurs when the velocity is low and consequently the viscous damping (which is velocity proportional) is also low.

From the foregoing discussions we note and emphasize again that spectral acceleration S_a is the total acceleration (the true acceleration of the structure) whereas S_d, the spectral displacement is a relative value, measured in relation to the ground movement (which is also moving) during an earthquake. This at first sight may seem confusing. Looking at it from another angle, the absolute acceleration of the system is governed by the total force on the structure (Equation 6.9). The force is determined by the relative deformation (imagine the compression/elongation in the spring in Figure 6.1). The velocity spectrum S_v is due to relative velocity.

Response spectra calculations are now routinely performed in engineering offices and are no longer considered to be a specialist area.

6.2 Design Response Spectra

The ensembles of spectra from different earthquakes have been studied and statistical analyses performed to produce a general spectrum for use in codes of practice. The designer is aware that the design spectrum embraces the spectra of many different earthquakes and is a good starting point for estimating the seismic loads.

The earliest attempt at defining a design response was in the form of *site independent design* spectra as proposed by Professor G.W. Housner (1959). He derived the spectra from four large US earthquakes (M ranging from 6.5 to 7.7). The recording sites were located on rock, stiff soil and deep cohesionless soil. Housner's method may be extended to include many other records and produce a smoothed design curve that bounds 84 % of the data. The United States Atomic Energy Commission introduced a site independent response spectra (1973)

The properties of the soil underneath a site can significantly modify the amplitude level and influence the spectral composition of the ground motion.

Buildings with a certain natural period are often damaged from ground shaking when located on a soil stratum having a similar natural period. Consider the following scenario shown in Figure 6.7. A multi-storey structure is located on a soil stratum of depth H and shear wave velocity V_s. The fundamental period of the site is given by $T_s = 4H/V_s$. If, for example, the soil depth is 75 m and soil deposit has a shear wave velocity of 300 m/sec,

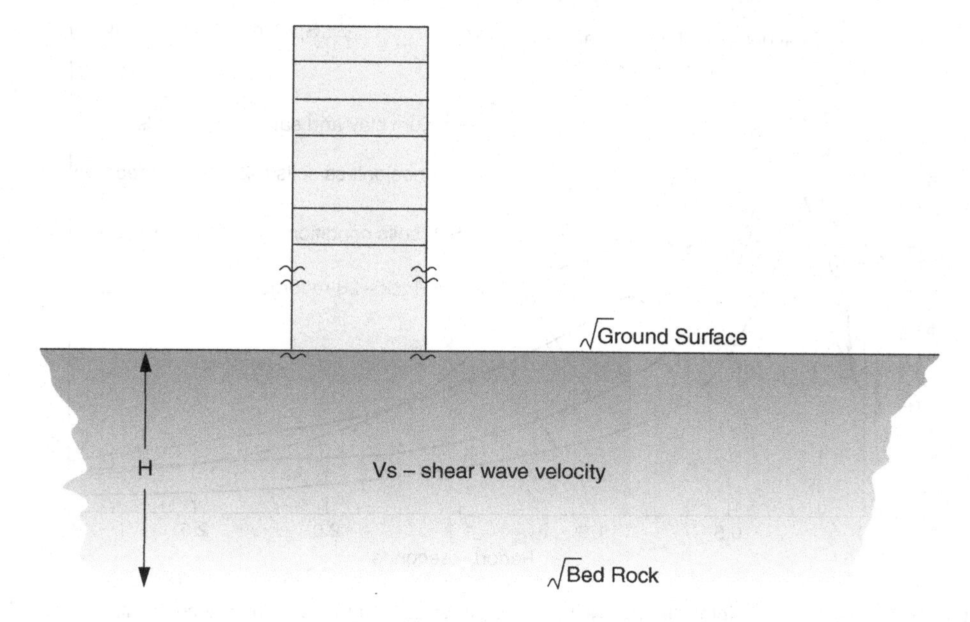

Figure 6.7 Multi-storey building on soil stratum.

then the fundamental period of the site, from the above relation, is 1.0 seconds. A rule of thumb often quoted for the purpose of fundamental period of buildings with N floors is N/10. Hence, buildings approximately 10 storeys high face possible resonance and severe damage as well. The damage is explained by the fact that the soil conditions in this type of situation amplify input ground motion in a period range coinciding with period of the structure. If the fundamental period of the structure is less than the fundamental period of the site, it is prudent to include soil-structure interaction (SSI) in the analyses, and if the fundamental period of the structure is longer than the fundamental period of the site then including SSI effects will reduce the response of the structure even if radiation damping effects are ignored. (Dowrick, 2003).

This phenomenon is most visibly demonstrated by the Mexico City earthquake (1985).

Seed *et al.* (1976) considered the influence of soil conditions on response spectra. The records were divided into four categories: rock, stiff soils, deep cohesionless soil and soft to medium clay. The plot of average acceleration spectra for different site conditions is shown in Figure 6.8.

Large earthquakes produce larger and *longer-period* ground motions than smaller earthquakes; consequently, the frequency content of a ground motion is related to earthquake magnitude. As seismic waves travel away from a fault, their higher-frequency components are scattered and absorbed more rapidly than their lower-frequency component. Hence, we would expect to see a change in frequency content with distance, as the higher-frequency components are absorbed. There is shift in the predominant period which is shown in Figure 6.9.

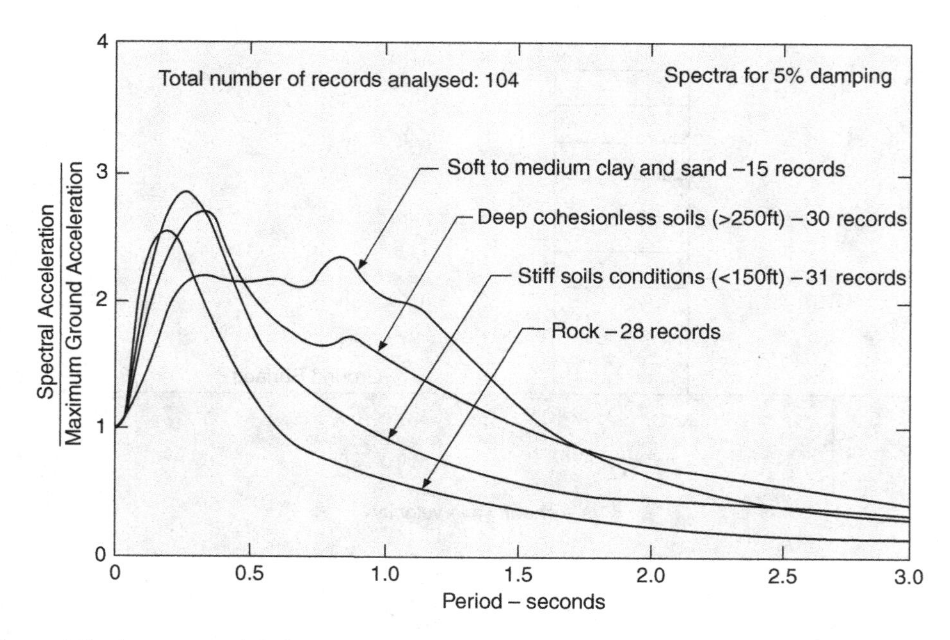

Figure 6.8 Average acceleration spectra for different site conditions. Reproduced from Seed *et al.* (1976), (c) the Seismological Society of America (SSA).

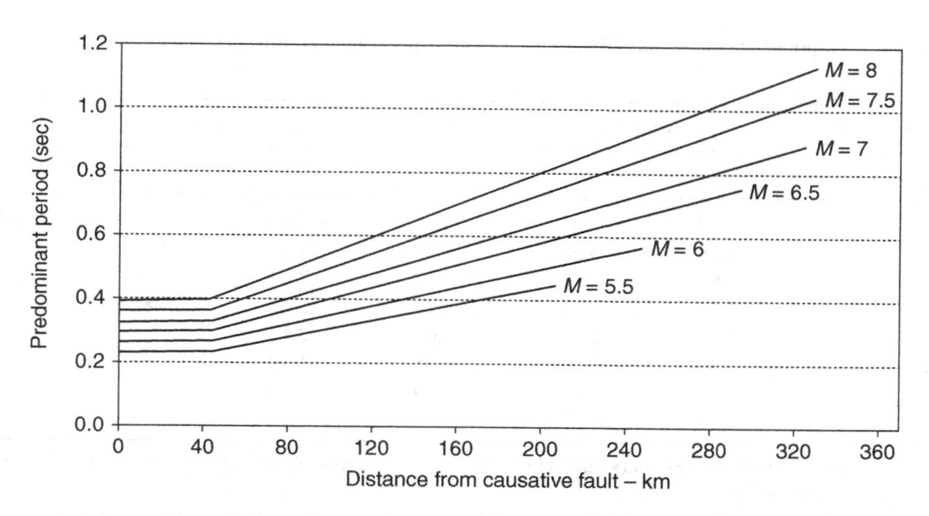

Figure 6.9 Variation of predominant period at rock outcrops with magnitude and distance. Reproduced from Seed and Idriss (1969) by permission of the Earthquake Engineering Research Center (EERC).

Amplification Factors

Referring back to Figure 6.3, it may be observed that the response spectrum over certain frequency ranges is related by an amplification factor to peak values of ground acceleration, velocity and displacement. Based on the above observation, Newmark and Hall (1969) proposed amplification factors for construction of the response spectrum on tripartite logarithmic, similar to Figure 6.3.

If the building is stiff (high natural frequency) the peak building acceleration will approach the peak ground acceleration. At low frequencies, the displacement will eventually approach the peak ground displacement. In between the low and high frequencies there is the intermediate range where most of the first mode natural frequencies of structures lie. The velocity is constant in this region. Thus separate amplification factors have been proposed for the high, intermediate and low frequency ranges with respect to peak ground acceleration, velocity and displacement. The response spectrum amplification factors are shown in Table 6.1.

The estimated peak ground acceleration, velocity and displacement are multiplied by the respective amplification factors for the level of damping. The response spectrum is constructed on tripartite paper. The values of the amplified acceleration, velocity and displacement are plotted along the respective axes. Lines of constant acceleration, velocity and displacement, are drawn through these points to produce the trapezoidal shaped response spectra similar to Figure 6.3.

The IBC (International Building Code) 2006 specifies detailed design spectra for a site in the USA. The general shape of the spectra specified may be found in ASCE 7.

The IBC specifies how to calculate the respective ordinates and its provisions are discussed in Chapter 8.

It may be seen that the response spectra method is a powerful design office tool and its wide acceptance in the field of earthquake engineering analysis is due to *the ease with which the base shears and base moments of practical structures can be evaluated.* A simple example

Table 6.1 Response spectrum amplification factors for peak acceleration (a), peak velocity (v), peak displacement (d).

Damping (%)	One sigma (84.1 %)			Median (50 %)		
	a	v	d	a	v	d
0.5	5.10	3.84	3.04	3.68	2.59	2.01
1.0	4.38	3.38	2.73	3.21	2.31	1.82
2.0	3.66	2.92	2.42	2.74	2.03	1.63
3.0	3.24	2.64	2.24	2.46	1.86	1.52
5.0	2.71	2.30	2.01	2.12	1.65	1.39
7.0	2.36	2.08	1.85	1.89	1.51	1.29
10.0	1.99	1.84	1.69	1.64	1.37	1.20
20.0	1.26	1.37	1.38	1.17	1.08	1.01

Source: Reproduced from Newmark and Hall (1982) by permission of the Earthquake Engineering Research Institute (EERI).

Note: The amplification factors quoted above are for firm ground. Amplification factors appropriate to soft soil sites are not available at present.

illustrates the use of the response spectra and the straightforward nature of the calculations involved.

6.2.1 A Worked Example: An MDOF System Subjected to Earthquake Loading

Earthquake Loading

Derivation of Equations for an MDOF System

Consider first the SDOF system shown in Figure 6.10. The only external loading is in the form of an applied motion at ground level.

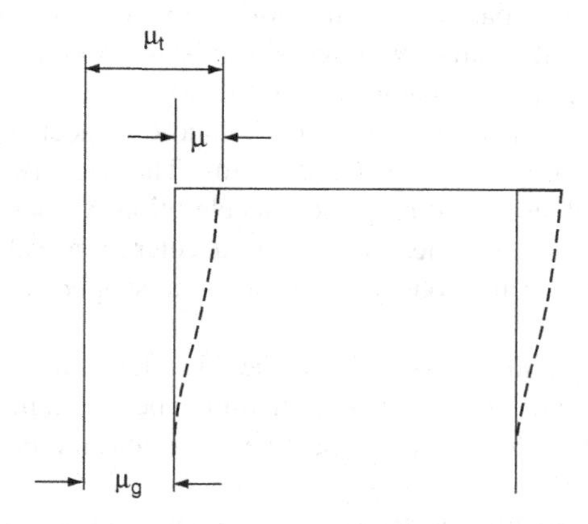

Figure 6.10 An SDOF Structure Subjected to Earthquake Loading.

The total acceleration of the mass 'm'

$\ddot{\mu}_t = \ddot{\mu} + \ddot{\mu}_g$ (Note: The dots denote differentiation with respect to time).

The equation of motion for the SDOF system is given by:

$$m\{\ddot{\mu} + \ddot{\mu}_g + c\{\dot{\mu}\} + k\{\mu\} = 0$$

or

$$m\ddot{\mu} + c\dot{\mu} + k\mu = -m\ddot{\mu}_g$$

Hence, extending the equations in the form for a lumped MDOF system we have:

$$[M]\{\ddot{\mu}\} + [C]\{\dot{\mu}\} + [K]\{\mu\} = F(t) \tag{6.11}$$

Based on stated assumptions

$$F(t) = -[M]\{I\}\{\ddot{\mu}_g(t)\} \tag{6.12}$$

Where $\{I\}$-resolving vector. The ith term in the column vector

$$\{I\} = 1$$

where the mass on the diagonal in the ith row of mass matrix corresponds with the direction of the ground acceleration.

$$\ddot{\mu}_g(t) = \text{absolute ground acceleration.}$$

Equation (6.11) can be solved directly using time history methods described in sections elsewhere. However, by transforming the equations to a normal mode coordinate system, the response of each mode of a multi-degree system can be treated as a single degree of freedom system. The objective is to demonstrate the use of the response spectrum for a multi degree freedom system.

We note, from earlier derivation in Chapters 2 and 3, the following relationships:

$$c_r = 2\sqrt{k}\sqrt{m}$$

$$\varsigma_r = \frac{c}{c_r} = \frac{\frac{c}{m}}{2\omega_r}$$

where

c_r = critical damping

At this stage, we need to make a coordinate transformation from the Cartesian to 'normal' coordinates so that the coupled Equations (6.11) may be uncoupled. The mode superposition technique may then be applied. By making the coordinate transformation:

$$\mu = [\phi]\eta,$$

where ϕ is the matrix of Eigen vectors and pre-multiplying each side of Equation (6.11) by $[\phi]^T$ we have,

$$[\phi]^T[M][\phi]\ddot{\eta} + [\phi]^T[C][\phi]\dot{\eta} + [\phi]^T[K][\phi]\eta = -[\phi]^T[M]\{I\}\ddot{\mu}_g(t) \tag{6.13}$$

The rth modal equation becomes

$$\ddot{\eta}_r + 2\varsigma_r\omega_r\dot{\eta}_r + \omega_r^2\eta_r = -\frac{\{\phi_r\}^T[M]\{I\}}{\{\phi_r\}^T[M]\{\phi_r\}}\ddot{\mu}_g(t) \tag{6.14a}$$

where

ω_r – frequency r

The term $\frac{\{\phi_r\}^T[M]\{I\}}{\{\phi_r\}^T[M]\{\phi_r\}}$ on the R.H.S. of equation (6.14a) is referred to as the *modal participation factor* ψ_r.

$$\psi_r = \frac{\{\phi_r\}^T[M]\{I\}}{\{\phi_r\}^T[M]\{\phi_r\}} \tag{6.14b}$$

The participation factor is indicative of how a particular mode responds to ground vibration in the rth direction. As may be seen from Equation (6.14b), the value ψ_r is dependent on the normalization scheme adopted for φ_r (see Chapter 3). A more detailed discussion may be found in Clough and Penzien (1993).

The term $\{\phi_r\}^T[M]\{\phi_r\}$ is the *modal mass* M_r (see Chapter 3).

The corresponding maximum response $\mu_r^{(0)}$ is obtained from the response spectrum because the response spectrum has been constructed from the maximum response and $\mu_r^{(0)}$ is assumed to be positive.

The maximum modal response is

$$\eta_{r\,(\max)} = \psi_r\mu_r^{(0)} \tag{6.15}$$

and the maximum displacement in relative coordinates

$$\{\mu_r\}_{\max} = \{\phi_r\}\eta_{r\,(\max)} \tag{6.16}$$

The corresponding joint inertia forces $\{F_r\}$ may be calculated from:

$$\begin{aligned}\{F_r\} &= \{I\}^T[K]\{\mu_r\}_{\max} \\ &= \{I\}^T[K]\{\phi_r\}\eta_{r(\max)} \\ &= \{I\}^T[K]\{\phi_r\}\psi_r\mu_r^{(0)} \\ &= \{I\}^T[K]\{\phi_r\}\frac{\{\phi_r\}^T[M]\{I\}}{M_r}\cdot S_{dr};\end{aligned}$$

with $S_a = \omega^2 S_{dr}$ and replacing K with $\omega^2 M$ we have

$$= \{I\}^T[M]\cdot\{\phi_r\}\frac{\{\phi_r\}^T[M]\{I\}}{M_r}S_a \tag{6.17}$$

Now, S_a may be read off the response spectrum plot.

Modal Equations for 3D Analysis

The formulation for a simple 2D system was outlined above. Buildings may qualify for 2D analyses. But even when a building qualifies for two separate 2D analyses in two orthogonal horizontal directions, X and Y, it is preferable to do the modal response spectrum analysis on a full 3D structural model. Then each mode shape, represented by vector ϕ_r for mode r, will in general, depending on modelling, have displacement and rotation components in all three directions, X, Y and Z, allowed for in the structural modelling.

The first step is the determination of the 3D modal shapes and natural frequencies of vibration (Eigen modes and Eigen values). Most finite element (FE) packages available will have this facility and it can be performed very reliably and efficiently.

From the Eigen-mode analysis the following need to be extracted for each mode for estimation of peak responses in all three directions:

- The natural circular frequency, ω_r, and the corresponding natural period, $T_r = 2\pi/\omega_r$
- The mode shape vector represented by ϕ_r
- The modal participation factors ψ_{Xr}, ψ_{Yr}, and ψ_{Zr}, in response to the seismic action in directions X, Y and Z computed as

$$\psi_{Xr} = \frac{\{\varphi_r\}^T[M]\{I_X\}}{\{\varphi_r\}^T[M]\{\varphi_r\}} = \frac{\sum_i \varphi_{Xi,r} m_{Xi}}{\sum_i (\varphi_{Xi,r}^2\cdot m_{Xi} + \varphi_{Yi,r}^2\cdot m_{Yi} + \varphi_{Zi,r}^2\cdot m_{Zi})} \tag{6.18a}$$

where i denotes the nodes of the structure associated with dynamic degrees of freedom, $[M]$ is the mass matrix, I_X is a vector with elements equal to 1 for the translational degrees of freedom parallel to direction X and with all other elements equal to 0, $\varphi_{Xi,r}$ is the element of ϕ_r corresponding to the translational degree of freedom of node i parallel to direction X and m_{Xi} is the associated element of the mass matrix (similarly for $\phi_{Yi,r}, \phi_{Zi,r}, m_{Yi}$ and m_{Zi}). If $[M]$ contains rotational mass moments of inertia, $\boldsymbol{I}_{\theta Xi}, \boldsymbol{I}_{\theta Yi}, \boldsymbol{I}_{\theta Zi}$, the associated terms also appear in the sum of the denominator. ψ_{Yr}, ψ_{Zr}, defined similarly.

- The effective modal masses in directions X, Y and Z, M_{Xr}, M_{Yr} and M_{Zr}, respectively, computed as shown below:

Effective Modal Mass

When considering the number of modes, the modal masses must account for more than 90 % of the total mass (ASCE 7, 2005, Cl.12.9). The effective modal masses may be computed as

$$M_{X,r} = \frac{\left(\{\varphi_r\}^T [M] \{I_X\}\right)^2}{\{[\phi_r\}^T[M]\{\phi_r\}} = \frac{\left(\sum_i \varphi_{Xi,r} m_{Xi}\right)^2}{\sum_i (\varphi^2_{Xi,r} \cdot m_{Xi} + \varphi^2_{Yi,r} \cdot m_{Yi} + \varphi^2_{Zi} \cdot m_{Zi})} \tag{6.18b}$$

and similarly for $M_{Y,r}$ and $M_{Z,r}$. These are essentially base-shear-effective modal masses, because the reaction forces (base shear) X, Y or Z due to mode r are equal to

$$F_{bX,r} = S_a(T_r) \cdot M_{Xr},$$

$$F_{bY,r} = S_a(T_r) \cdot M_{Yr}$$

and

$$F_{bZ,r} = S_a(T_r) \cdot M_{Zr} \text{ respectively.}$$

T_r – as defined above

The effective weight, for mode r (effective modal weight) may now be defined as

$$W_r = \frac{\left(\{\varphi_r\}^T [M] \{I\}\right)^2}{\{\phi_r\}^T[M]\{\phi_r\}} \cdot g \tag{6.19a}$$

$$= \frac{L_r^2}{M_r} \cdot g \tag{6.19b}$$

By computing the effective modal weights participating for each mode, the percentage of the participating mass may be calculated. This is illustrated in an example in Chapter 8, Section 8.5.

Example Problem – MDOF system

The structure to be analysed is shown in Figure 6.11. The structure is subjected to earthquake loading and the loading is represented by the response spectrum (simplified for the purpose of illustration) shown in Figure 6.12.

The first step is to perform an Eigen value analysis for the structure (refer to Chapter 3).

Mode	$\omega(rad)$	T(sec)	$S_a(m/\sec^2)$
1	7.9067	0.7947	5.00
2	40.7280	0.1543	3.85

$S_a(m/\sec^2)$ for the two frequencies are read from the response spectrum plot for the example problem (Figure. 6.12).

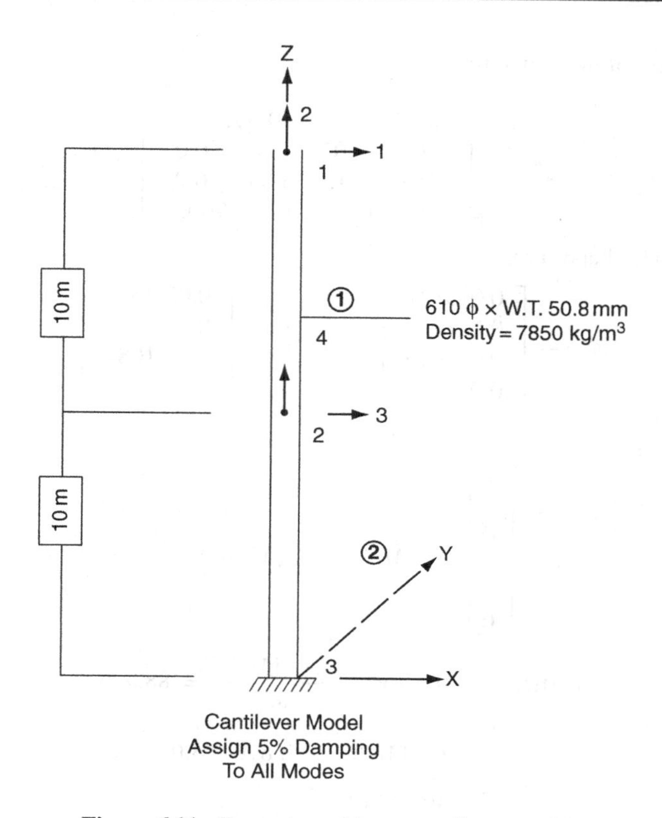

Figure 6.11 Example problem – cantilever model.

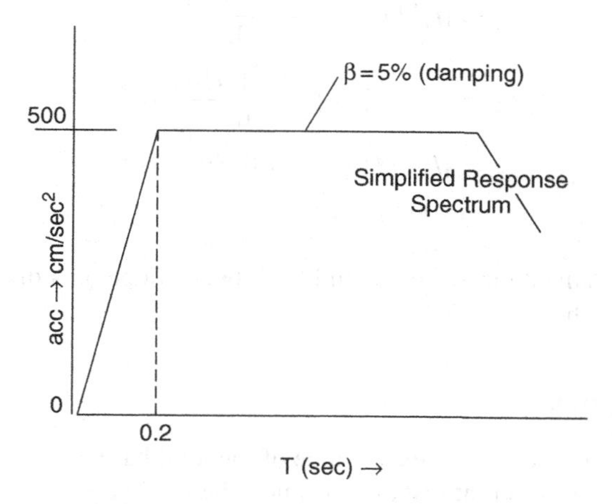

Figure 6.12 Example problem – simplified response spectrum.

The mass matrix of the structure is:

$$[M] = \begin{bmatrix} 3503\ldots0.0\ldots0.0\ldots0.0 \\ 0.0\ldots3503\ldots0.0\ldots0.0 \\ 0.0\ldots0.0\ldots7006\ldots0.0 \\ 0.0\ldots0.0\ldots0.0\ldots7006. \end{bmatrix}$$

The first two mode shapes are:

$$\{\phi_1\} = \begin{bmatrix} 0.01533 \\ 0.0 \\ 0.005019 \\ 0.0 \end{bmatrix} \qquad \{\phi_2\} = \begin{bmatrix} 0.007098 \\ 0.0 \\ -0.01084 \\ 0.0 \end{bmatrix}$$

Calculation: Mode 1

$$\{I\} = \begin{bmatrix} 1 \\ 0 \\ 1 \\ 0 \end{bmatrix} \qquad M_1 = \{\phi_1\}^T[M]\{\phi_1\}$$

$$\text{Participation Factor} = \frac{\{\phi_1\}^T[M]\{I\}}{M_1} = 88.876$$

$$\begin{aligned} F_1 &= \{I\}^T[M]\{\phi_1\} \cdot 88.876 \times 5.0 \\ &= 39490.4 \text{ N} \end{aligned}$$

Calculation: Mode 2

Referring to above, we note that

$$\begin{aligned} F_n &= \{I\}^T\,[M]\,\{\phi_n\}\frac{\{\phi_n\}^T[M]\{I\}}{M_n}S_a \\ F_2 &= \{I\}^T[M]\{\phi_2\}\frac{\{\phi_2\}^T[M]\{I\}}{M_2}S_a \\ &= \{I\}^T[M]\{\phi_2\} \cdot 51.0789 \times 3.85 \\ &= 10041.3 \text{ N} \end{aligned}$$

Note: A different normalization scheme could have been adopted as discussed in Chapter 3 but the final results are the same.

Modal Combination Rules

It will not be possible to find out the exact value of the total base shear, as the peak values of the modes are attained at different instants of time. The total base shear may be found from modal combination rules.

The ABS (Absolute Sum) Rule

The absolute sum (ABS) rule gives a very conservative value.
Total (ABS) = 39490 + 10041 = 49531 N

The Square Root of Sum of Squares (SRSS) rule

A more realistic estimate may be obtained from the square root of sum of squares (SRSS) rule. The peak response in each mode is squared and then summed. The square root of the sum provides an estimate of the response.

$$\text{Total (SRSS)} = \sqrt{39490^2 + 10041^2} = 40747\text{N} = 40.747 \text{ kN}$$

Moment at the Base

Referring to Figure 6.11 the moment at the base would be contributed by the horizontal shear at DOF (degree of freedom) 1 and 3.

Calculation: Mode 1

$$F_{\text{mode 1}} = \{I\}^T[M]\{\phi_1\}\frac{\{\phi_1\}^T[M]\{I\}}{M_1}S_{a(\text{mode 1})}$$

$$F_{dof1} = \{I\}^T[M]\{\phi_1\} \times 88.876 \times 5.0$$

$$= [1..0..0..0]\begin{bmatrix} 23865 \\ 0.0 \\ 15624 \\ 0.0 \end{bmatrix}$$

$$= 23865 \text{ N}$$

$$F_{dof2} = [0..0..1..0]\begin{bmatrix} 23865 \\ 0.0 \\ 15624 \\ 0.0 \end{bmatrix}$$

$$= 15624 \text{ N}$$

Calculation: Mode 2

$$F_{\text{mode 2}} = \{I\}^T[M]\{\phi_2\}\frac{\{\phi_2\}^T[M]\{I\}}{M_2}S_{a(\text{mode 2})}$$

$$F_{dof1} = [1..0..0..0]\begin{bmatrix} 4888.7 \\ 0.0 \\ -14934.5 \\ 0.0 \end{bmatrix}$$

$$= 4888.7 \text{ N}$$

$$
\begin{aligned}
F_{dof2} &= [0..0..1..0]\begin{bmatrix} 4888.7 \\ 0.0 \\ -14934.5 \\ 0.0 \end{bmatrix} \\
&= -14934.5 \text{ N}
\end{aligned}
$$

Therefore,

$$
\begin{aligned}
M_{\text{mode } 1} &= 23.865 \times 20.0 + 15.624 \times 10.0 \text{ kN} \cdot \text{m} \\
&= 633.54 \text{ kN} \cdot \text{m}
\end{aligned}
$$

$$
\begin{aligned}
M_{\text{mode } 2} &= 4.8887 \times 20.0 - 14.935 \times 10.0 \text{ kN.} \\
&= -51.59 \text{ kN} \cdot \text{m} \\
\text{Total (SRSS)} &= \sqrt{633.54^2 + (-51.59)^2} = 635.64 \text{ kN} \cdot \text{m}
\end{aligned}
$$

Deflection Calculations

The maximum modal response is

$$\eta_{r(\max)} = \{\phi_r\}\mu_r^{(0)}$$

and

$$\{\mu_r\}_{\max} = \{\phi_r\}\eta_{r(\max)}$$

where $\mu_r^{(0)}$ (max. response: defl., vel., acc.)

$$
\begin{aligned}
\eta_{1(\max)} &= \psi_1 \cdot \frac{5.0}{7.9067^2} \qquad \left[S_d = \frac{S_a}{\omega^2} = \frac{5.0}{7.9067^2}\right] \\
&= 7.108
\end{aligned}
$$

Mode 1 Deflections $(\mu_{1(\max)})$

$$
\begin{aligned}
\begin{Bmatrix} dof_1 \\ dof_2 \\ dof_3 \\ dof_4 \end{Bmatrix} &= \begin{Bmatrix} 0.01533 \\ 0.0 \\ 0.005019 \\ 0.0 \end{Bmatrix} 7.108 \\
&= \begin{Bmatrix} 0.10896 \\ 0.0 \\ 0.03568 \\ 0.0 \end{Bmatrix}
\end{aligned}
$$

Mode 2 Deflections ($\mu_{2(\max)}$)

$$\psi_2 = 51.0789 \cdot \frac{3.85}{40.728^2}$$
$$= 0.118$$

$$\begin{Bmatrix} dof_1 \\ dof_2 \\ dof_3 \\ dof_4 \end{Bmatrix} = \begin{Bmatrix} 0.007098 \\ 0.0 \\ -0.01084 \\ 0.0 \end{Bmatrix} 0.118$$
$$= \begin{Bmatrix} 0.00084 \\ 0.0 \\ -0.0012853 \\ 0.0 \end{Bmatrix}$$

SRSS Deflections

$$dof_1 = \sqrt{0.10896^2 + (0.00084)^2}$$
$$= 0.10896 \text{ m}$$
$$dof_3 = \sqrt{0.03568^2 + (-0.0012853)^2}$$
$$= 0.0357 \text{ m}$$

Complete Quadratics Combination (CQC) Rule

It has been recognized that the SRSS rule provides very good estimates for systems with separated frequencies, but may not yield the desired result when the frequencies are closely spaced. This situation is most frequently encountered in the piping systems of many industrial facilities and was not apparent when the SRSS rule was initially proposed. The CQC rule overcomes this problem and is now accepted as the standard way of combining modes. The codes of practice recommend the CQC rule and the software packages have implemented this combination rule. According to the CQC rule:

$$r_{\max} \cong \left(\sum_{i=1}^{i=N} \cdot \sum_{j=1}^{j=N} \rho_{ij} r_{i\max} r_{j\max}\right)^{1/2} \tag{6.20}$$

where $r_{\max}$ is the maximum estimate from the combination rule
ρ_{ij} – a correlation coefficient to be determined
$r_{i\max}, r_{j\max}$ – peak responses in modes i and j.

The correlation coefficient ρ_{ij} for the two modes varies between 0 and 1 and $\rho_{ij} = 1$ for $i = j$. One of the earliest formulations of ρ_{ij} is due to Newmark and Rosenbleuth (1971).

The correlation coefficient most widely in use is due to Der Kiureghian (1981). The suggested equation for computing the correlation coefficient takes the form

$$\rho_{ij} = \frac{8\sqrt{\zeta_i\zeta_j}\,(\lambda_{ij}\zeta_i + \zeta_j)\,\lambda_{ij}^{3/2}}{(1-\lambda_{ij}^2)^2 + 4\zeta_i\zeta_j\lambda_{ij}(1+\lambda_{ij}^2) + 4(\zeta_i^2+\zeta_j^2)\lambda_{ij}^2} \tag{6.21}$$

where $\lambda_{ij} = \omega_i/\omega_j$
ω_i, ω_j - frequency of mode i and j
ζ_i, ζ_j – modal damping ratios in modes i and j.

The above formulation is for modal damping ratio being different for the two modes i and j. If the modal damping ratios are the same for the two modes then the equation reduces to:

$$\rho_{ij} = \frac{8\zeta^2(1+\lambda_{ij})\lambda_{ij}^{3/2}}{(1-\lambda_{ij}^2)^2 + 4\zeta^2\lambda_{ij}(1+\lambda^2_{ij}+) + 4(2\zeta^2\lambda_{ij}^2)} \tag{6.22}$$

The above combination rule is recommended in the SEAOC Code (1999).

It has been shown (Chopra, 2001) that the correlation coefficient ρ_{ij} diminishes rapidly as the two natural frequencies ω_i, ω_j move further apart. As a result all cross terms in the CQC rule become very small and may be neglected and the estimate reduces to the SRSS rule.

Referring to the above problem, let us now compute the base shear V_{base} according to CQC rules.

The frequencies are:

Mode	$\omega(rad)$
1	7.9067
2	40.7280

Peak modal responses

Mode	V_{base}
1	39.490 kN
2	10.041 kN

Natural frequency ratios

Mode i	$j = 1$	$j = 2$
1	1.00	0.194
2	5.151	1.00

Correlation coefficients ρ_{ij}

Mode i	$j = 1$	$j = 2$
1	1.00	0.00233
2	0.00233	1.00

Components in CQC Rule

Base Shear V_{base}

Mode i	$j = 1$	$j = 2$
1	1559.46 kN	0.91202 kN
2	0.91202 kN	100.821 kN

Base Shear V_{base}
Adding the individual terms and taking the square root gives us:

$$V_{base} = (1559.46 + 0.91202 + 0.91202 + 100.821)^{1/2} \quad \text{(see Equation 6.20)}$$
$$= 40.769 \text{ kN}$$

Note: Since the frequencies are well separated the estimate of base shear is close to the base shear obtained from the SRSS rule.

6.3 Site Dependent Response Spectra

Establishing a design response spectrum is an activity otherwise covered under the topic seismic hazard analysis. Seismic hazard analysis usually covers a fairly wide range of activities. Attention here will be focused on design response spectra. The two approaches followed are the deterministic and the probabilistic. The deterministic approach is simpler and is discussed in this chapter. The probabilistic approach is discussed in Chapter 7.

Deterministic ground motion plays an important part in planning, preparedness and loss estimation due to earthquakes in urban areas with valuable assets. In many situations seismic hazard may be readily related to a specific earthquake which has happened before.

Deterministic ground motion hazard estimates are sometimes made for evaluating design ground motions at sites potentially near an active fault capable of producing a significant earthquake. Such an event may have a sufficiently high frequency of occurrence or high public profile necessitating the requirement for inclusion in design.

Another role for site dependent deterministic ground motion arises when it is required by a regulatory agency. However, the trend is shifting towards probabilistic approaches.

The various seismological inputs required have been discussed earlier. These quantities form the various inputs for deterministic seismic hazard analysis.

From the foregoing discussions it emerges that for strong earthquakes on firm ground at intermediate to medium focal distances ($7.5 < R < 60$ km), the attenuation relationships permit the estimation of ground motion parameters, a, v, d at the given site. In developing a response spectrum at the given site, three basic quantities must be established.

The attenuation relationships are in terms of magnitude M and some form of distance measure to the source. Hence these two quantities must be known. The other quantity, which also needs to be established, is the peak ground acceleration at the site.

Two levels of earthquake activity need to be defined. There are pressing economic reasons for this. A strong earthquake imposes the most severe kind of loading the structure is ever likely to experience, but the probability of this level of earthquake at the site is very low. A pragmatic approach is to design the structure for mild earthquakes so that the stresses are within *elastic* limits. This is often referred to as the Operating Basis Earthquake (OBE). For strong earthquakes the approach dictates that the structure should deform but should not collapse, thus ensuring the safety of the occupants. This is referred to as the Safe Shutdown Earthquake (SSE). As its name implies the design must cater for orderly shutdown and prevent loss of life. The design is *inelastic* and concepts of *Performance Based Earthquake Engineering (PBEE)* are slowly being introduced into the design process. PBEE concepts are briefly discussed in Chapter 12.

The Operating Basis Earthquake represents an earthquake which is most likely to occur during the lifetime of the structure (e.g. 40 years). This level of earthquake would in all likelihood have occurred in the past.

The Safe Shutdown Earthquake, on the other hand, is thought to represent the severest earthquake in terms of magnitude at the site. An earthquake of this magnitude may not have occurred in the past, but this value is arrived at after very detailed investigation (part of the pre-activity) of past earthquakes, local geology (fault structure, distance from the fault etc.), and local underlying soil characteristics. The nature of pre-activity usually undertaken is shown in Figure 6.13.

6.3.1 The Design Process

The design process due to seismic loading is a tortuous one. Judgements made must be conservative from the design point of view. There is also the dual design philosophy to contend with – to produce an economically designed structure as well.

The controversial, or rather the uncertain part is, the prediction of peak ground acceleration. This stems from the fact that we are dealing with a *future* earthquake about which we know very little. It is possible to figure out from the previous discussions that the seismic

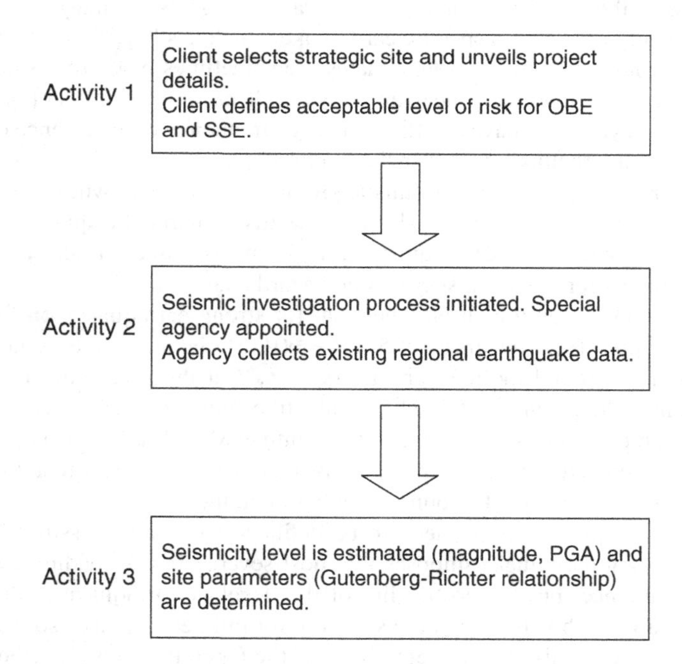

Figure 6.13 Site investigation process (pre-activity).

loading evaluation process has been an evolutionary one. New facts and data are being made available for analyses all the time. For example, only after instrumentation was installed *fortuitously* before the San Fernando earthquake (1971), did a clearer picture emerge regarding peak ground acceleration with causative faults less than 10 km away.

Establishing the peak ground acceleration (PGA) at a site, where very little historical data or no attenuation relationships are available, makes the task more difficult.

Loading Definition Phase

For the practical example given in the next section, we may now identify the next set of activities summarized as shown in Figure 6.14.

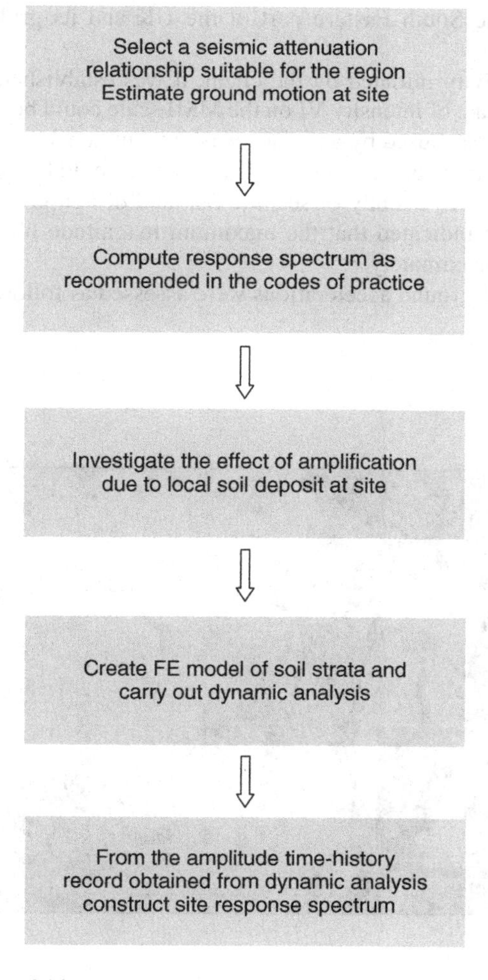

Figure 6.14 Flow chart for site dependent response spectra.

The use of local seismological records to develop site-dependent response spectra is preferable and certainly more desirable than site independent spectra. The site-dependent spectrum will account for the local soil conditions, which may not be reflected in the site independent spectra. However there are difficulties in the way. First, in estimating the magnitude and peak ground acceleration at the site, the source mechanism and local transmission path characteristics may not be fully explored or thoroughly evaluated. Secondly, available seismogram records may not be adequate for statistical analyses. Given these difficulties, caution and more importantly, engineering judgement must be exercised to ensure that a reasonable set of approaches has been adopted.

6.3.2 Practical Example: Construction of a Site Dependent Response Spectrum

The site is located in the South Eastern part of the UK and its grid location is shown in Figure 6.15.

As part of the pre-activity initiated by the client, it was established that for the 100 year seismic event an earthquake of intensity VI on the MMI scale could be considered appropriate and it would be likely to be caused by a surface wave magnitude $M_s \approx 5.4$.

For the 1 in 10,000-year event a magnitude 7 earthquake could be considered appropriate.

Dowrick (1981) conducted a study of seismic risk and design ground motions in the UK offshore area which also indicated that the maximum magnitude for the 10,000-year event earthquake would be approximately 7.

Subsequently, the peak ground accelerations were assessed as follows:

100 yr event : 0.05g
10,000 yr event: 0.25g

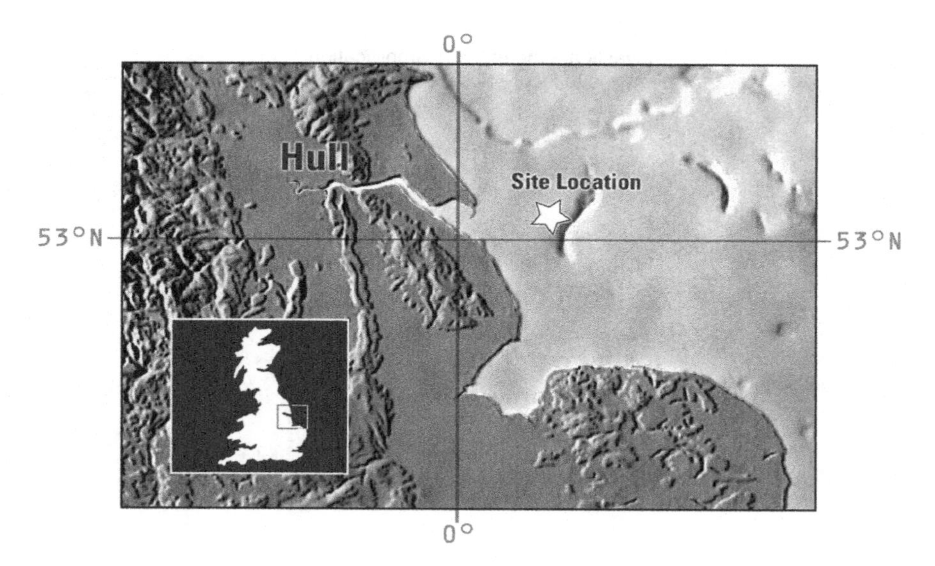

Figure 6.15 UK grid location of project site.

As we have seen in the previous discussions, a considerable amount of effort has been devoted by seismologists towards developing various attenuation equations, which predict ground motion in terms of magnitude, distance and site geology for the seismically active *inter-plate* regions.

The UK is a part of an *intra-plate* region of low seismic activity but is nevertheless seismic and an earthquake is considered a possibility. Hence a brief seismic study was carried out. It so happens that for most *intra-plate* regions of the world including the UK and Northwest Europe, no attenuation relationships exist. Strong motion data available for the region are sparse and there are few empirical studies to differentiate between *inter-plate* and *intra-plate* regions. Dahle *et al.* (1990) presented attenuation relationships for *intra-plate* regions. However, in order to obtain a sufficiently large database, more than a third of the events were from *inter-plate* regions. Hence the database was diluted and as such studies on *intra-plate* ground motions have been inconclusive.

A study of seismic ground motions for UK design (Ambraseys *et al.*, 1992) showed that for previous seismic designs within the UK *inter-plate* region attenuation relationships were used.

Preliminary Data

The soil profile is shown in Figure 6.16. Bedrock has been assumed at 17 m where chalk is encountered. For the deposits up to 17 m the measurements at site showed the shear wave velocity $((V)_s$ to be in the range of 250–350 m/sec.

Detailed geological maps of the region showed major faulting associated with a geological formation occurring at a distance of approximately 53 km.

Preamble

The expected ground motion at a site is notoriously difficult to predict. Local residents in San Francisco Bay area will perhaps testify to this. The Loma Prieta earthquake hit San Francisco Bay area in 1989. Its magnitude was 7.1 and the epicentre 120 km distant. The severity of ground motion was considerably more than expected and consequently the damage much greater than would have been predicted. As an example of how predictions may go wrong, the Loma Prieta earthquake was an eye opener in many senses. One effect that was not predicted was the complete loss of bearing strength of loose, saturated, sandy soils (liquefaction). Why did the Bay area suffer so much devastation? Were the predictions of seismic effects adequate? Structural engineers, for example, need reliable predictions of ground motion before they can design critical structures. Experience since suggests that we need to know more about the fault rupture process associated with a predicted earthquake; a more reliable characterization of the geological make-up of the soil along the seismic wave's travel path and the liquefaction potential of the underlying soil at a site. This aspect has been further explored in the section on soil-structure interaction in Chapter 10.

Estimate of Ground Motion

A preliminary estimate of seismic loading, which may be carried out based on available information, is outlined next. These preliminary estimates of seismic loading enable the designer

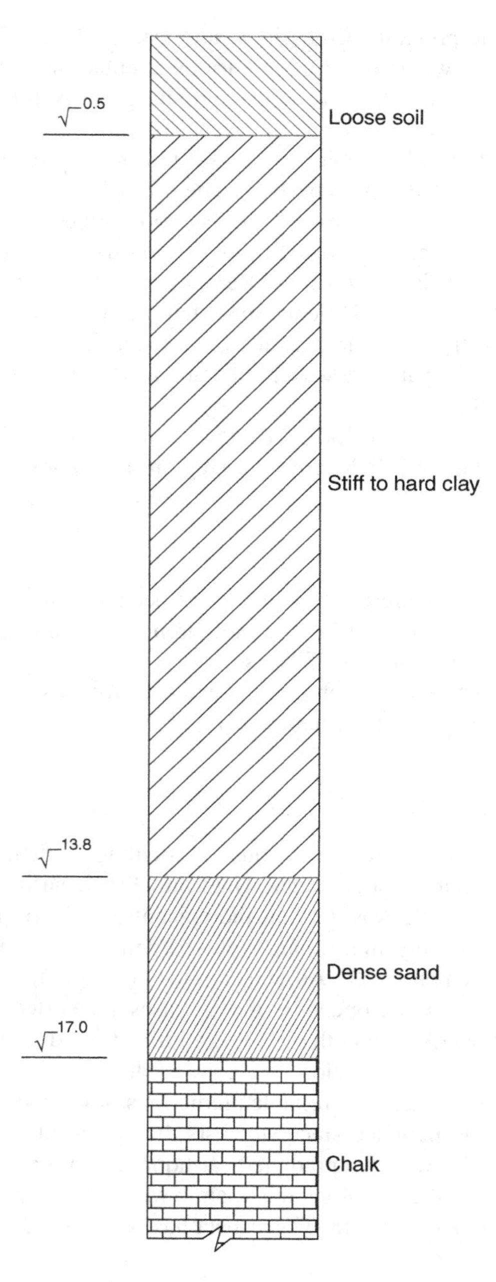

Figure 6.16 Soil profile at site.

to assess the severity of loading. These are later supported by more detailed seismological investigations by specialists.

In view of the nature of soil deposit ($V_s \approx 250 - 350m/\text{sec}$) several different approaches may be adopted for the site. Here we restrict ourselves to *four* pragmatic, design office oriented, approaches for this problem:

1. As a first pass evaluation, adopt the amplification factor method (Newmark and Hall, 1982). The database for amplification factors was from *inter-plate* records.
2. Use the most recent attenuation relationship for the European region based on weighted regression analyses on spectral ordinates. Coefficients are included to model the impact of local site effects (Ambraseys *et al.*, 2005).
3. Compute response spectrum as recommended in the codes of practice. The codes of practice have recommendations for low activity seismic zones.
4. The soft soil deposit will produce its own brand of ground amplification, hence carry out a dynamic analysis of the local soil strata subjected to recorded ground motion from representative earthquakes of approximately similar magnitude.

Finite Element (FE) modelling is most appropriate and the *amplified ground motion* at the surface may be used as input to develop the response spectra (see Section 6.1).

Approach 1

Referring to Step 1 (Figure 6.13), McGuire's (1974) attenuation relationship (Equations (5.12 (a and b)) will be used for the first pass assessment. McGuire's relationship is as follows:

$$y = b_1' 10^{b_2 M} (R + 25)^{-b_3} \qquad [5.12]$$

With $R = 53\,\text{km}$ and

$M = 5.4$

$a = 51.73\ cm/s^2$ (see Table 5.1). This, fortuitously, is close to 0.05g ($49.25\ cm/s^2$) assessed earlier.

$v = 4.39\ cm/s$.

Esteva and Villaverde's (1973) equation (Equation 5.10 (a and b)) gives:

$$a = 48.68\ cm/s^2$$

$$v = 4.30\ cm/s.$$

Ground displacement may be estimated from the empirical relationship proposed by Newmark and Rosenbleuth (1971) (Equation 5.11). Using the relationship (for R = 53 km)

$$d \approx 12v^2/a$$

$d = 4.47\ cm$ (values due to McGuire)

$d = 4.56\ cm$ (values due to Esteva and Villaverde).

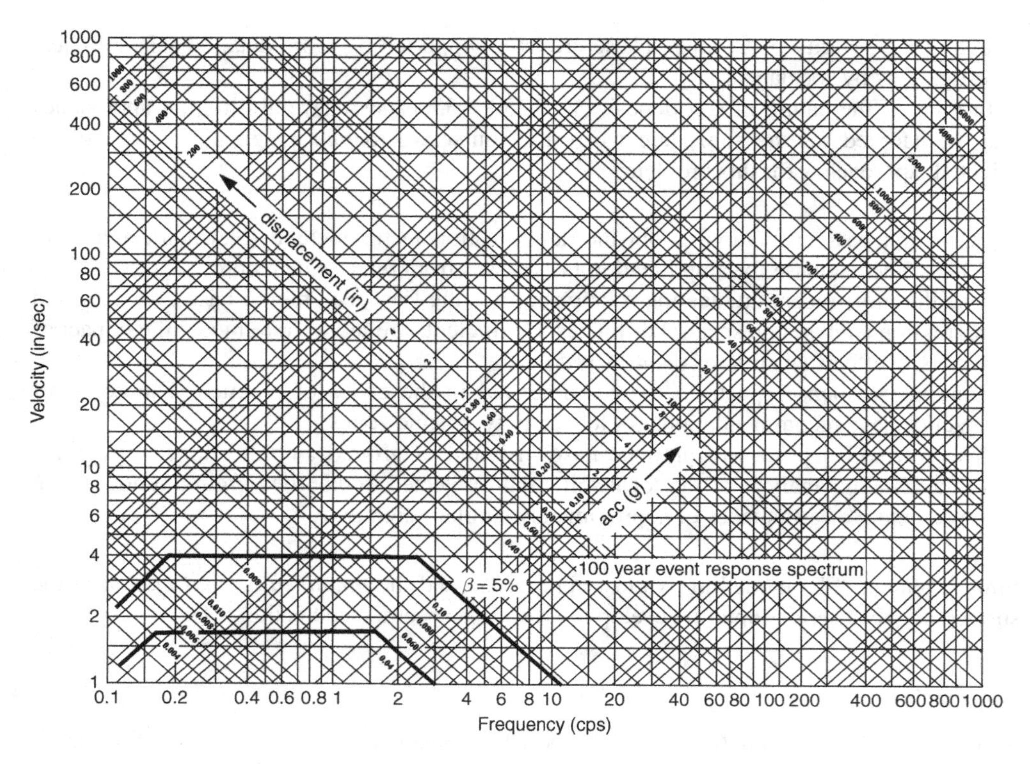

Figure 6.17 Tripartite plot.

The values of acceleration, velocity and displacement due to McGuire ($51.73\,cm/s^2$, 4.39 cm/sec and 4.47 cm are plotted along the respective axes. Lines of constant acceleration, velocity and displacement are drawn through these points to produce the trapezoidal shaped response spectrum shown in Figure 6.17. These are the values at bedrock level. Amplification factors (84 percentile values for design purposes – Table 6.1) on acceleration, velocity and displacement proposed by Newmark and Hall (1982) are applied next to produce a site dependent design response spectrum (Figure 6.17). The amplification factors used have been suggested for firm ground.

The design acceleration response spectrum is shown in Figure 6.17.

Approach 2

Attenuation Relationship with Soil Model (Ambraseys et al., 2005)

The idea is to use the most recent attenuation relationship developed for the region, which also includes factors for soft soil. The relationship proposed by Ambraseys *et al.* (2005), may be used. The recommended relationship is shown in Equation 5.21. The soil may be characterized as soft ($V_s \approx 250 - 350m/\text{sec}$); site category D, IBC 2003.

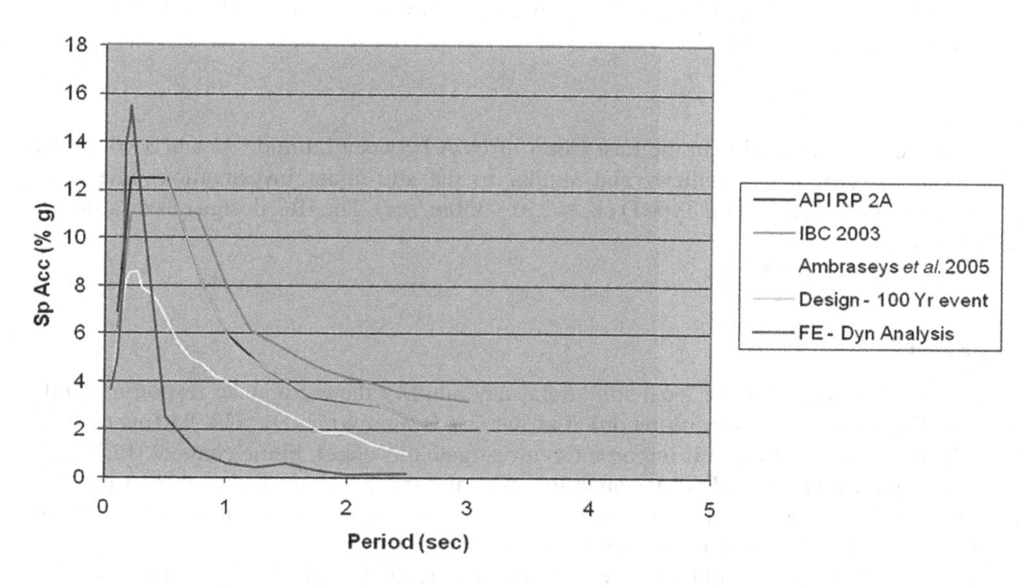

Figure 6.18 Site-dependent response spectra (5 % damping).

Referring to Section 5.2.2 and setting $S_S = 1$ for soft soil and 0 otherwise, $S_A = 1$ for stiff soil sites and 0 otherwise, S_A and S_S are set to 0 for rock sites. $F_N = 1$ for normal faulting earthquakes and 0 otherwise and $F_0 = 1$ for odd faulting earthquakes and 0 otherwise, the response spectra are shown in Figure 6.18.

Approach 3

Code Provisions

The site is located in a low seismic zone of Europe.

The codes of practice provide guidance on constructing response spectra. Several categories of seismic zones are identified on the map. Also identified is the soil category of the site. The response spectra according to provisions of the following codes are constructed:

1. API RP 2A-LRFD (1993 R 2003)
2. API RP 2A-WSD (2000)
3. IBC 2003.

API RP 2A

Response spectra for design of offshore structures for specific seismic zones in the USA are specified in API RP 2A, sec. 2.3.6. The peak ground acceleration 0.05g specified for the

100-year event, corresponds to Zone 1 of a seismic risk map of USA. The response spectrum for soil type B (description of which closely matches that with the soil profile in the code) has been reproduced in Figure 6.18.

IBC 2003

The seismic zone selected is in the East Coast of USA between Latitude 34 and 35 degrees. This is an area of low seismicity and similar to the site under investigation. The code parameters chosen are for Soil Type D ($V_s \approx 250 - 350 m/\text{sec}$). The IBC design curve is shown in Figure 6.18.

Approach 4

It is well established that the local soil strata may amplify the earthquake response significantly. The shear wave velocity recorded at the site is approximately 250–350 m/sec (soil Type D, IBC 2003). Hence, it is important to investigate this effect. Finite element (FE) modelling is a good way forward. The availability of quality FE packages makes this easier. The soil strata at the site are composed of stiff clay up to a depth of 13.8 m followed by 3.2 m of sand (Figure 6.16). The width of the FE model is 17 m.

The FE package ABAQUS (v. 6.5) was used to create the model. The FE soil is modelled as shown in Figure 6.19. The loose topsoil of 0.5m has been ignored. The details of the FE model are shown below:

- Element : CPE4R (2D continuum, 4noded Reduced Integration Element)
- (Variables: U_x, U_y at the nodes) The time varying earthquake excitation was applied to the bottom nodes in the model. The nodes on the two sides were constrained to move in-phase. The input excitation at the bottom and the output excitation at the top are also shown in Figure 6.19.

With reference to Figure 6.16, the properties of the stiff clay and sand used in the model were as follows:

Clay
Young's Modulus, 'E': 250 MPa increasing @ 10 Kpa/m
Poisson's Ratio, 'ν' : 0.3
Yield Stress : 10 Kpa
Sand
Friction Angle : 34°
Dilation Angle : 12°

The earthquake record details selected from the European Site are shown in Table 6.2. The input excitation at the bottom and the output excitation at the top are also shown in Figure 6.19.

The response spectrum obtained from the excitation at the top for 5 % damping is also shown in Figure 6.18.

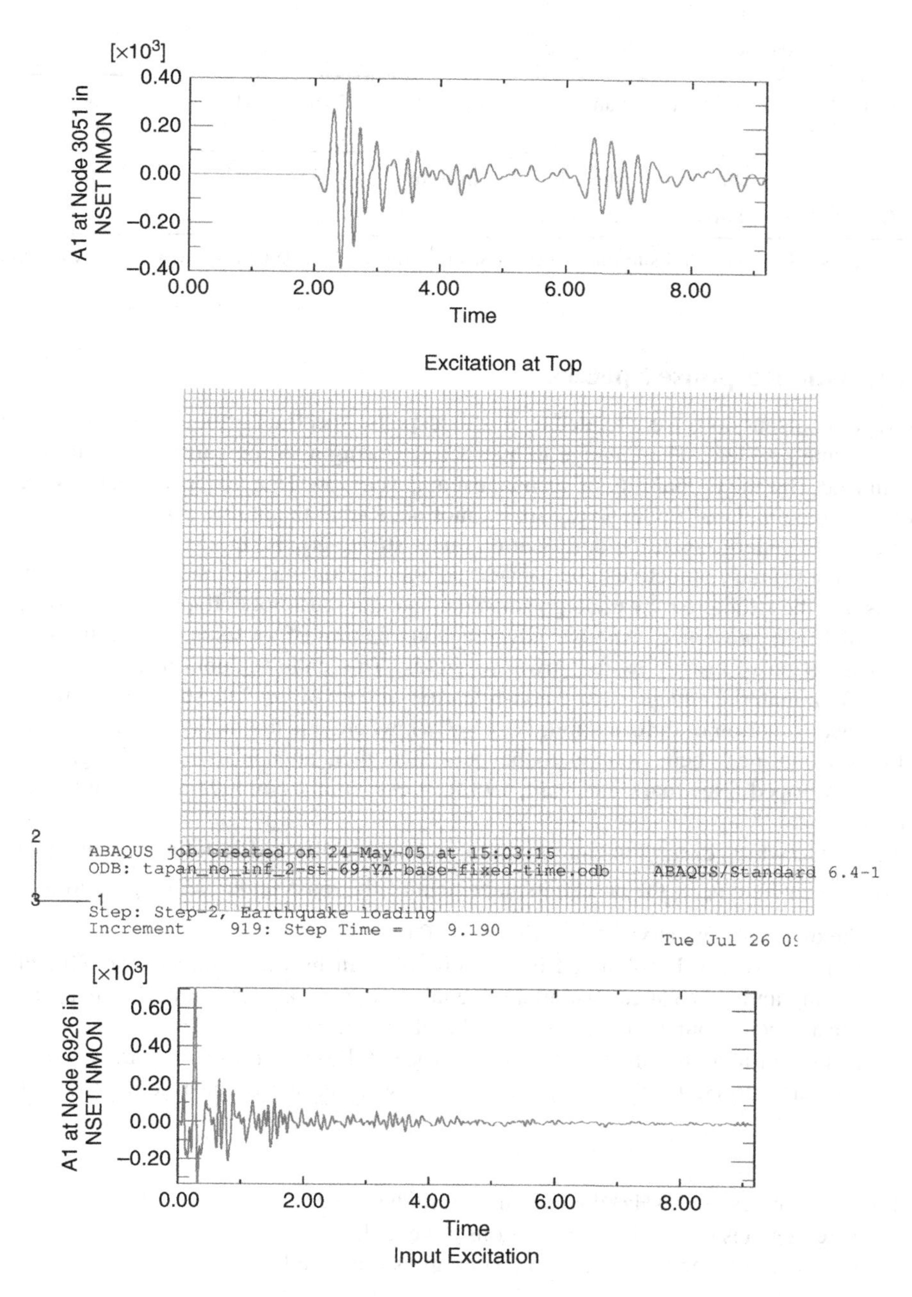

Figure 6.19 2D FE model of the soil and input and output excitation.

Table 6.2 The earthquake time history record.

Waveform ID	Earthquake name	Region	Year	Ms	Soil Cond. Sh.Wave Vel
69	Fruili (Aftershock)	North'n Italy	1976	4.72 Rock	847 m/sec
Peak Ground Acceleration: 0.706 m/sec^2					

Source: accessed from the website for European Strong Motion Data (ISESD) (www.isesd.cv.ic.ac.uk 20/07/08).

6.4 Inelastic Response Spectra

The foregoing discussions were related to elastic response spectra. Earthquake design presents another unique problem. The damage to a structure during a strong earthquake can be very severe indeed. But the probability of such an event is very low. Hence, most structures are not expected to remain elastic under a very strong motion. Rather the philosophy is that structures are expected to undergo inelastic deformation under strong ground motion.

The main or primary objective to be considered with the known excursions into the inelastic regime is that the structure may undergo some damage but must not collapse. This is to ensure that loss of life is prevented. Another advantageous feature of inelastic behaviour is that it allows dissipation of energy as the structure yields. This calls for nonlinear, elasto-plastic analyses. Both material and geometric non-linearities are involved. The structural stiffness (k) is a non-linear function of deformation, i.e. $k = k(u)$, where u is the inelastic deformation.

If limited nonlinear behaviour is to be permitted then an approximate design method has been developed (Newmark and Hall, 1982). It essentially employs a modified response spectrum.

The extent to which excursions into inelastic regime are allowed is defined by the ductility factor $\mu = \frac{u_m}{u_y}$, where u_m is the actual displacement of the mass under actual ground motions and u_y is the displacement at yield (see Figure 6.20a).

Inelastic response can be obtained in the time domain by direct integration. But this is time consuming and also a specialist area to be taken up after a round of preliminary design. If elasto-plastic behaviour is assumed, then the elastic response can be modified to reflect inelastic behaviour and preliminary design can proceed. Based on research carried out in the 1970s, Newmark and Hall (1982) proposed the following basis for constructing the inelastic spectra:

1. at low frequencies (< 0.3 Hz) displacements are the same;
2. at high frequencies (> 33 Hz) accelerations are equal;
3. at intermediate frequencies the absorbed energy is preserved.

The ductility modified response spectra thus obtained is shown in Figure 6.20b.

Construction of the inelastic response spectrum has been discussed thoroughly by others (Newmark and Hall, 1982; Chopra, 2001). Studies suggest that the method is reasonably accurate if ductility is limited to 5 or 6 or less (Newmark and Hall, 1982).

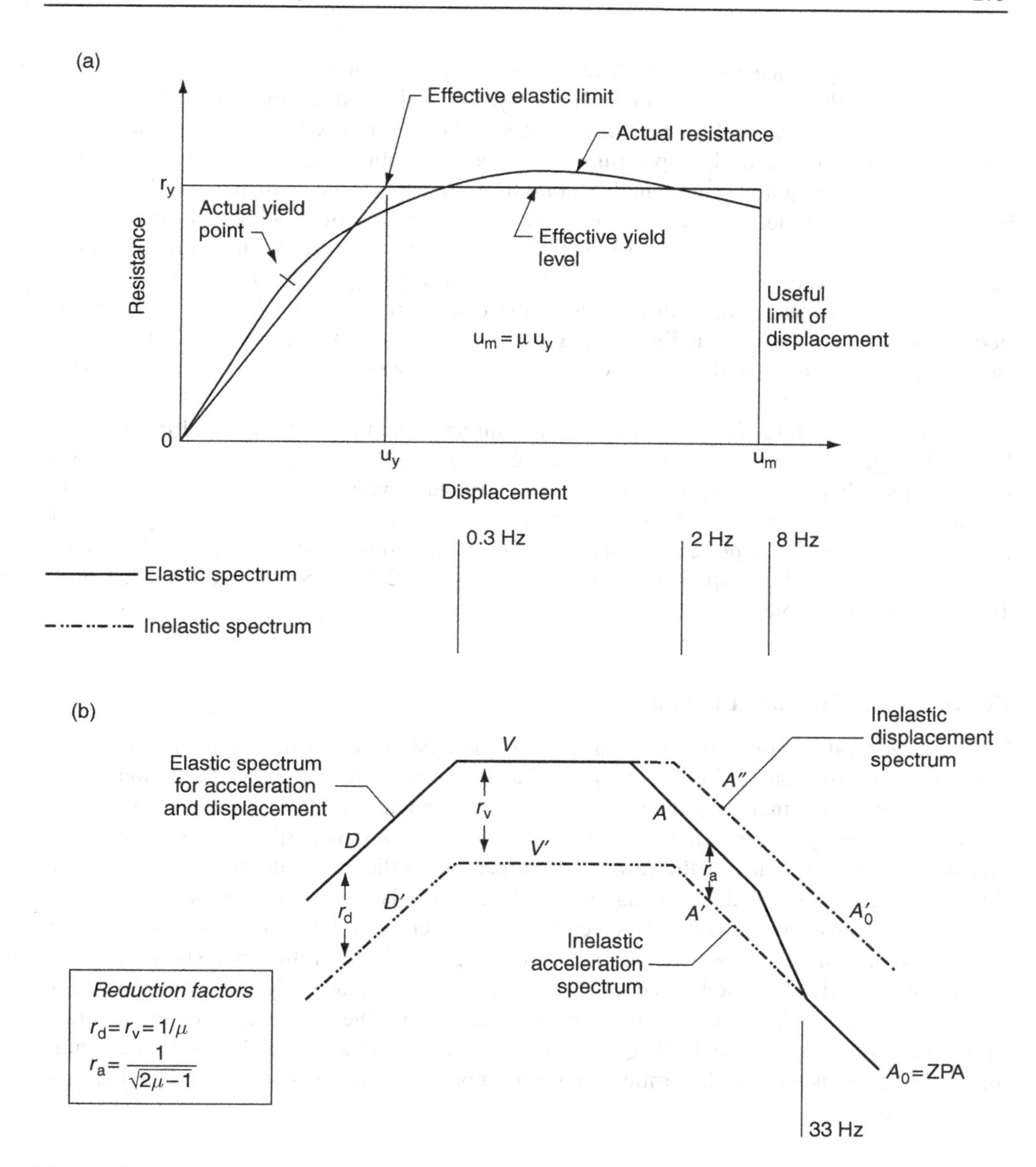

Figure 6.20 Inelastic design response spectra. Reproduced from Newmark and Hall (1982) by permission of the Earthquake Engineering Research Institute (EERI).

Carrying out the analysis as linear elastic we may now derive peak acceleration and internal forces with the modified ductility spectrum used exactly in the same way as normal elastic spectrum. The defections, however, derived from this method of analysis, must be multiplied by μ to obtain the inelastic deformation that has been allowed for in the design. It is now standard practice to analyse MDOF systems in the same way. In other words the ductility is introduced in the yielding MDOF system and the ductility modified spectrum used in the same way as the linear elastic. Ductility factors used in this way involve a general reduction in the design spectrum. Hence, a reasonable assessment of the allowable ductility factor is required. Here another complication arises. One must make a distinction between the various ductility factors involved in the response of a building to earthquake excitation. The ductility factor, for example, of a member, is defined by the rotational hinge capacity at a joint of a flexural member. The storey ductility factor is defined in terms of the relative displacement with respect to the displacement of the floor above and the floor beneath. The overall ductility is in general a weighted average of the member and storey ductility factors. Guidance on evaluation of ductility factors may be found in Chopra (2001).

Acceleration and force responses are obtained directly and the defections multiplied by μ. While for regular structures it may provide acceptable results, the method is to be used with great caution when dealing with irregular shaped structures with pronounced torsional effects or for structures with weak storeys. For structures with weak storeys the ductility demand is not evenly spread (Chopra 2001). In spite of the limitations, most codes allow the use of ductility modified inelastic spectra (Eurocode 8, Sec. 4, UBC 1997, Sec.1631, IBC 2003, Sec. 1617, ASCE 7-05 Table 12.4-1).

FE Analysis in the Time Domain

The most accurate form of this kind of dynamic analysis involves time history integration methods. The advantage of this method is that it can provide detailed information about nonlinear behaviour, including displacement, velocity and acceleration, the number of reversals of cycles and phase information of response between various parts. The disadvantage is that it is a complex task and the results are sometimes difficult to interpret. It is also very time consuming. Another disadvantage is that the analysis is for a single record. As we have seen in our discussions on response spectra, it is impractical to base the design for a future earthquake on a single record. Hence, it is not enough to carry out the analysis for a single earthquake record. Most modern codes (Eurocode 8) require that analyses be performed for at least five records. This is to account for, to some extent, the randomness of the records. Specialist software packages (ABAQUS, DYNA etc.) are available which take into account material and geometric non-linearities. In general non-linear time history analyses remains a complex task.

6.4.1 Response Spectra: A Cautionary Note

Referring to Figure 6.18, from the four different approaches there is now a reasonable spread of response spectra that may be expected at the site and on which to base a preliminary design, though only one earthquake record (Fruili Aftershock, Italy, 1976, Ms = 4.72) has been analysed. The magnitude is less than the estimated magnitude at the site (Ms = 5.4). More representative earthquakes need to be analysed. The steps outlined are indicative of what would usually be carried out in practical situations. These are followed by further specialist investigations by seismologists.

The seismic wave fields that are generated by an earthquake are complex and have large random and spatial variations. This is often manifested in the irregular pattern of damage that is observed in an earthquake. The irregularity is attributed to the characteristics of the earthquake source, seismic wave propagation, structure type and soil conditions.

The prediction of the occurrence of an earthquake at any specific location and time remains an elusive goal even today. The most challenging unknown factor for an earthquake prediction is the timing of the earthquake. Although a great deal is known about where earthquakes are likely to occur, there is currently no reliable way to predict when an event will occur at any specific location.

Earthquake magnitude and timing are controlled by the size of a fault segment, the stiffness of the rocks and the amount of accumulated stress. Where faults and plate motions are well known, the fault segments most likely to break can de identified. If a fault segment is known to have broken in a past large earthquake, recurrence time and probable magnitude can be estimated based on fault segment size, rupture history, and strain accumulation.

It is to be borne in mind that we are dealing with a future earthquake. The forecasting technique gives only an estimate and the ground motion characteristic may not be repeated for a future earthquake. Forecasts for poorly understood faults, hidden or small fault systems would carry a greater amount of caution. However, lesser known faults have caused some major recent earthquakes including the 1994 Northridge, California earthquake, and the 1995 Kobe, Japan and 1999 Ji-Ji, Taiwan earthquakes. As predicting the timing of future earthquakes is not possible it is better to aim for engineering solutions that result in earthquake resistant structures under a given magnitude of earthquake.

For engineering purposes the characteristics of the earthquake motion that are of primary significance include the magnitude, frequency content, duration of motion and amplitude. The magnitude of the earthquake is sometimes related to the fault length. At present the magnitude of maximum credible earthquake that might occur at a given location is estimated by fault mapping. This is the basic requirement for earthquake hazard maps all over the world. The frequency content is an entirely different matter being almost entirely unpredictable. It often depends on which part of the fault triggers the earthquake and on distance of the site from the fracture zone. The role of site conditions in controlling the amplitude and frequency content of the surface motions is still the subject of intense research.

The emphasis is on a safe structure taking into consideration the socio-economic factors. Usually the minimum level of loading that would be acceptable is that which the code provides. In addition, it may so happen that the soil condition at the site needs further investigation. FE analysis procedures are a valuable aid to investigate further, the recommendations obtained from seismologists.

Field inspection and analyses of structures due to shaking during earthquake reminds us that design concepts embodied in the codes are still evolving.

6.5 References

Ambraseys, N.N., Bommer, J.J., and Sarma, S.K. (1992) *A Review of Seismic Ground Motions for UK Design*, ESEE Research Report No: 92–8. Imperial College, London.

Ambraseys, N.N., Douglas, J. and Smit, P.M. (2005) 'Equations for estimation of strong ground motions from shallow crustal earthquakes using data from Europe and the Middle East: horizontal peak ground acceleration and spectral acceleration'. *Bull. Earthquake Eng.* **3**(1): 11–53.

American Petroleum Institute (1993 R 2003) API RP 2A-LRFD *Recommended Practice for Planning, Designing and Constructing Fixed Offshore Platforms – Load and Resistance Factor Design*, 1st edn; Errata – October 1993; Supplement 1 – February 1997. New York: API.

American Petroleum Institute (2000) API RP 2A-WSD *Recommended Practice for Planning, Designing and Constructing Fixed Offshore Platforms – Working Stress Design* –21st edn; Errata and Supplement 3: October 2007. New York: API.

American Society of Civil Engineers (2005) ASCE 7-05 *Minimum Design Loads for Buildings and Other Structures – Including Supplement 1*. Reston, Virginia: ASCE.

Benioff, H. (1934) 'The physical evaluation of seismic destructiveness'. *Seismol. Soc. America Bull.* **24**: 398–403.

Biot, M.A. (1943) 'Analytical and experimental methods in engineering seismology'. *Transactions of the Am. Soc. Civil Engineers*, ASCE, Vol. 180, 365–375.

British Standards Institution (2005) BS EN 1998-1: 2005 *Eurocode 8: Design of Structures for Earthquake Resistance Part 1: General Rules, Seismic Actions and Rules for Buildings*. Milton Keynes: BSI.

Chopra, A.K. (2001) *Dynamics of Structures: Theory and Applications to Earthquake Engineering*. New Jersey: Prentice Hall.

Clough, R.W. and Penzien, J. (1993) *Dynamics of Structures*. New York: McGraw Hill Inc.

Dahle, A., Bungum, H. and Kvamme, L.B. (1990) 'Attenuation models inferred from intraplate earthquake recordings'. *Earthq. Eng. Struct. Dyn.* **19**: 1125–1141.

Der Kiureghian, A. (1981) 'A response spectrum method for random vibration analysis of MDOF systems'. *Earthq. Eng. Struct. Dyn.* **9**: 419–435.

Dowrick, D.J. (1981) 'Earthquake risk and design ground motions in the UK'. *Proc. Inst. Civil Engrs*, Vol. 77 Part. 2, 305–321.

Dowrick, D.J. (2003) *Earthquake Risk Reduction*. Chichester: John Wiley & Sons, Ltd.

Esteva, L. and Villaverde, R. (1973) 'Seismic risk, design spectra and structural reliability'. Proceedings of the 5th World Conference on Earthquake Engineering, Rome, 2586–2596.

European Strong Motion Data (ISESD) website (www.isesd.cv.ic.ac.uk 20/07/08), maintained by Imperial College, London.

Housner, G.W. (1959) 'Behaviour of structures during earthquakes'. *J. Eng. Mech. Div., Am. Soc. Civ. Eng.* **85**: 109–129.

International Code Council (1997) *Uniform Building Code (UBC) 1997*. Washington, DC: ICC.

International Code Council (2003) *International Building Code (IBC) 2003*. Washington, DC: ICC.

Levy, S. and Wilkinson, J.P.D. (1976) *The component element method in dynamics*. McGraw Hill Inc., New York.

McGuire, R.K. (1974) *Seismic structural response risk analysis, incorporating peak response regressions on earthquake magnitude and distance*. Research Report R74-51, Dept of Civil Engineering, Massachusetts Institute of Technology (MIT), Cambridge, Massachusetts.

Newmark, N.M. and Hall, W.J. (1969) 'Seismic design criteria for nuclear reactor facilities'. Proceedings of the 4th World Conference on Earthquake Engineering, Santiago, Chile, 2, B-4, 37–50.

Newmark, N.M. and Hall, W.J. (1982) *Earthquake Engineering and Spectra*. Berkeley, California: Earthquake Engineering Research Institute.

Newmark, N.M. and Rosenbleuth, E. (1971) *Fundamentals of earthquake engineering*. Prentice Hall, New Jersey.

Seed, H.B. and Idriss, I.M. (1969) 'Influence of soil conditions on ground motions during earthquakes'. *J. Soil Mechanics and Foundations Division*, ASCE, **95**(1): 1199–1218.

Seed H.B. and Idriss, I.M. (1969) *Rock Motion Accelerograms for High Magnitude Earthquakes*. EERC Report 69-7. Earthquake Engineering Research Centre, University of California, Berkeley.

Seed H.B., Murarka R., Lysmer J. and Idriss I.M. (1976) 'Relationships of maximum acceleration, maximum velocity, distance from source, and local site conditions for moderately strong earthquakes'. *Bull. Seism. Soc. Am.* **66**(4): 1323–1342.

Seed, H.B., Ugas, C. and Lysmer, J. (1976) 'Site dependent spectra for earthquake-resistance design'. *Bull. Seism. Soc. Am.* **66**(1): 221–243.
Structural Engineers Association of California (1999) *Recommended Lateral Force Requirements and Commentary (SEAOC Blue Book).* Washington, D.C. International Code Council.
US Atomic Energy Commission. (1973) Regulatory Guide 1.60 *Design Response Spectra for Seismic Design of Nuclear Power Plants*. Washington, DC: US Atomic Energy Commission.

7

Probabilistic Seismic Hazard Analysis

7.1 Introduction

It should be apparent to the reader by now that uncertainties are a fact of life when it comes to defining seismic loading at a site. These uncertainties are reflected in the discussions on seismic source, travel path, local site conditions and also in dealing with the example problem in Chapter 5.

Earthquakes cause widespread disruption of human activity. A major earthquake is a disaster. It is a prime requirement in our society that all forms of manmade structures be protected during an earthquake. If we think about it, it is the collapse of the structure that causes the damage and loss of human life and not the earthquake itself.

The *deterministic approach* towards seismic hazard analysis (SHA) provides an easy-to-follow and transparent method for defining seismic loading at a site. The analysis makes use of discrete, single valued events or models, as we have seen in our example in the preceding chapter, to arrive at the required design level of ground motion. The maximum possible earthquake was identified for the 100 and 10,000 year earthquakes. How likely or unlikely are these earthquakes? In a deterministic approach, the chance of exceeding the design ground motion is not addressed directly.

The question that is posed before us is:

> '*Is this approach (seismic hazard analysis) satisfactory in dealing with uncertainties associated with practically all the elements that go towards defining the loading*?'

It is not possible to be certain about the exact magnitude of an earthquake, its location, let alone the complexities inherent in ground motion make-up arising from source, travel path, local site conditions etc. We can design for the worst possible scenario. However, designing for this scenario may not be economically feasible and some projects may consequently never be constructed.

Fundamentals of Seismic Loading on Structures Tapan Sen

The designer has to think about moderate earthquakes as well as large earthquakes. What are the probabilities of both large and moderate earthquakes occurring? In its simple form it is easy to comprehend that the likelihood of a large earthquake is low compared to that of a moderate one. In the socio-economic context it makes sense to design for a large earthquake so that the structure does not collapse (preventing valuable loss of life), accepting the fact that it will be damaged, but designing for moderate earthquakes to prevent damage. This dual philosophy has been widely adopted by the engineering community. It involves making a fundamental trade-off (as in other walks of life) between providing very costly higher resistance of the building to seismic loading during a large earthquake and the risk of damage to the building with a lower resistance, but avoiding collapse. It also requires in-depth assessment of performance of various utilities and the more important economic implications thereof.

The economic implications due to the expected damage are part of 'Seismic Risk Analysis' (SRA). Typical probabilistic analysis will provide the information for the structure to be designed for ground motion that will be exceeded once during its lifetime with a *given probability*. Depending on the importance of the structure a *higher or lower probability* may be chosen.

Thus, it may be seen, that the above scenario calls for well-informed and orderly decision making. This is precisely what *probabilistic seismic hazard analysis* helps to achieve.

All pertinent data relating to past earthquakes, ground motion intensity, location and geological make-up must be made available to the engineer responsible in a form suitable for informed decision making. A higher ground motion will have a lower probability. The engineer must know, for example, how the probability level drops with increasing ground motion. Probabilistic seismic hazard analysis (PSHA) provides a way forward. It may be noticed that several terms in relation to hazard and risk analyses have been used. It is helpful to clarify the usage of these terms. Figure 7.1 shows a general plan of hazard analysis with a brief explanation of the terms.

Seismic Hazard Analysis (SHA)
(Evaluating design parameters of earthquake ground motion at site.)

Deterministic Approach
Deterministic Design Code Provisions
Seismic Hazard Analysis (DSHA)
(Quantitative estimate of earthquake hazard based on single earthquake magnitude assumed to occur at a fixed distance from site and a specified ground motion probability level – which could be either 0 or 1 standard deviation above the median by tradition.)

Probabilistic Seismic Hazard Analysis (PSHA)
(Quantitative estimate of earthquake hazard considering all possible earthquakes from all possible sources and probability of these occurrences; an integrated approach covering all uncertainties.)

Seismic Risk Analysis (SRA)
(The estimation of damage arising from earthquake hazard and evaluating its socio-economic impact; SRA would be preceded by PSHA.)

Figure 7.1 General plan of seismic hazard analysis (SHA).

7.1.1 Seismic Hazard Analysis (SHA)

(Evaluating design parameters of earthquake ground motion at site.)

Yes, there are differences between the two approaches, DSHA and PSHA. At the heart of both these methods is the very fundamental and basic issue of identifying the potential sources of an earthquake and its magnitude. Quantitative assessment follows thereafter. When viewed with this in the background, it will slowly become clear in the following sections of this chapter, that there are many features which are common to both the methods. In fact they have much more in common than not. An element of probability of an earthquake scenario occurring (see example in Section 5.1.1) in DSHA, is implied. The split between the two approaches is not as pronounced as is often implied (Bommer, 2001). A good introduction to PSHA is provided by McGuire and Arabasz (1990).

7.1.2 Features of PSHA

The PSHA provides a sound basis with an engineering bias for representing a fairly wide range of natural variability and allows treatment of uncertainties arising from incomplete knowledge.

In the probabilistic approach, all possible and relevant deterministic earthquake scenarios (all possible magnitude and location combinations) are considered as well as possible ground motion probability levels, i.e. a range of the number of standard deviations above or below the median.

This approach paves the way for risk benefit analyses, which are required before making a major investment. It is also useful for code writers, who rather than defining the loading, can use it for establishing probabilistic seismic zone maps of a country (as in UBC, 1997, for example). Some of the frequently asked questions are: what are the scenarios for widespread disruption of public services? What is the vulnerability to financial loss? Answers to these questions are required and that is why this approach has proved so useful and has been widely adopted in contemporary PSHA.

A Disadvantage

A disadvantage of the PSHA proposed is that the concept of a 'design earthquake' is lost. Compare this situation with the 'deterministic approach' discussed in Chapter 5. In the PSHA there is no single event that represents the earthquake hazard at the site. This disadvantage is a direct outcome of the mathematical formulation adopted in this approach, where the variables are treated as random variables.

7.2 Basic Steps in Probabilistic Seismic Hazard Analysis (PSHA)

The methodology being used in most contemporary probabilistic seismic hazard analysis was first proposed by Cornell (1968) in a landmark paper. It has now come to be accepted as the preferred approach to the problem of determining the design ground motion at site. The key PHSA steps as outlined by Reiter (1990) are shown in Figure 7.2.

The following aspects need careful consideration:

1. Historical seismicity of the region.
2. Geology, tectonics, identification of earthquake source and the geometry.
3. Establishment of a recurrence law which specifies the *average rate* at which an earthquake of some size will be exceeded.
4. The ground motion parameter (peak acceleration, velocity etc.) produced at a site with respect to earthquake size and distance expressed in terms of attenuation relationships (see sec. 5.3).
5. Ground motion probability levels (a range of number of standard deviations above or below the median).

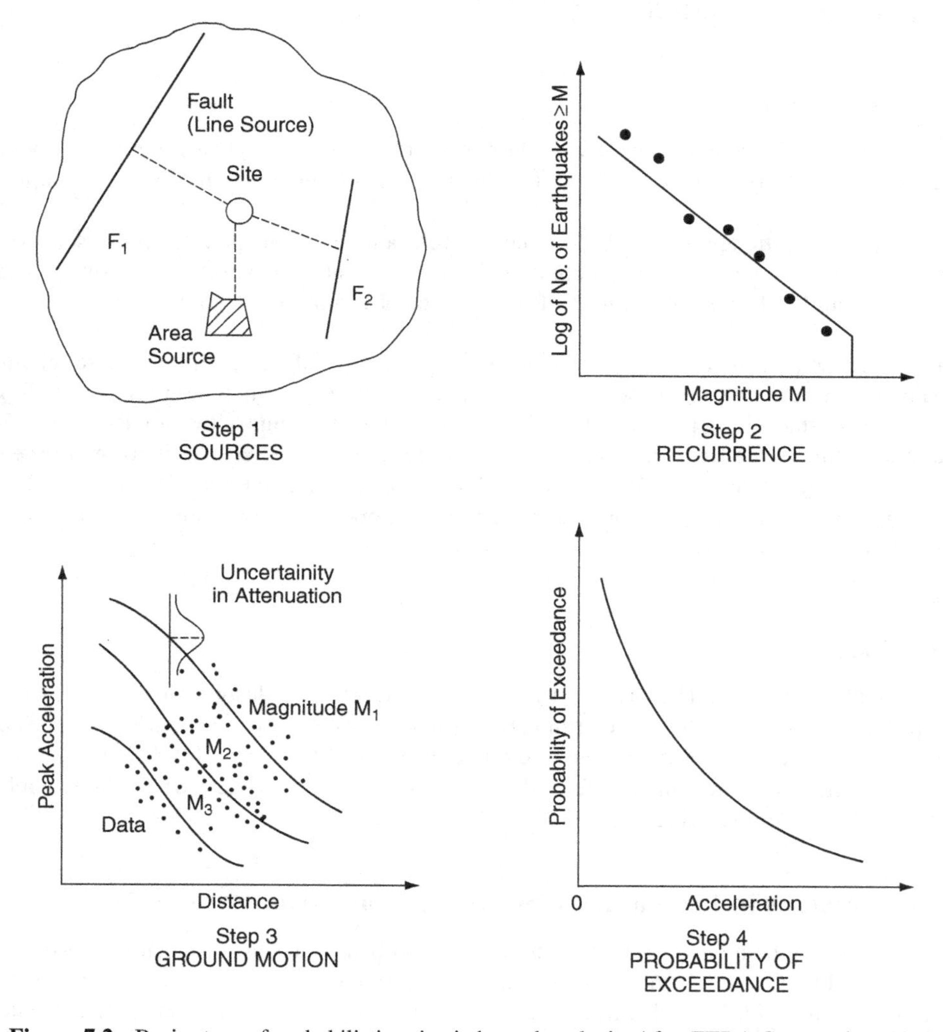

Figure 7.2 Basic steps of probabilistic seismic hazard analysis. After TERA Corporation, 1978.

6. Local soil conditions – it has been seen in many earthquakes that soft soils can amplify earthquake ground motion significantly.
7. Finally, the temporal model. All the uncertainties of source geometry, distance, magnitude and ground motion parameter of interest are combined in this final step of probability calculations (probability of exceeding a ground motion parameter at a site). This step is achieved by incorporating a temporal model (distribution of earthquake occurrence with respect to time, sometimes simply referred to as the arrival process).

A General Note on Engineering Modelling

As a matter of pragmatism, there should be a balance in the detail (of engineering significance) of these inputs. The detail that significantly influences the outcome should be captured in relatively more detail as far as practicable. Complicated models of earthquake occurrence that require data, which cannot easily be obtained with reasonable accuracy and reliability should be avoided. The same general comment is also applicable for identifying the recurrence law and ground motion modelling. We only have available at our disposal strong motion data. Only parameter details, which can be extracted with statistical significance, should be included.

7.2.1 Historical Seismicity

In general the region of interest around the site could be large, perhaps $5^\circ \times 5^\circ$ on the regional map or even larger. However, the idea is to establish and demarcate regions which could be termed 'homogeneous' meaning belonging to the same tectonic makeup. Data regarding earthquake size (magnitude, moment), location (epicentre), focal depth if available are collected for earthquakes that have happened in the past. Instrumental recordings are of recent origin and thus the data will normally consist of historical data and instrumental recordings. National institutions such as the International Seismological Centre (ISC) based in the UK, and the National Earthquake Information Center and National Geophysical Data Center (NEIC and NGDC) based in the USA, can supply instrumental data from 1906 to present day. The data collected in the early stages when seismographs were first introduced are not that accurate. Errors associated with the epicentre could be of the order of $\pm 25km$. A few instruments with limited sensitivity, unevenly located were deployed. Further to this small earthquakes were not recorded and one may say that the data is incomplete in that respect. Current seismographs are much more sensitive and their network spread around the world.

The locations of pre-instrumental period earthquakes are obtained from historical records extending as far back as possible in time and these vary from region to region. Historical records in most regions date back only a few hundred years, with the exception of China and the Middle East, where meaningful records of earthquakes date back more than 2000 years.

There is a problem with magnitude determination as well. Magnitude determination in the instrumental period is confusing as different formulae were used in different periods. It is often necessary to recalculate the magnitude of all data in a uniform manner. There were also errors associated with focal depth, which was virtually unknown. A normal depth of 33 km was assigned to most shallow focus earthquakes with unknown depths. Focal depth corrections to attenuation equations were introduced later.

7.2.2 Geology, Tectonics, Identification of Earthquake Source and the Geometry

The idea is to map the locations of all known active faults and if possible the depth at which they are located. Some areas are better documented than others. For example, along the Californian belt *inter-plate* boundaries, either active crustal faults or large scale seismological features such as zones of plate subduction have been mapped. These are the possible locations where future earthquake energy release may occur. In the case of *intra-plate* regions it may not be that straightforward as sources of energy release may not be readily associated with identifiable faults. Moreover, some faults are located below the surface. It is customary to designate these regions as area sources. Even in Western North America where faults are well mapped, unpredictable events have taken place, including Coalinga (1983). This event happened on a hitherto unknown fault.

Categorization of source and identifying the geometry for the purpose of PSHA is an important step, which is still very subjective at present, depending to a great extent on the tectonic process involved. A volcanic source, for example, would be concentrated in a small region and could be identified as a small volumetric source or simpler still as a point source. Being close to the site of area or volumetric source models could add to the accuracy. The vast majority of global earthquakes occur along *inter-plate* boundaries where active faults are usually located. Faults may be conveniently be modelled as a line source and can be located at a certain depth below the ground surface. If faulting in an area is extensive and if it is difficult to locate individual faults, the source should be modelled as a volumetric source or an area source. Distant source zones can be modelled as a line or point source.

The size of the rupture is significant in estimating the distance for the attenuation relation (Der Kiureghian and Ang, 1977).

7.2.3 Modelling Recurrence Laws

The next step in the process is to establish the recurrence relationship which establishes the *average rate* at which an earthquake of some size will be exceeded, or rather the sizes of earthquakes the region is likely to exhibit.

How often do earthquakes occur? Does some kind of pattern exist in the time scale at which earthquakes are known to occur? This has been a quest of seismologists for a very long time. In a seismically active region earthquakes occur at irregular intervals of time. It is fairly obvious that in order to extract a meaningful pattern, the length of the record must be reasonably large. The longer the record the better it is. Historical records are dated and hence it is possible for seismologists to analyse and assess the recurrence relationship.

7.2.4 Gutenberg-Richter Recurrence Law

Gutenberg and Richter (1954) first described the general underlying pattern of magnitudes of earthquakes and their occurrence. The data (spread over a certain length of time) were organized in a manner to reflect the number of earthquakes that exceeded a certain magnitude. Also from the organized data, the *mean annual rate* λ_M of an earthquake exceeding magnitude M was defined as the number of occurrences greater than M divided by the length of the time period. In other words the average rate at which an earthquake of some size

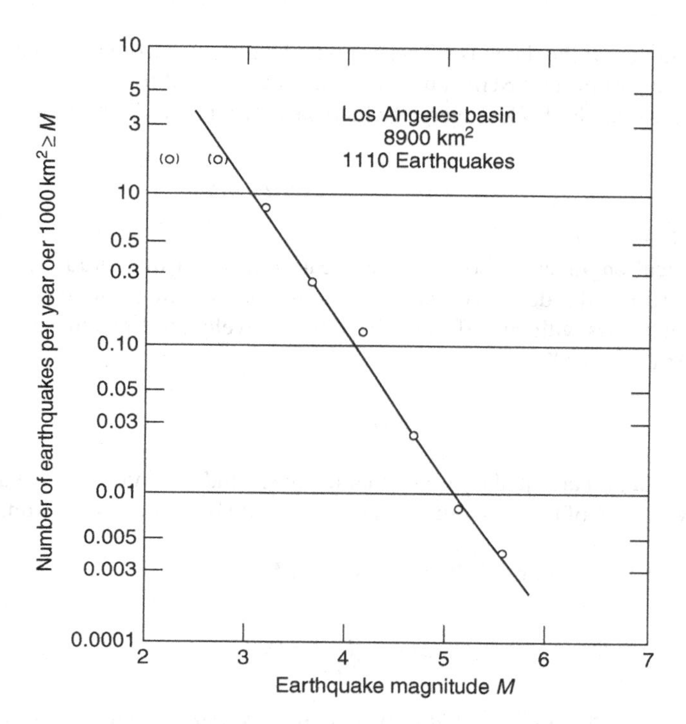

Figure 7.3 Earthquake recurrence plot. Reproduced from Allen *et al.* (1965), © Seismological Research Society of America (SSA).

will be exceeded. The reciprocal of the mean annual rate of being exceeded was referred to as the *mean return period* of the earthquake. An empirical relation between magnitude and frequency was proposed:

$$\log \lambda_M = A - bM \tag{7.1}$$

where A and b are seismic constants for a region.
An example of this relationship obtained using the Californian data is shown in Figure 7.3.

The term 'A' varies from region to region and 'b', denotes the relative likelihood of large and small earthquakes of the region and varies between 0.5 to 1.5 (Dowrick, 1987). Note that a decrease in the value of 'b' (the slope of the line) would mean an increase in the likelihood of larger earthquakes. Esteva (1967) applied the Gutenberg and Richter law to worldwide seismicity data. Another well documented, in-depth study of magnitude frequency relationships in various parts of the world is due to Everenden (1970).

For the kind of exercise to produce the plot shown in Figure 7.3, regression techniques (numerical techniques applied to curve fitting) are usually resorted to. The regression model is applied to the database of seismicity for the source zone of interest. The database is more than likely to be sparse and use of instrumentally recorded (past sixty years or so) and historical databases of the region is required. In some regions the presence of before shocks and aftershocks could distort the database. These should be removed from the database as they are not intended to be part of the PSHA, and their effects accounted for separately.

Expressing the standard Gutenberg-Richter law (Equation 7.1) in exponential form we have:

$$\lambda_M = 10^{A-bM} = \exp(\alpha - \beta M) \tag{7.2}$$

where $\alpha = 2.303A$ and $\beta = 2.303b$.

From a practical engineering point of view, earthquakes of magnitude four or below are not of great interest to the designer since they do not cause major damage. Hence, bounded Gutenberg-Richter laws with cut-off limits have been developed (Epstein and Lomnitz, 1966; McGuire and Arabasz, 1990).

$$M_o = \text{lower cut-off limit}$$

Let us suppose that an earthquake of the smallest magnitude of interest, M_o has taken place. The probability density of the magnitudes in excess would be a diminishing function as shown in Figure 7.4.

The probability of its magnitude between M_o and M is

$$F(M) = \int_{M_O}^{M} f(M)dM \tag{7.3}$$

A suitable expression for the *Cumulative Distribution Function* (CDF), $F(M)$ as suggested by Epstein and Lomnitz (1966) and McGuire and Arabasz (1990) is of the form:

$$F(M) = 1 - e^{-\beta(M-M_O)} \tag{7.4}$$

and the probability density function (PDF):

$$f(M) = \beta e^{-\beta(M-M_O)} \tag{7.5}$$

$$M_u = \text{upper cut-off limit}$$

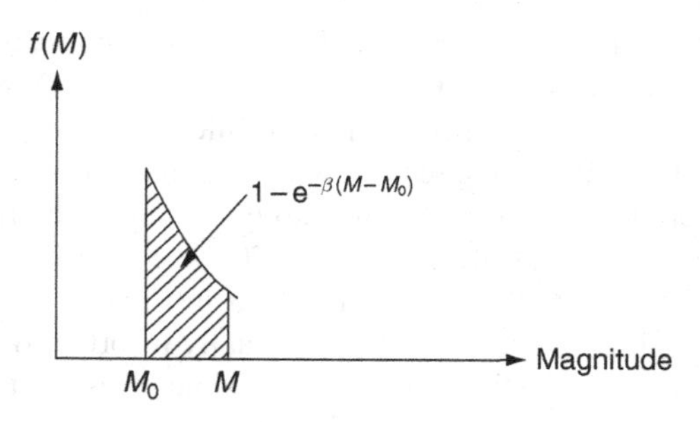

Figure 7.4 Probability density function for earthquake magnitudes.

Again, the Gutenberg-Richter expression (Equation 7.2) covers an infinite range of magnitudes and mathematically may predict much higher magnitudes than may be possible at the source zone (see Figure 7.2b). An upper limit of magnitude is associated with all source zones. Seismologists are able to assign some maximum magnitude M_u (upper cut-off) for each source zone from the geological make-up, fault rupture length and other features.

The mean annual rate of an earthquake exceeding magnitude M, with $M_o \leq M \leq M_u$, may be found in Kramer (1996).

The CDF and PDF for the Gutenberg-Richter relationship with lower and upper bounds may be expressed as (Kramer, 1996):

$$F(M) = \frac{1 - \exp[-\beta(M - M_0)]}{1 - \exp[-\beta(M_u - M_0)]} \tag{7.6a}$$

and the PDF as

$$f(M) = \frac{\beta \exp[-\beta(M - M_0)]}{1 - \exp[-\beta(M_u - M_0)]} \tag{7.6b}$$

7.2.5 Alternative Models

An alternative method of obtaining recurrence rate is modelling on the basis of seismic moment. As mentioned earlier, the seismic moment of an earthquake M_{mom} is a measure of the strength of the earthquake. Though the seismic moment can be evaluated from the recordings from some of the seismographs in use, it is related to the fault slip rate through the well known theoretical result, which states that

$$M_{mom} = \mu AD \tag{7.6c}$$

where

μ – shear modulus of the fault material
A – area of the fault involved in the seismic slip
D – average slip over the rupture

The annual slip across the fault is linked to the seismic moment rate and methods are available for determining the recurrence rate. Descriptions of the method may be found in some of the related references (Youngs and Coppersmith, 1985; Anderson, 1979; and Schwartz and Coppersmith (1984).

Another physical phenomenon needs to be pointed out very briefly. It has been observed that certain faults can exhibit a trend of generating similar magnitude earthquakes over time. Seismologists have referred to this phenomenon in the literature as '*characteristic earthquakes*'. The reader is referred to works by Wesnousky *et al.* (1984), and Wu *et al.* (1995).

7.2.6 Why the Gutenberg-Richter Model?

The validity of the Gutenberg-Richter law has been questioned by various investigators (Schwartz and Coppersmith, 1984; Schwartz, 1988; Youngs and Coppersmith, 1985).

Although the Gutenberg-Richter Law may have been questioned, available worldwide seismicity data *does not* support other models being adopted.

Because of its simplicity and as it fits the data reasonably well in the lower magnitude range, this model has been found to be convenient and is currently in general use.

The Gutenberg-Richter law will be used in the worked examples that follow.

7.2.7 Ground Motion Parameter (peak acceleration, velocity etc.)

The uncertainty associated with predictive models of ground motion parameters needs to be addressed within PSHA. The predictive relationships have been discussed in Chapter 5, Section 5.2.1. The ground motion parameters' variation with M and R (magnitude and hypocentral distance) are log normally distributed (the logarithmic of the parameter are normally distributed). The regression laws themselves are not complicated but it is worth noting that all authors have reported considerable scatter in the data (Boore *et al.*, 1997; Campbell, 2003a,b; Ambraseys, 1996a,b, 2005a,b). A certain amount of scatter is inevitable as not enough is known about the source, travel path and local site conditions. More sophisticated models of attenuation relationships are awaited.

The assumption implicit in the above formulation is that energy is released at the hypocentre. We have noted earlier that energy is released from the entire fault rupture surface. It has been noted that the rupture surface of a large earthquake with a distant hypocentre could release energy much closer to the site than assumed above (Der Kiureghian and Ang, 1977).

7.2.8 Local Soil Conditions

Local soil conditions are treated separately as a special case. Recent attenuation equations (Ambraseys, 2005a,b and others) have introduced parameters to account for different types of soil conditions when predicting ground accelerations. However finite element (FE) modelling of the actual soil strata at site still remains the best alternative for investigating local soil conditions (see example Chapter 5).

7.2.9 Temporal Model (or the arrival process)

As noted earlier, the final step in our calculations is the incorporation of a mathematical model to account for distribution of earthquake occurrence with respect to time. The incorporation of this model within PSHA explicitly or implicitly assumes that the occurrence of one earthquake is not related to the previous one (the process is memory less). In other words, the time of occurrence and the magnitude of the next earthquake are probabilistically independent. The memory less model is inconsistent with the physical concept of *elastic rebound theory* – of continuous crustal strain accumulation and intermittent release of strain when the build-up exceeds the strain limits the faults can sustain. The release mechanism of this energy build up is through movements along the faults.

Cornell (1968) first introduced the Poisson model as the mathematical model. Since then all contemporary PSHA has been based on the Poisson model. From an engineering perspective, the reasons are not far to seek and are outlined in the EPRI (Cornell and Winterstein, 1986) Report:

1. Some successful comparison of its predictions with observations (Gardiner and Knopoff, 1974, McGuire and Barnhard, 1981)
2. The broad acceptance by the engineering community of the fact, that lacking detailed and substantiated evidence to the contrary, the model is not unreasonable physically (especially for the smaller to medium events that may dominate the hazard scenario).
3. Also, the fact from a mathematical angle that the sum of non-Poissonian processes may be approximately Poissonian (Cornell and Winterstein (1986, 1988)).
4. Perhaps, the most important reason is that *it is the simplest model that captures the basic elements of the entire concept.*

7.2.10 Poisson Model

Earthquake arrival or the recurrence rate of earthquakes is represented mathematically by the Poisson distribution model.

If the *average number* of random occurrences of earthquakes in a *given time interval* is $= \xi$, the probability 'p' of a random variable 'N' denoting the number of earthquakes in that time interval is given by

$$p\,[N = n] = e^{-\xi}\left[\frac{\xi^n}{n!}\right] \tag{7.7}$$

where

ξ = average number of earthquakes in the time interval
n = number of occurrences in that time interval

The time interval between events in a Poisson process can be shown to have an exponential distribution. The process can otherwise also be represented in the form

$$p\,[N = n] = \frac{e^{-\lambda t}(\lambda t)^n}{n!} \tag{7.8}$$

where

λ = *average rate* of occurrence of the event within the time interval of interest (e.g. inverse of nominal return period)
t = time interval of interest

Suppose that 25 earthquakes ($\geq m$) have occurred in the past 100 years. What is the probability that '*just one*' such ($\geq m$) earthquake will occur in the next year?

$$\lambda = 25/100 = 0.25, t = 1, n = 1, p = \lambda e^{-\lambda} = 0.194$$

The probability that '*at least one*' such earthquake will happen in the next year (any one year) is:

$$p = 1 - \text{'}\textit{probability that no such earthquake will happen}\text{'}$$
$$\text{i.e.}\quad n = 0,\ t = 1,\ \lambda = 0.25\ \text{(remember that } 0! = 1)$$
$$= 1 - e^{-\lambda}$$
$$= 0.22\ [\text{if } \lambda = 1,\ \text{then } p = 0.63]$$

Probability of Occurrence

The probability of occurrence *of at least one favourable event* (earthquake) in the time interval of interest is given by

$$p\,[N \geq 1] = 1 - e^{-\lambda t} \tag{7.9}$$

Example

The design life of a manufacturing plant is 40 years. The return period for the operating basis earthquake (OBE) is 100 years. What is the probability of an earthquake happening within the design life?

The average rate of occurrence in this case is $\lambda = 1/100$. The time interval of interest is 40 years. Hence the probability of an earthquake

$$p = 1 - e^{-40^{*}1/100}, = 0.3296$$

Applicability of the Poisson Model

On the applicability of the Poisson model, Cornell and Winterstein (1988) point out that cases in which the Poisson estimate is insufficient are restricted only to those situations where the hazard is controlled by a single feature for which the elapsed time since the last significant event exceeds the average time between such events. However, such situations create a problem only if the fault displays strong 'characteristic time' behaviour.

What are the engineering implications of including the Poison process within the PSHA framework? The EPRI (Cornell and Winterstein, 1986) report concluded that 'the choice of earthquake occurrence model is a not a major source of uncertainty'.

7.3 Guide to Analytical Steps

7.3.1 Multiple Point Sources

Figure 7.5 shows several point sources. The site would be affected by ground shaking from these sources. In addition the attenuation constants for each source may be different.

Thus λ_i for each source is known. There will be at least one favourable event (earthquake) from any of the sources.

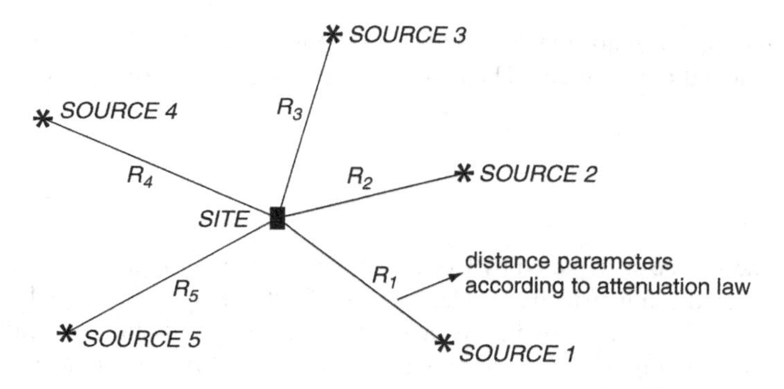

Figure 7.5 'r' Independent point sources at varying distances.

The probability that '*at least one*' such earthquake from any of the sources will happen in 1 year is:

$$p_i = 1 - \text{'probability that } no \text{ such earthquake of magnitude } (\geq m_i) \text{ happen in 1 year in } any \text{ of the sources'}$$

$$= 1 - \prod_{i=1}^{r} e^{-\lambda_i}$$

$$= 1 - e^{-\Sigma\lambda_i}$$

(The symbol $\prod$ represents product similar to Σ representing sum.)

A worked example of a single point source is illustrated in Section 7.5.2.

7.3.2 Area Source

The location of the site and area source is shown in Figure 7.6. An 'area source' can be considered as 'multiple point sources' of elemental area dA having the same A and b values of the recurrence relationship:

$$\log \lambda_M = A - bM.$$

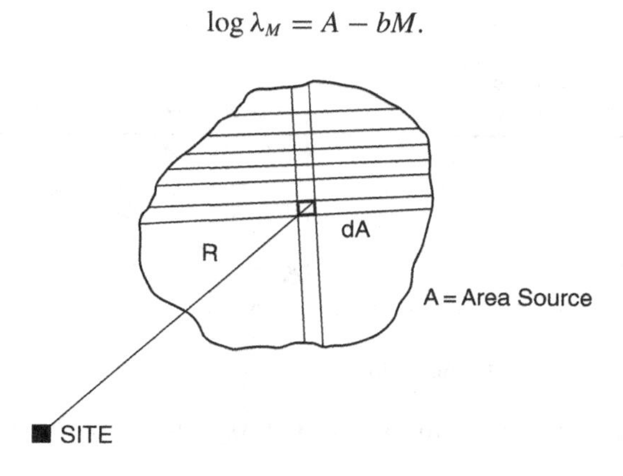

Figure 7.6 Area source.

Also the attenuation constants may be assumed to be the same but with different distances for each of the elemental area sources. The activity is defined in terms of numbers per unit area per unit time.

7.3.3 Line Source

Consider the line source as depicted by Cornell (1968) shown in Figure 7.7.

The site is situated symmetrically in relation to the fault (a line source) at a depth 'h' below the ground surface. The perpendicular distance from the site to the line vertically above the fault is 'Δ' as shown in Figure 7.7. Assuming that events are likely to occur at any point on the line, the location variable 'X' is uniformly distributed along the interval $(-l/2, +l/2)$. The shortest inclined distance from the site to the fault line is $d = \sqrt{h^2 + \Delta^2}$.

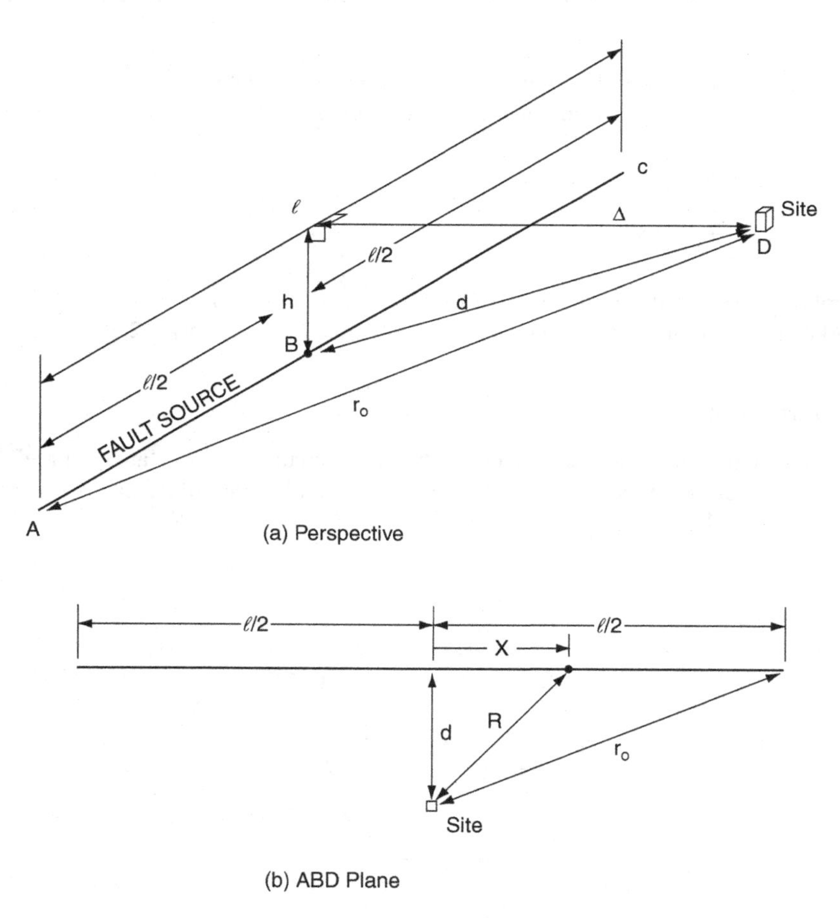

Figure 7.7 Line source. Reproduced from Cornell (1968), © the Seismological Research Society of America (SSA).

The focal distance, R, to any future earthquake source at distance X from the point B is $R = \sqrt{d^2 + X^2}$. In general, since the size and location of future earthquakes are uncertain, the distance X and R are to be treated as random variables.

The cumulative probability of R may therefore be expressed as:

$$F_R(r) = P[R \leq r] = P[R^2 \leq r^2]$$
$$= P[X^2 + d^2 \leq r^2] = P[|X| \leq \sqrt{(r^2 - d^2)}].$$
$$F_R(r) = 2\sqrt{(r^2 - d^2)}/l \qquad d \leq r \leq r_o$$

Therefore the probability distribution function of R is

$$f_R(r) = \frac{dF_R(r)}{dr}$$
$$= \frac{2r}{l\sqrt{(r^2 - d^2)}} \qquad d \leq r \leq r_o$$

7.4 PSHA as Introduced by Cornell

Cornell (1968) first introduces this method of seismic risk evaluation at a site where an engineering project is to be located. As pointed out earlier it is vitally important that the engineer should have all the information and relevant data to opt for best possible solution – the informed decision making mentioned earlier in this chapter. Cornell's assessment of the requirements for seismic hazard analysis may be summed up as 'seismologists have long recognized the need to provide engineers with their best estimates of seismic risk'. The formulation proposed by Cornell (1968) is outlined first in this section and reproduced with minor annotations only. The formulation is set in terms of Modified Mercalli Intensity, I. It may be noted, that the early development of PSHA by Cornell (1968) did not account for ground motion variability.

Models that include ground motion variability have not been discussed here. The details may be found in Esteva (1970). Further observations on the selection of the range of the standard deviation parameter (ε) to be included may be found in Bommer and Abrahamson (2006).

First the conditional distribution of Modified Mercalli Intensity, I, at the site is sought, given that an earthquake occurs at a focal distance $R = r$. A simple expression for I with common assumptions at that time is chosen (Ipek *et al.*, 1965; Esteva and Rosenbleuth, 1964; Wiggins, 1964; Kanai, 1961). The intensity is assumed to have the following expression in the range of our engineering interest. Note that it is dependent on the magnitude and focal distance R:

$$I = c_1 + c_2 M - c_3 \ln R \tag{7.10}$$

in which ln denotes natural logarithm and c_1, c_2 and c_3 are semi-empirical constants with values of the order of 8, 1.5 and 2.5 respectively for firm ground in Southern California (Esteva and Rosenbleuth, 1964).

We next express the probability of I being greater than any value i given that an earthquake occurs at a focal distance of $R = r$. This may be expressed in the form

$$P[I \geq i|R = r]$$

which represents the probability of I given r. Thus from Equation 7.10

$$P[I \geq i|R = r] = P[c_1 + c_2 M - c_3 \ln r \geq i|R = r] \tag{7.11}$$

Assuming probabilistic independence of M and R we may rearrange Equation (7.11)

$$P[I \geq i|R = r] = P[M \geq \frac{i + c_3 \ln r - c_1}{c_2}] \tag{7.12}$$

$$= 1 - F_M[\frac{i + c_3 \ln r - c_1}{c_2}] \tag{7.13}$$

where $F_M(m)$ is the cumulative distribution function of earthquake magnitudes in the region.

Usually, earthquake magnitudes below a certain value M_o (for example 3 or 4) are not of any engineering significance. Let us recall from our earlier discussions the bounded Gutenberg-Richter relationship with a lower cut-off limit. The expression for cumulative distribution function with a lower cut off limit is given by Equation (7.4):

$$F(M) = 1 - e^{-\beta(M - M_O)} \qquad [7.4]$$

Combining Equations 7.4 and 7.13 we have

$$P[I \geq i|R = r] = \exp\left[-\beta\left\{\frac{i + c_3 \ln r - c_1}{c_2} - M_o\right\}\right] \tag{7.14}$$

In order to consider the influence of all possible values of the focal distance (note: focal distance restricted to $d \leq R \leq r_o$) and their relative likelihood, we must integrate over the focal distance. We seek the cumulative distribution of I, $F_I(i)$, given the occurrence of an earthquake of magnitude $M \geq m_o$. Hence, the probability that a certain value of I will be exceeded is given by:

$$1 - F_I(i) = P[I \geq i] = \int_{source}^{r_o} P[I \geq i'|R = r] f_R(r) dr$$

In this case the source is the line source, hence

$$1 - F_I(i) = P[I \geq i] = \int_{d}^{r_o} P[I \geq i'|R = r] f_R(r) dr \tag{7.15}$$

in which $f_R(r)$ is the probability density function of R (see above).

With this step a key stage of the analysis has been reached. This determines the probability of the intensity I at the site being greater than a certain value, which we may call i, given that an event of interest $M \geq M_o$ has occurred somewhere along the fault. What is missing at this stage is the number of random occurrences that might take place. This is introduced through the temporal model.

The solution of Equation (7.15) is obtained by numerical methods. Cornell (1968) has provided the solution in the following form:

$$\begin{aligned} 1 - F_I(i) &= P[I \geq i] \\ &= \frac{1}{l} CG \exp[-\frac{\beta}{c_2} i] \qquad i \geq i' \end{aligned} \tag{7.16}$$

where i' is the lower limit of validity of i (from Equation 7.10) and is given by (remember we specified a lower cut off limit M_o for magnitude):

$$i' = c_1 + c_2 M_o - c_3 \ln d \tag{7.17}$$

C and G are constants. The constant C is related to parameters in the various relationships used above:

$$C = \exp\left[\beta\left(\frac{c_1}{c_2} + M_0\right)\right] \tag{7.18}$$

The constant G is a geometric parameter specific for the site geometry and is solved by numerical methods. Cornell (1968) reports the following:

$$G = 2\int_d^{r_0} \frac{dr}{r^{\gamma}\sqrt{r^2 - d^2}}$$

in which

$$\gamma = \beta\frac{c_3}{c_2} - 1$$

For the line source chosen for illustration, charts based on numerical methods have been provided in the paper for quantifying the intensity risk at the site.

The question of random occurrences in any time period is now considered. It is the next key step.

It is assumed that occurrences of these major events follow a Poisson arrival process (discussed in the previous section) with average occurrence rate (along the entire fault in our illustration) of λ per year. If certain events are Poisson arrivals with average arrival rate λ and each of these events are independent with probability, p, a special event, then these special events are Poisson arrivals with average rate $p\lambda$. In our case the special events are those which cause an intensity at the site in excess of some value i. The probability, p_i, that any event of interest ($M \geq M_o$) will be a special event is given by Equation 7.16:

$$p_i = P[I \geq i] = \frac{1}{l} CG \exp\left[-\frac{\beta}{c_2} i\right] \tag{7.19}$$

Thus, the number of times N that the intensity at the site will exceed i in an interval of length t (see Equation (7.8) for Poisson process) is

$$p_N(n) = P[N = n] = \frac{e^{-p_i\lambda t}(p_i\lambda t)^n}{n!} \qquad n = 0, 1, 2, \ldots \tag{7.20}$$

We should note that such probabilities are useful in studying losses due to succession of moderate intensities or cumulative damage due to two or more major ground motions as part of SRA discussed above.

Of particular interest to the engineering community is the probability distribution of $I_{\max}^{(i)}$ the maximum intensity over an interval of time t. Let $t = 1$ (i.e. time interval year). Note that $P[I_{\max}^i \leq i] = P$ [exactly zero special events in excess of i occur in time interval of 1 year] which from Equation (7.20) is

$$P[I_{\max}^i \leq i] = P[N = 0] = e^{-p_i\lambda} \qquad (t = 1) \tag{7.21}$$

$$F_{I\max}^i = e^{-p_i\lambda} = \exp[-\frac{\lambda}{l}CG\exp(-\frac{\beta}{c_2}i)] \qquad i \geq i' \tag{7.22}$$

$$(p_i = \frac{1}{l}CG\exp[-\frac{\beta}{c_2}i]) \qquad \text{(see above)}$$

We see from the above equation that for the larger intensities the annual maximum intensity has a distribution of the double exponential or Gumbel type. This distribution is widely used in engineering studies of extreme events. Now, if the annual probability of the intensity being exceeded is small (say ≤ 0.05) the distribution of $I_{\max}$ can be approximated by:

$$1 - F_{I\max}^i = 1 - e^{-p_i\lambda} \cong 1 - (1 - p_i\lambda)$$

$$\cong \frac{\lambda}{l}CG\exp(-\frac{\beta}{c_2}i) \quad i \geq i' \tag{7.23}$$

We also note that the average return period, T_i, of an intensity equal to or greater than i is defined as the reciprocal of $1 - F_{I\max}^{(i)}$ or $(\rho = \lambda/l)$

$$T_i \cong \frac{1}{\rho CG}\exp(\frac{\beta}{c_2}i) \qquad i \geq i' \tag{7.24}$$

The term $\rho = \lambda/l$ has been introduced and it is the average number of earthquake occurrences per unit length per year.

Expressing intensity in terms of, T_i, we have:

$$i \cong \frac{c_2}{\beta}\ln(\rho CGT_i) \qquad i \geq i' \tag{7.25}$$

7.4.1 *Line Source Illustration Problem (Cornell, 1968)*

The site is located in Turkey where in one region it was found that in 1953 years the following relationship was valid:

$$\log_{10}\lambda_{ne} = A - bM$$
$$= 5.51 - 0.644M \tag{7.26}$$

in which λ_{ne} is the number of earthquakes greater than M in magnitude. If we set $\lambda_{ne} = 5$, then the number of earthquakes above magnitude 5,

$$\log_{10}\lambda_{ne} = 5.51 - 0.644x5$$
$$\lambda_{ne} = 194 \tag{7.27}$$

Hence

$$\lambda = \frac{194}{1953} = 0.098$$

The major fault line is 650 km long. Assuming these earthquakes all occur along the 650 km of the fault, the average number of earthquakes in excess of magnitude 5 (i.e. $M_o = 5$) per year per unit length is

$$\rho = \frac{194}{(1953)(650)} = 1.5 \times 10^{-4}(year)^{-1}(kilometre)^{-1} \tag{7.28}$$

Also

$$\beta = 2.303b = 0.644(2.303) = 1.48$$

Attenuation constants evaluated for Californian data (Esteva and Rosenbleuth, 1964) were used.

$$c_1 = 8.16$$
$$c_2 = 1.45$$
$$c_3 = 2.46$$

The line source is shown in Figure 7.7. The site is located at a distance Δ, 40 km from the line source of earthquakes, at a depth of $h = 20km$.

Referring to Fig. 7.7

$$d = \sqrt{h^2 + \Delta^2} = 44.6km$$
$$\gamma = \beta\frac{c_3}{c_2} - 1 = 1.52$$
$$C = \exp[\beta(\frac{c_1}{c_2} + m_0)] = 6.85 \times 10^6 \tag{7.29}$$

Numerical integration gives $G = 6.58 \times 10^{-3}$

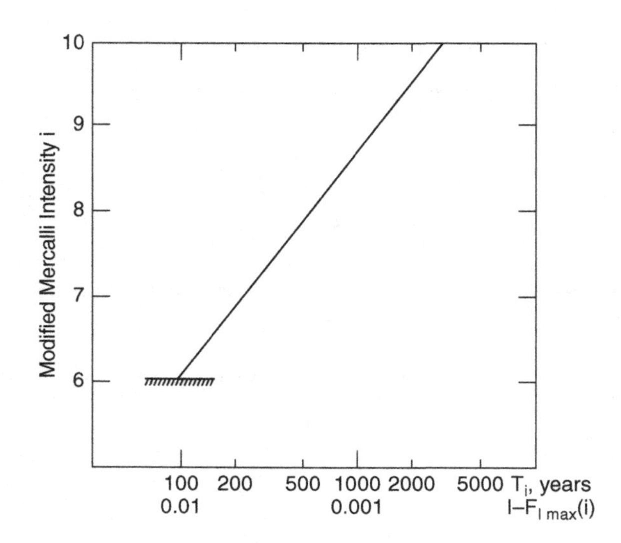

Figure 7.8 Numerical Example: Intensity versus return period. Reproduced from Cornell (1968), (c) the Seismological Research Society of America (SSA).

Thus the intensity at this site with a return period of T_i is

$$i \cong \frac{c_2}{\beta} \ln(\rho CGT_i) \tag{7.30}$$

$$\cong 0.98 \ln(6.9T_i) \tag{7.31}$$

Note the logarithmic relationship between i and T_i.

The plot of this equation is shown in Figure 7.8.

7.5 Monte Carlo Simulation Techniques

Monte Carlo simulation is an established technique for solving probabilistic models. In this section we shall explore how the line source illustrative problem worked out by Cornell (1968) may be solved by the Monte Carlo simulation process. The approach is different as we shall see, but it is a transparent process which the reader can follow easily. It mirrors real life events. The computational scheme is shown in the following flow chart (Figure 7.9).

Monte Carlo simulation is a powerful technique being used in many fields of science and engineering, notably nuclear science (used successfully on the Manhattan Project,1945 and as a matter of digression the process derives its name from the scientists who wanted to protect the atomic secrets from the enemy with a seemingly innocuous name). All the examples have been worked out with the established mathematical software *Mathcad 11*, *Enterprise Ed 1*. The software has a random number generator incorporated within. It may be noted that similar software packages (e.g. MATLAB, Mathsoft) could also have been used.

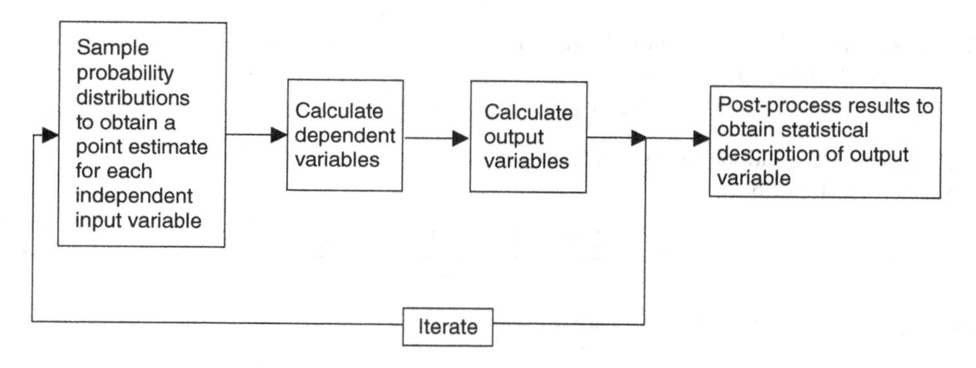

Figure 7.9 General scheme of a Monte Carlo simulation.

Preliminary Steps

We review the problem at hand again and refer to Figure 7.7. The intensity at the site is governed by the relationship (Equation 7.10):

$$I = c_1 + c_2 M - c_3 \ln R \tag{7.10}$$

The constants, c_1, c_2 and c_3 have been established for the site and these constants do not change. In the above intensity relationship R is related to the position where the earthquake originates which happens to be randomly distributed.

The preliminary steps are as follows:

1. Fault Line

The fault line is 650 km long and earthquakes can occur anywhere along the fault line. Thus, uniform random distribution for R may be assumed.

2. Gutenberg-Richter Recurrence Law

It has been assumed that the magnitude M in the above intensity relationship is governed by the Gutenberg-Richter Law and we will adopt the bounded model with a lower cut-off limit. In this instance the lower cut off limit happens to be $M_o = 5$.

We need to dwell upon the Gutenberg-Richter relationship further in order to set up the simulation process with random numbers. In order to facilitate programming the probabilistic relationship (cumulative distribution function). The probabilistic expression in Mathcad needs to be expressed in such a form that the independent variable is expressed in terms of a random number.

We begin with the cumulative distribution function. The CDF for the bounded Gutenberg-Richter relationship is of the form Equation (7.4):

$$F(M) = 1 - e^{-\beta(M - M_o)} \tag{7.4}$$

$$\beta = 2.303b \text{ (see section 7.2)}$$

We equate the CDF expression on the right hand side to the random number and solve for the independent variable M.

$$rnd(1) = 1 - e^{-2.303b(M-M_o)} \tag{7.32}$$

($rnd(1)$ generates random numbers between $0 \leq 1$)

Solving for M we have

$$M = \frac{[2.303bM_o - \ln(1 - rnd(1))]}{2.303b} \tag{7.33}$$

This expression is used in the software *Mathcad*.

Setting out the Problem in *Mathcad*

The *Mathcad* steps as programmed are outlined below:

$$\text{GUTENBURG3}(c, d) := \frac{(\ln(10) \cdot d \cdot c - \ln(1.0 - \text{rnd}(1)))}{(\ln(10) \cdot c)}$$

$$\text{UNIF}(A, B) := \text{rnd}(B - A) + A$$

$$\text{SIM}(N) := \left| \begin{array}{l} \text{for } i \in 1..N \\ \quad \left| \begin{array}{l} x \leftarrow \text{UNIF}(-325, 325) \\ M \leftarrow \text{GUTENBURG3}(0.644, 5) \\ r \leftarrow \sqrt{x^2 + 44.6^2} \\ I_{i-1} \leftarrow 8.16 + 1.45\,M - 2.46\ln(r) \end{array} \right. \\ \text{sort}(I) \end{array} \right.$$

$$\text{NO} := 1.5 \cdot 10^{-4} \cdot 650$$

$$\text{NO} = 0.098$$

$$N := 200000$$

$$I := \text{SIM}(N)$$

$$i := 0..N - 1$$

$$F_i := \frac{i + 0.5}{N}$$

Several values from Cornell's solution were calculated from Equation (7.31) for comparison and are shown on the plot below.

$$\text{int} := \begin{pmatrix} 7.98 \\ 8.66 \\ 9.05 \\ 9.34 \\ 10.0 \end{pmatrix} \qquad R := \begin{pmatrix} 500 \\ 1000 \\ 1500 \\ 2000 \\ 3915 \end{pmatrix}$$

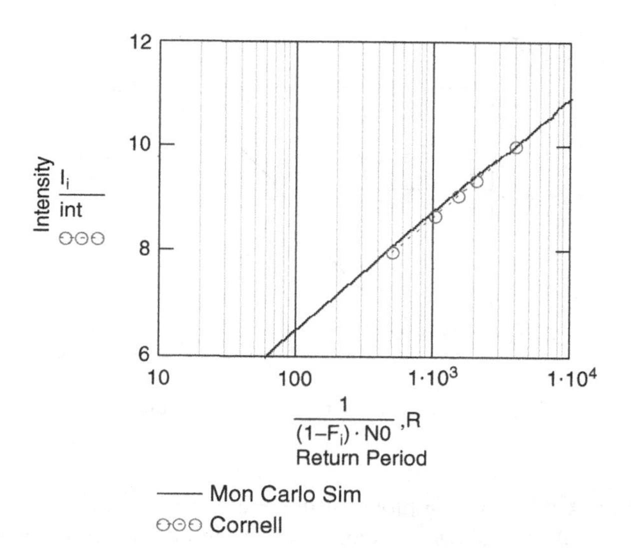

Figure 7.10 Plot of Monte Carlo Simulation and Cornell's solution (Equation 7.31).

The arguments (c,d) in the Gutenberg-Richter relationship correspond to b and M_o. The values entered for uniform distribution along the fault line are self-explanatory.

The simulation is carried for 200,000 random iterations and sorted i.e. arranged in ascending order. The best estimate of probability of ith serial number is $\frac{i+0.5}{N}$ as shown above. 'N0' corresponds to λ.

The plot of Monte Carlo Simulation and points from Cornell's solution Equation (7.31) are shown in Figure 7.10.

7.5.1 *Monte Carlo Simulation Process – An Insight*

An insight in to the workings of Monte Carlo simulation may be gained from following the simulation process. The simulation process is particularly helpful in complex situations when the outcome of a random variable is dependent on two or more independent random variables, and an analytical solution does not exist.

Consider the worked example in the previous section. We had peak ground acceleration (PGA), Z, dependent on two independent variables of magnitude (M) and distance (R). Referring to the flow chart below, we

- sample two random values from the uniform distribution [0, 1];
- transform these values using the probability distributions of M and R to obtain point estimates, m and r;
- calculate a point estimate z of the peak ground acceleration Z from the above values using the relationship between Z, M and R;
- repeat this process thousands of times to generate corresponding values of Z;
- post process results to obtain the probability distribution function of Z.

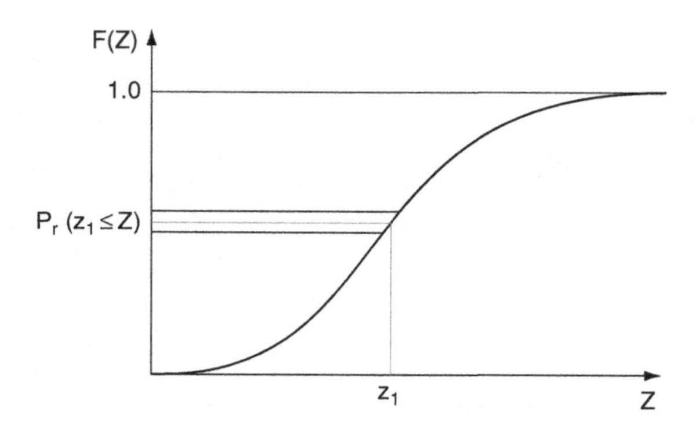

Figure 7.11 Cumulative distribution function.

If plotted against PGA the resulting plot will in general have a continuous curve and in the present case will look like that shown schematically in Figure 7.11 (Z representing the random variable peak ground acceleration).

We can therefore define

$$P(z) = \Pr[Z \leq z]$$

without difficulty and plot it with precision if a large enough number of trials are made. Sampling for various values of magnitudes and distance we will obtain the distribution function of the peak ground acceleration. The plot is the cumulative distribution function of the peak ground acceleration.

The distribution will also conform to the physical conditions: the probability that any acceleration will have a negative value is zero; the probability of not exceeding a small positive value is small; the probability of any given value z_1 is less than 1.

The probability that the acceleration will have a value between limits z_1 and $z_2 > z_1$ will be

$$\Pr[z_1 \leq Z \leq z_2] = \Pr[Z \leq z_2] - \Pr[Z \leq z_1]$$

The Monte Carlo simulation exploits this situation. We can physically conceive that with a very large number of random simulations (sampled 200,000 times in this case) we will have a good approximation of the cumulative distribution function. The process is transparent and easy to follow.

7.5.2 Example: Point Source (extracted from Cornell, 1968)

We will only consider the parameters for the point source. Though a hypothetical situation, it is instructive to go through the site description given in the paper (reproduced below from Cornell, 1968), as more often than not we come across situations where little is known about past seismicity. Consider the regional layout and the point source shown in Figure 7.12.

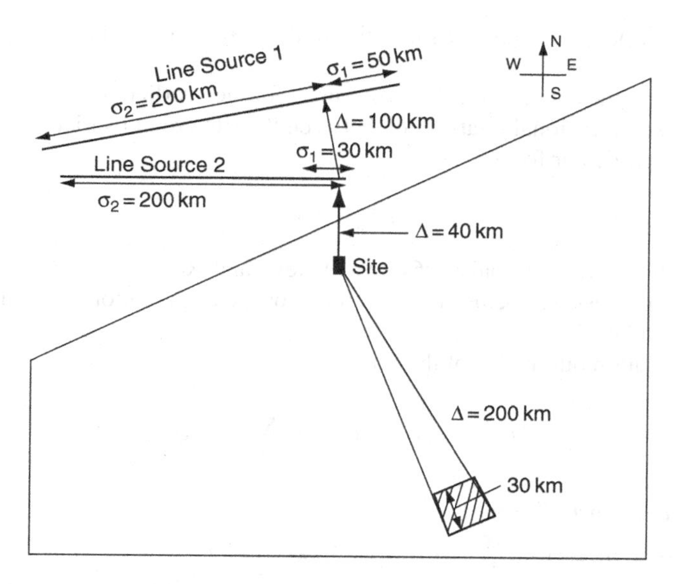

Figure 7.12 Point source model. Reproduced from Cornell (1968), (c) the Seismological Research Society of America (SSA).

The site is located on a deep alluvial plane (shaded) such that the geological structure below the site and to the South and East is not known in detail. Historically, earthquakes have occurred throughout this plane, but not often enough to determine fault patterns. The engineer chooses to treat the region as if the next earthquake were equally likely to occur in any unit area. The average rate (ρ) was estimated by dividing the region's total number of earthquakes (with magnitudes in excess of 4) by the total area. The exception, historically, is a small area some 200 km southeast. It is also below the alluvial plane. The frequency of all sizes of earthquake there has been relatively high, including several of higher magnitude. Although the engineer can easily account for any suspected local difference in the parameter β (Note that a decrease in the value of 'β' (the slope of the line) would mean an increase in the likelihood of larger earthquakes), he chooses to use the same β value, 1.6, for the entire region. In other words, he chooses to attribute the small area's observed larger magnitudes to the same population $f_M(m)$. The justification is that the larger the average arrival rate, the larger the number of observations and more likely it is that larger magnitudes will be included among the observations in a given period of time. Exactly what area (here shown as 30 by 30 km) is used to estimate the aerial occurrence rate, $\rho = 1.0 \times 10^{-4}$ per km^2, is not critical in this case since the area is small enough and far enough from the site that the entire source will be treated as a point with rate $\rho = 0.09$. Finally, to the Northwest where the geological structure is exposed, two faults have been located. Neither can be assumed inactive. Past activity on the first (and other geologically similar faults) suggests an average rate occurrence of $\rho = 1.0 \times 10^{-4}$ per km. No earthquake on the second, but closer fault has been recorded, but its geological similarity to the first suggests that it be given a similar activity level.

With reference to the solution for the line source outlined earlier, a similar exercise can be carried out with any functional relationship between the site ground variable Y, and M and R. For example, the particular form

$$Y = b_1 e^{b_2 M} R^{-b_3} \tag{7.34}$$

has been recommended by Kanai (1961) and Esteva and Rosenbleuth (1964). The Monte Carlo simulation process will be used on this form for the construction of the uniform hazard spectra as outlined later.

The general solution outlined is of the form:

$$1 - F_{Y_{\max}}(i) \cong C y^{-\beta/b_2} \sum_{j=1}^{n} \rho_j G_j \tag{7.35}$$

Point Source Data (Cornell, 1968)

$r_0 = 216.0; m_0 = 4; b_2 = 0.8 \quad \beta = 1.6, b_1 = 2000, b_3 = 2$

r_0 – distance from point source to site; m_0 – lower cut off limit

$\rho G = 4.3 \times 10^{-11} \quad C = 2.4 \times 10^9$ ...... peak acceleration (numerical values extracted from Cornell (1968))

$$\begin{aligned} 1 - F_a(a) &= C\rho G y^{-\beta/b_2} \\ &= 2.4 \times 10^9 \times 4.3 \times 10^{-11} a^{-1.6/0.8} \\ &= 10.32 \times 10^{-2} a^{-2} \end{aligned}$$

Return Period

$$\begin{aligned} T_a(a) &= \frac{1}{10.32 \times 10^{-2}} a^2 \\ &= 9.68 a^2 \end{aligned}$$

Therefore,

$$a = 0.32 T_a^{0.5}$$

Thus, if $T_a = 200$

$$a = 0.32 \times \sqrt{200} = 4.54 \text{ cm/sec}^2$$

The plot for return period vs. acceleration is shown in Figure 7.13.

Compare the value of acceleration, how it fits on the plot of return period vs. acceleration obtained from Monte Carlo simulation which follows next.

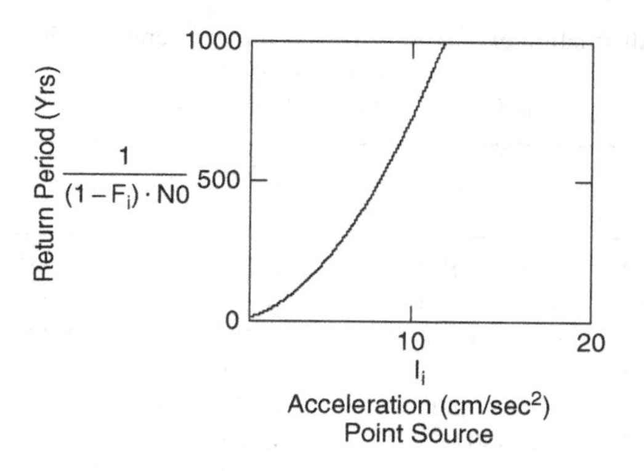

Figure 7.13 Plot for return period vs. acceleration.

Mathcad Sheet

Note the change in program steps.

$$b3 := 2$$

$$b1 := 2000.0$$

$$r := 216.0$$

$$b2 := 0.8$$

$$\mathrm{SIM}(N) := \left| \begin{array}{l} \text{for } i \in 1..N \\ \quad \left| \begin{array}{l} m \leftarrow \mathrm{GUTENBURG3}(0.644, 4) \\ I_{i-1} \leftarrow b1 \cdot \exp(b2 \cdot M) \cdot r^{-b3} \end{array} \right. \\ \mathrm{sort}(I) \end{array} \right.$$

$$NO := 1.5 \cdot 10^{-4} \cdot 600$$

$$NO = 90 \times 10^{-3}$$

$$N := 200000$$

$$I := \mathrm{SIM}(N)$$

$$i := 0..N - 1$$

$$F_i := \frac{i + 0.5}{N}$$

7.6 Construction of Uniform Hazard Spectrum

Before we consider constructing the uniform hazard spectrum for a site, we must know the probability of peak ground motion (peak acceleration, velocity etc.) being exceeded.

Cornell's probabilistic methodology outlined above can be applied to any functional relationship between site ground motion variable Y (peak acceleration, velocity and displacement), and M and R.

Table 7.1 McGuire's Attenuation expressions for spectral acceleration with 5% damping (1974).

$$S_a = b_1' 10^{b_2 M}(R+25)^{-b_3}$$
$$\log S_a = b_1 + b_2 M - b_3 \log(R+25)$$

Period(s)	b_1'	b_1	b_2	b_3	Coeff. of var. of S_a
0.1	1610	3.173	0.233	1.341	0.651
0.2	2510	3.373	0.226	1.323	0.577
0.3	1478	3.144	0.290	1.416	0.560
0.5	183.2	2.234	0.356	1.197	0.591
1.0	6.894	0.801	0.399	0.704	0.703
2.0	0.974	−0.071	0.466	0.675	0.941
3.0	0.497	−0.370	0.485	0.709	1.007
4.0	0.291	−0.620	0.520	0.788	1.191

Source: (Reproduced from Dowrick (1987) by permission of John Wiley and Sons Ltd)

We could, for example adopt the functional relationship provided by McGuire (1974) working on data from sites in Western USA, as shown:

$$S_a = b_1' 10^{b_2 M}(R+25)^{-b_3} \quad \text{cm/sec}^2 \text{ or}$$
$$\log S_a = b_1 + b_2 M - b_3 \log(R+25) \tag{7.36}$$

The constants for the above equations are shown in Table 7.1.

7.6.1 Monte Carlo Simulation Plots for Peak Ground Acceleration

We can apply the Monte Carlo process to derive the plot for peak ground acceleration vs. return period, R, (or probability of being exceeded, $1 - F_{Y_{max}}(y)$). Remember,

$$R \cong \frac{1}{1 - F_{Y_{max}}(y)}$$

For application of the steps, let us also assume same source region and site distance shown in Fig.7.7. Assume the same seismic data. Hence,

$$\log_{10} \lambda_m = A - bM$$
$$= 5.51 - 0.644M$$
$$M_o = 5$$

The program steps are on similar lines and are as follows:

$$\text{GUTENBURG3}(c,\ d) := \frac{(\ln(10)\cdot d\cdot c - \ln(1.0 - \text{rnd}(1)))}{(\ln(10)\cdot c)}$$

$$\text{UNIF}(A,\ B) := \text{rnd}(B - A) + A$$

$$\text{SIM}(N) := \left| \begin{array}{l} \text{for } i \in 1..N \\ \quad \left| \begin{array}{l} x \leftarrow \text{UNIF}(-325, 325) \\ M \leftarrow \text{GUTENBURG3}(0.644, 5) \\ r \leftarrow \sqrt{x^2 + 44.6^2} \\ S_{i-1} \leftarrow 161010^{0.233\,M}\cdot(r+25)^{-1.341} \end{array} \right. \\ \text{sort}(S) \end{array} \right.$$

The plot of return period versus acceleration for period of 0.1 second is shown in Figure 7.14. The plots for period 0.2 and 0.3 sec are shown in Figures 7.15 and 7.16.

7.6.2 Procedure for Construction of Uniform Hazard Response Spectrum

As we have seen in Chapter 5, a response spectrum is particularly useful as it gives a direct indication of the peak response of a structure when subjected to an earthquake ground motion. Response spectra are usually plotted as a function of peak acceleration response against structural period.

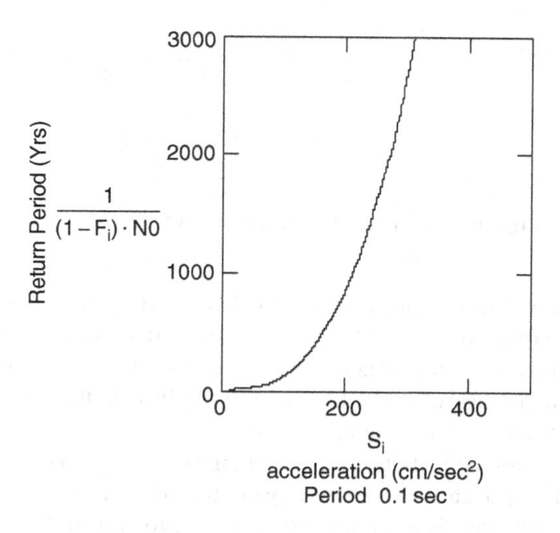

Figure 7.14 Return period versus acceleration for period of 0.1 seconds.

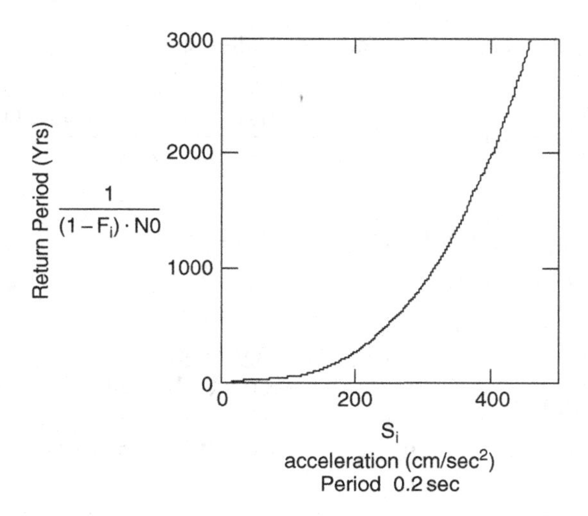

Figure 7.15 Plot for return period vs. acceleration.

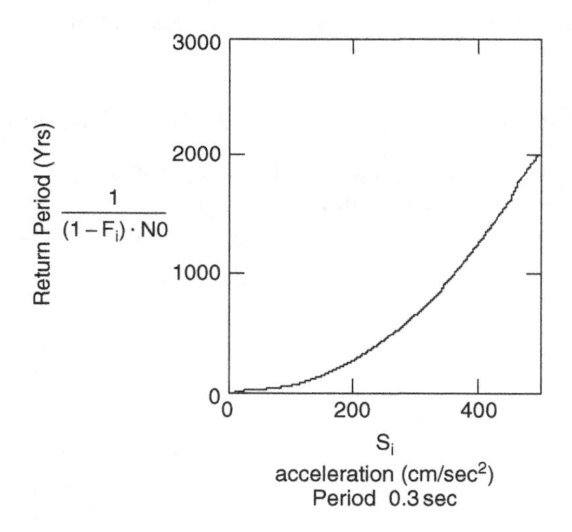

Figure 7.16 Plot for return period vs. acceleration.

The difference between the response spectrum discussed earlier and the uniform hazard response spectrum (UHRS), is that the latter is constructed so that all points on the spectrum have an equal probability of being exceeded within a pre-determined return period.

If we think about the processes embedded within the PSHA, the uniform hazard spectrum is a multi-parameter description of ground motion.

The UHRS envelopes are contributions from a range of earthquakes. For example, the short period motion of UHRS is usually dominated by contributions from small nearby earthquakes while the long period response is dominated by contributions from large distant earthquakes. This is taken care of within the PSHA. Hence the plan for generating the UHRS:

Step 1

Calculation of peak ground accelerations for the remaining periods 0.4–4.0 sec (as shown in Table 7.1) are carried out first (plots not shown here) by modifying the *Mathcad* sheets.

Step 2

The concept and methodology are shown in Figure 7.17.

Finally, we follow the plan outlined in Figure 7.17 for constructing the UHRS. With the return period vs. peak ground acceleration plots at various periods being generated, we can now draw the uniform hazard response spectrum. The plot for the UHRS for a predetermined return period of 1000 years is shown in Figure 7.18.

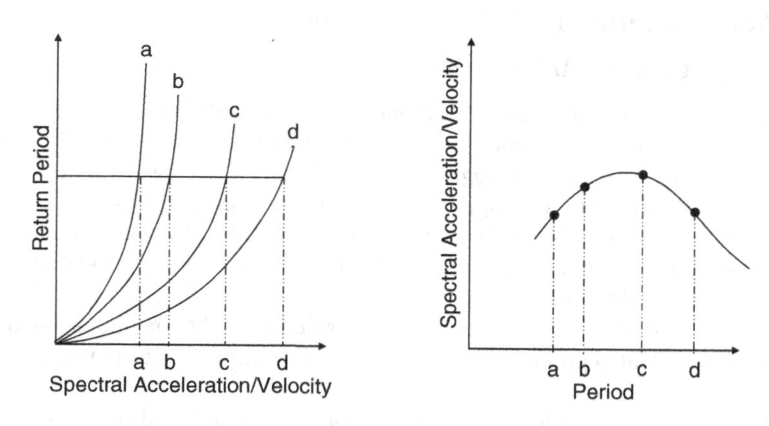

Figure 7.17 Construction of UHRS.

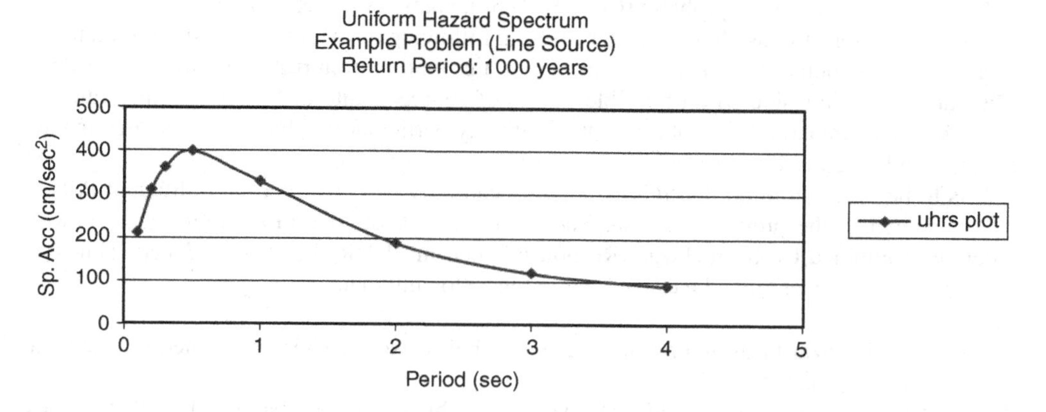

Figure 7.18 Plot of UHRS (return period: 1000 years).

7.6.3 Further Attenuation Equations

The computations above have been carried out with attenuation equations proposed by McGuire (1974). Other attenuation equations can easily be incorporated into the simulation model.

For example, the attenuation equation proposed by Boore *et al.* (1997) referred to in Chapter 5 is of the form:

$$\ln Y = b_1 + b_2(M-6) + b_3(M-6)^2 + b_5 \ln r + b_v \ln {}^{V_S}\!/\!{}_{V_A}$$

All the coefficients in this equation are defined in the paper. Again, it is a simple proposition to incorporate this equation in the *Mathcad* sheet for simulation.

7.7 Further Computational Considerations

7.7.1 De-aggregation – An Introduction

There is usually a project requirement to estimate the most likely combination of earthquake magnitude and source distance contributing most to the mean annual rate of the parameter being exceeded. In the process of *de-aggregation*, we further analyse *individual* PSHA results to discover which combination of magnitude and distance contributes most. The reader may be wondering why do we need *de-aggregation*? Sooner or later there will be a requirement of dynamic analyses (at least for all large projects) for detailed assessment of local soil conditions and amplifications of seismic excitation.

The development of time histories requires a knowledge of the magnitudes and distances that dominate the calculated ground-shaking hazard at the return periods and structural periods of interest.

These estimates may be used to select existing ground motion records for dynamic analyses (see procedure in Chapter 6 where European records were searched). Candidate time histories may be selected, for example, within say $\pm 0.3M$ (consistent with magnitude scale for PSHA) and $\pm 10km$.

Consider the line source problem (Cornell, 1968), analysed in Section 7.6.1.

The calculations that we have been performing following the Monte Carlo simulation technique allow computation of mean annual rate of a ground motion parameter to be exceeded. The calculations considered all possible values of source to site distances and magnitudes. Thus the mean annual rate is not associated with any particular combination of source to site distance and magnitude.

It is to be borne in mind that *de-aggregation* can only be performed on individual PSHA results shown in the previous section. The information sought from *de-aggregation* is lost when the combined Uniform Hazard Response Spectrum (UHRS) plot is produced. Thus the *de-aggregation* process may be carried out in the following manner:

1. The seismic hazard (ground motion parameter being exceeded) is partitioned into several magnitude-distance bins.
2. A return period is selected (say 1000 years). At this point we have a further decision to make. We must decide on the most likely estimate of the natural period of the structure

(if not known at this stage). The natural period must be away from the dominant period of seismic excitation (refer to Chapter 4 for frequency domain analyses of time history records).

3. Hence, a target period is selected (e.g. 2 seconds).

The bins with the largest relative contribution identify those earthquakes that contribute the most towards the total hazard. Time histories (from existing records) are then selected which are consistent with the magnitude and distances of these design earthquakes.

The idea is similar to defining the dominant earthquake and has been explored by others (Chapman, 1995; McGuire, 1995; Bazzuro and Cornell, 1999).

7.7.2 Computational Scheme for De-aggregation

Consider the Monte Carlo simulation plot for peak ground acceleration against return period for 100 years for a structural period of 2 seconds as shown in Figure 7.19.

The Monte Carlo simulation was carried out for 200,000 samples. Each of the random number sampled produced *a pair of magnitude and distance*. Thus we have 200,000 pairs of magnitude and distance data.

The return period of interest is 100 years. The acceleration corresponding to a return period of 100 years is approximately 30 cm/sec^2. From the samples, a subset with ground motion parameter (acceleration) corresponding to 100 years and above is created. De-aggregation is carried out on the subset.

The number of hits in each of the bins was counted. The relative contribution of each bin was calculated by dividing the number of hits in each bin by the total number in the subset.

The resulting plot with relative percentage contribution is shown in Figure 7.20.

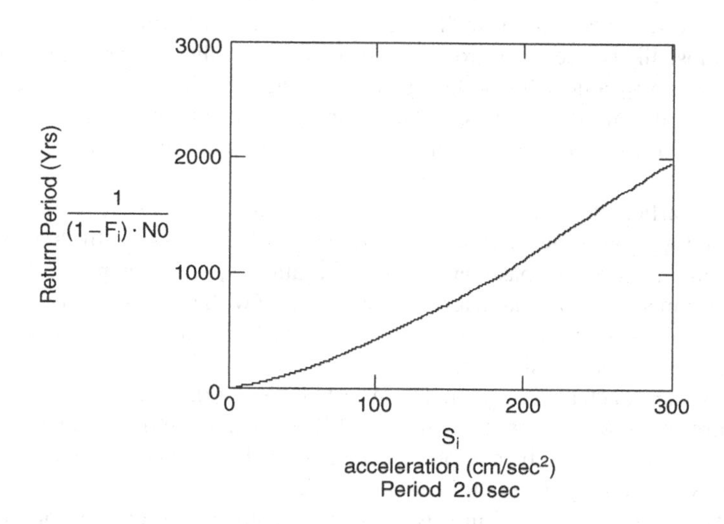

Figure 7.19 Return period vs. acceleration (period: 2 sec) (attenuation relationship: McGuire, 1974).

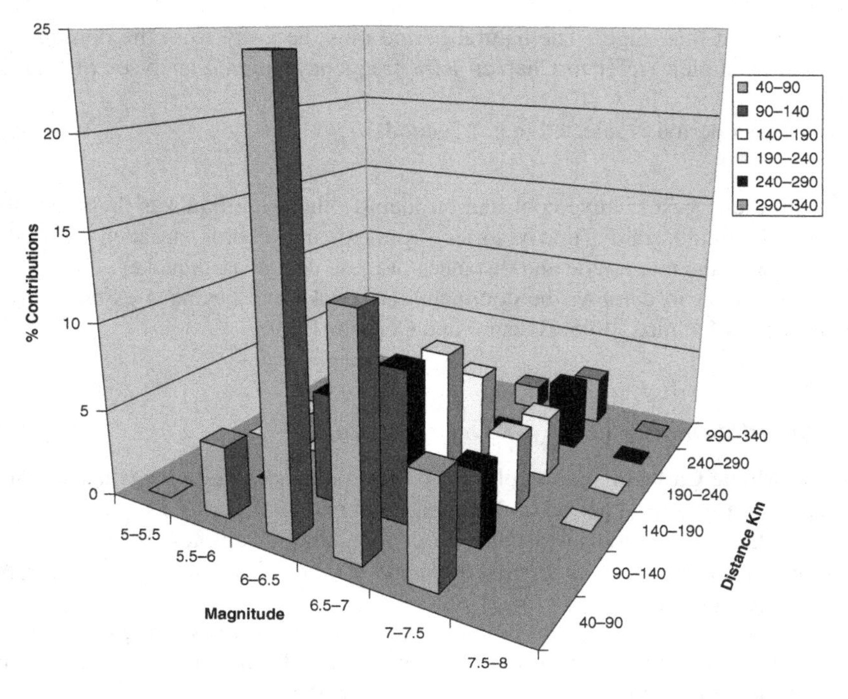

Figure 7.20 De-aggregation plot (return period 100 yrs).

De-aggregation Bin Size

Intuitively we are able to work out that bin sizes will affect the de-aggregation plot. In most hazard studies carried out for projects, there is little or no discussion on the selection of magnitude and distance bins. Usually, de-aggregation is associated with the 'mode' and refers to the most likely scenario group. It corresponds to the scenario group that has the largest de-aggregation value. The mode has the advantage that it will correspond to a realistic source. However, the mode will depend on grouping of scenarios. It has been observed by Abrahamson (2006) that with the grouping of the scenarios, the mode will change if the bin sizes are changed.

As observed earlier, it is often the case that the main use of de-aggregation analysis is selecting the appropriate time histories for further detailed analysis . In this case the characteristics of time histories of past earthquakes (duration, spectral shape etc.) will influence the selection of bin sizes. The characteristics will change with magnitude and distance. Thus, the selection of bin sizes will be project specific. Abrahamson (2006) has provided a very informative discussion on bin sizes.

An alternative approach also suggested by Abrahamson (2006) is to use fine de-aggregation bins. The controlling scenarios are then identified. The mean magnitude and distance for each controlling scenario may then be computed. The approach is robust and will not be sensitive to bin size. However, it may get complicated if the scenarios overlap.

There are two other terms used in seismic hazard analysis, which are not so common in other fields. They are 'aleatory variability' and 'epistemic uncertainty'. Aleatory variability is the natural randomness in a process. Epistemic uncertainty represents the scientific uncertainty associated with the modelling process. It is the result of lack of knowledge and simplified modelling of a complex process. Both sources of uncertainty are represented through a probability density function (PDF) / a cumulative distributive function (CDF); it is common to combine these into a single analysis. However, it is important to understand the dominant source of uncertainty in interpreting the results of probabilistic analysis. If the result is dominated by epistemic uncertainty, perhaps more effort should be made in reducing this source of uncertainty through further data collection or refined modelling. Aleatory variability cannot be reduced by collecting more data. Again Abrahamson (2006) provides a useful discussion of these issues.

7.7.3 Logic-Tree Simulation – An Introduction

Let us refer to the hazard analysis steps shown in Figure 7.2. In the simulations carried out in the previous section, only one model-set was used. In the real world we cannot rely on one model-set. There are various options on recurrence laws, attenuation models etc., all well documented in the literature. A computational framework is needed to provide a systematic, clear and logical way of assimilation of various elements of hazard. Otherwise, there is a risk of mixing the effects of various options and the consequence of not being able to provide the in-depth decision-making. The *logic-tree* concept provides a systematic way forward.

The concept of a flow chart of logical sequences of activities is not new to engineering. The logic tree is essentially a decision flow chart with branches and sub-branches. Each branch represents a discrete set of choices for the input parameter. For example, whether the attenuation model should be according to Boore, Joyner and Fumal (1997), or Campbell and Bozorgnia (2003). Similarly whether the magnitude, $M = 6.5, 7.0$ or 7.5. Each branch is also assigned the likelihood of being correct (weighting factor). Figure 7.21 shows a simple

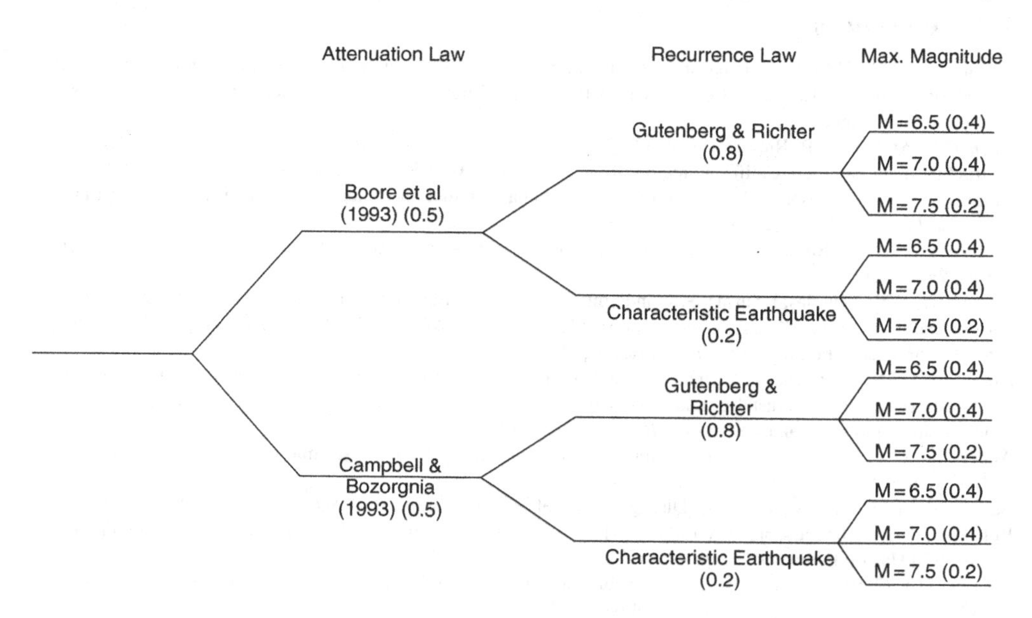

Figure 7.21 *Logic-tree* layout for model uncertainty.

logic-tree. Uncertainties pertaining to attenuation model, recurrence law and magnitude are shown.

In this logic-tree both the attenuation models are considered equally likely. Therefore, each is assigned a weighting factor of 0.5. The Gutenberg-Richter model is the most widely used recurrence model, hence it has a much higher weighting factor than the characteristic earthquake recurrence model, which is also likely according to local seismic reports. Similarly there are different weighting factors for the magnitudes. The weighting factors reflect the degree of belief of the scientific community in the alternative models.

Note that at each node the sum of probabilities must add up to 1. It is easy to comprehend that as more uncertainties are incorporated, the number of branches in the logic-tree might rapidly increase. Some projects may require a much more detailed logic-tree. A very detailed logic-tree for a project in Norway is reported in Coppersmith and Youngs (1985).

7.7.4 Lest We Forget . . .

Probability calculations are important. However, it should be emphasized that seismic risk is not all about calculating probabilities. For instance, the input parameters such as location of faults, seismic activity levels along the faults, travel path, predicted magnitude etc. are all currently very uncertain quantities. Moreover, seismic probabilities have to be considered in the overall context of what is acceptable to society at large. What are the economic consequences of reducing risk? Can the society sustain this? The acceptable level of risk will depend on the importance of the structure. Probability calculations should be seen as one of the inputs to the wider decision-making process. The debate over nuclear safety continues as the disastrous consequences of a possible earthquake close to a nuclear site loom forbiddingly in the future.

7.8 References

Abrahamson, N. (2006) 'Seismic hazard assessment: problems with current practice and future developments'. Proceedings of the 1st European Conference on Earthquake Engineering and Seismology. Keynote Address K2; Geneva, Switzerland.

Allen, C.R., St Amand, P., Richter, C.F. and Nordquist, J.M. (1965) 'Relationship between seismicity and geological structure in the Southern California region'. *Bull. Seism. Soc. Am.* **55**: 753–97.

Ambraseys, N.N., Simpson K.A. and Bommer, J.J. (1996a) 'Prediction of horizontal response spectra in Europe'. *Earthq. Eng. Struct. Dyn.* **25**: 371–400.

Ambraseys, N.N. and Simpson K.A. (1996b) 'Prediction of vertical response spectra in Europe'. *Earthq. Eng. Struct. Dyn.* **25**: 401–12.

Ambraseys, N.N., Douglas, J. Sarma, S.K. and Smit, P.M. (2005a) 'Equations for estimation of strong ground motions from shallow crustal earthquakes using data from Europe and the Middle East: horizontal peak ground acceleration and spectral acceleration'. *Bull. Earthquake Eng.* **3**(1): 11–53.

Ambraseys, N.N., Douglas, J., Sarma, S.K. and Smit, P.M. (2005b) 'Equations for estimation of strong ground motions from shallow crustal earthquakes using data from Europe and the Middle East: vertical peak ground acceleration and spectral acceleration'. *Bull. Earthquake Eng.* **3**(1): 55–73.

Anderson, J.G. (1979) 'Estimating the seismicity from geological structure for seismic-risk studies'. *Bull. Seism. Soc. Am.* **69**: 135–58.

Bazzuro, P. and Cornell, C.A. (1999) 'Disaggregation of seismic hazard'. *Bull. Seism. Soc. Am.* **89**(2): 501–20.

Bommer, J.J. and Abrahamson, N.A. (2006) 'Why Do Modern Probabilistic Seismic-Hazard Analyses Often Lead to Increased Hazard Estimates?', *Bull. Seism. Soc. Am.* **96**: 1967–1977.

Bommer, J.J. (2001) 'Deterministic vs. probabilistic seismic hazard assessment: An exaggerated and obstructive dichotomy'. *J. Earthquake Eng.* **6**, Special Issue, 43–73.

Boore, D.M., Joyner, W.B. and Fumal, T.E. (1993) *Estimation of Response Spectra and Peak Acceleration from Western North American Earthquakes: An Interim Report.* Open-File Report 93-509, United States Geological Survey.

Boore, D.M., Joyner, W.B. and Fumal, T.E. (1997) 'Equations for estimating horizontal response spectra and peak acceleration from western North American Earthquakes: a summary of recent work'. *Seism. Res. Lett.* **68**: 128–53.

Campbell, K.W. (2003a) 'Engineering models of strong ground motion', in Chen, W-F., Scawthorn, C.S. (eds), *Earthquake Engineering Handbook.* Boca Raton, Florida: CRC Press LLC.

Campbell, K.W. (2003b) 'Prediction of strong ground motion using the hybrid empirical method and its use in the development of ground motion (attenuation) relations in eastern North America'. *Bull. Seism. Soc. Am.* **93**(3): 1012–33.

Campbell, K.W. and Bozorgnia, Y. (1994) 'Near-source attenuation of peak horizontal acceleration from world-wide accelerograms recorded from 1957 to 1993'. *Proceedings of the 5th US National Conference on Earthquake Engineering*, Chicago, Illinois, 283–92.

Campbell, K.W. and Bozorgnia, Y. (2003) 'Updated near-source ground motion (attenuation) relations for the horizontal and vertical components of peak ground acceleration and acceleration response spectra'. *Bull. Seism. Soc. Am.* **93**(1): 314–31.

Chapman, M.C. (1995) 'A probabilistic approach to ground motion selection for engineering design', *Bull. Seism. Soc. Am.* **85**(3) 937–42.

Coppersmith, K.J. and Youngs, R.R. (1986), 'Capturing uncertainty in probalistic seismic hazard assessments with intraplate tectonic environments'. *Proc. 3rd U.S. National Conf on Earthquake Engineering, Charleston, South Carolina*, Vol. 1, pp. 301–312.

Cornell, C.A. (1968) 'Engineering seismic risk analysis'. *Bull. Seism. Soc. Am.* **58**: 1583–1606.

Cornell, C.A. and Winterstein, S.R. (1986) 'Applicability of the Poisson earthquake occurrence model'. Technical Report NP-4770, Palo Alto, CA: Electric Power Research Institute (EPRI).

Cornell, C.A. and Winterstein, S.R. (1988) 'Temporal and magnitude dependence in earthquake recurrence models'. *Bull. Seism. Soc. Am.* **78**(4): 1522–37.

Der Kiureghian, A. and Ang, A.H.S. (1977) 'A fault-rupture model for seismic risk analysis'. *Bull. Seism. Soc. Am.* **67**(4): 1173–94.

Dowrick, D.J. (1987) *Earthquake Resistant Design: For Engineers and Architects.* Chichester: John Wiley & Sons, Ltd.

Epstein, B. and Lomnitz, C. (1966) 'A model for the occurrence of large earthquakes'. *Nature* **211**: 954–6.

Esteva, L. (1967) 'Criteria for the construction of spectra for seismic design'. *3rd Panamerican Symposium*, Caracus, Venezuela.

Esteva, L. (1970), 'Seismic risk and seismic design decisions' in Seismic Design for Nuclear Power Plants, Ed R.J. Hansen, MIT Press, Cambridge, Mass., 142–182.

Esteva, L. and Rosenblueth, E. (1964) 'Spectra of earthquakes at moderate and large distances'. *Soc. Mex. De Ing. Sismica* **11**: 1–18.

Everenden, J.F. (1970) 'Study of regional seismicity and associated problems'. *Bull. Seism. Soc. Amer.* **60**(2): 393–446.

Gardiner, J.K. and Knopoff, L. (1974) 'Is the sequence of earthquakes in Southern California, with aftershocks removed, Poissonian?'. *Bull. Seism. Soc. Amer.* **64**: 1363–67.

Gutenburg, B. and Richter, C.F. (1954) *Seismicity of the Earth and Associated Phenomena.* Princeton University Press, reprinted 1965, New York: Stechert-Haffner.

Ipek, M. *et al.* (1965) 'Earthquake zones of Turkey according to seismological data'. *Prof. Conf. Earthquake Resistant construction Regulations* (in Turkish), Ankara, Turkey.

Kanai, K. (1961). 'An empirical formula for the spectrum of strong earthquake motions', *Bull. Eq. Res. Inst.* **39**: 85–95.

Kramer, S.L. (1996) *Geotechnical Earthquake Engineering.* New Jersey: Prentice Hall.

McGuire, R.K. (1974) 'Seismic structural response risk analysis incorporating peak response regressions on earthquake magnitude and distance', *Res. Report R74-51*, Dept of Civil Engineering, Massachusetts Institute of Technology.

McGuire, R.K. (1995) 'Probabilistic seismic hazard analysis and design earthquakes: closing the loop'. *Bull. Seism. Soc. Amer.* **85**: 1275–84.

McGuire R.K. and Arabasz, W.J. (1990) 'An introduction to probabilistic seismic hazard analysis', in S.H. Ward (ed.), *Geotechnical and Environmental Geophysics*, Vol. 1, Review and Tutorial, 333–53.

McGuire, R.K. and Barnhard, P. (1981) 'Effects of temporal variations in seismicity on seismic hazard'. *Bull. Seism. Soc. Amer.* **71**: 321–34.

Reiter, L. (1990) *Earthquake Hazard Analysis: Issues and insights*. New York: Columbia University Press.

Schwartz, D.P. (1988) 'Geology and seismic hazards: moving into 1990s'. *Proceedings of Earthquake Engineering and Soil Dynamics II, Recent Advances in Ground Motion Evaluation*, Geotechnical Special Publication 20, ASCE, New York, pp 1–42.

Schwartz, D.P. and Coppersmith, K.J. (1984) 'Seismic hazards: new trends in analysis using geologic data', in *Active Tectonics*. Orlando, Florida: National Academy Press, 215–30.

TERA Corporation (1978) *Bayesian Seismic Hazard Analysis; A Methodology*.

Wesnousky, S.G., Scholz, C.H., Shimazaki, K. and Matsude, T. (1984) 'Integration of geological and seismological data for analysis of seismic hazard: A case study of Japan'. *Bull. Seism. Soc. Amer.* **74**: 667–708.

Wiggins, J.H. (1964) 'Effect of site conditions on earthquake intensity'. *J. Structural Division*, ASCE, **90**, 279–312.

Wu, S.C., Cornell, C.A. and Winterstein, S.R. (1995) 'A hybrid recurrence model and its implication on seismic hazard results'. *Bull. Seism. Soc. Amer.* **85**(1): 1–16.

Youngs, R.R. and Coppersmith, K.J. (1985) 'Implications of fault slip rates and earthquake recurrence models to probabilistic seismic hazard estimates'. *Bull. Seism. Soc. Amer.* **75**: 939–64.

8

Code Provisions

8.1 Introduction

The consequences of earthquake damage are well documented and dealing with earthquakes has always been a challenge. Design procedures for building safer structures have evolved over the years and in most countries codes of practice have developed in a similar way. The primary purpose of the building codes is to protect the human lives and the buildings that are being designed.

The codes ensure that safe design and construction practices are being followed and that structures important for civilian protection and safety remain operational. This is in the interest of public welfare and the safety of individual citizens. The codes enable this by allowing the designer to estimate the design ground motion parameters when site specific analyses are not available.

Building codes set minimum standards for construction materials, for the type and importance of the structure, the type of detailing to be achieved at the joints and also indicate the amount of deformation that may be allowed under a particular loading condition. The government ensures that they become legal standards. These national building codes differ in content and detail from one country to another, but essentially strive towards the same goal of achieving a safe structure to meet potential seismic demands.

The environmental loads (wind, snow, earthquake) set out are against a particular return period and have a low probability of being exceeded during the lifetime of the structure. Thus wind speeds, for example, may be specified for a return period of 100 years and the specified speed would depend on the return period and the region. Similarly the seismicity could well vary from one part of the country to another. This information is presented in codes in map form.

Fundamentals of Seismic Loading on Structures Tapan Sen

Most codes will usually be classified in to three broad areas:

- a preamble on the general philosophy, purpose, the scope of application and the goals to be achieved;
- computational aspects of structural behaviour, for example, methods of evaluation of global and local forces, effective design loading and/or deformation/ductility limits;
- earthquake resistant constructional details.

Defining seismic loading has always been an important step in the design process, in view of the uncertainties associated with any future earthquake. The design philosophy, which has developed over the years, recognizes that large magnitude earthquakes are rare. This follows from observations that even in regions of frequent seismic activity, such as those around the Pacific Rim, intense earthquakes seldom occur. The frequency of occurrence would range from a few hundred years to a few thousand years. Most buildings would never experience this kind of intense load. To design buildings to resist such loads without some damage would be economically impractical. The codes make provision for this problem in a rather novel way, which at the same time is simple in implementation and will be discussed further in later sections of this chapter.

Early observation of damage to buildings during major earthquakes (including Lisbon, 1755), led building regulation authorities to conclude that certain types of construction fared badly and ultimately collapsed. Early focus, therefore, was on the use of certain construction details, which were known to perform better. Construction details are as important today as they were then.

The seismic provisions currently in use throughout the world will generally follow one of the following basic code models:

- The NEHRP (2000) (National Earthquake Hazard Reduction Program) – recommended provisions developed by the Building Safety Council in the USA.
- The UBC (Uniform Building Code).
- The IBC (International Building Code).
- Eurocode 8 (EN 1998-1-6).

The first three codes were developed in the USA. The 2000 NEHRP Provisions and their later editions reflect the state-of-the art and are followed in many other countries throughout the world. Eurocode 8 differs in detail, but based on similar earthquake resistant design philosophy. In addition to the above, in the California region, the Structural Engineers Association of California Code (SEAOC, 1999) published as the 'Blue Book' is followed widely in the USA. The SEAOC code (Recommended Lateral Force Requirements and Commentary) is revised every few years and has had an important influence on other codes internationally. The UBC (Uniform Building Code) has traditionally been linked to the guidelines published by the SEAOC.

Code development in other countries such as China, Japan, New Zealand, Canada, and India, has followed a similar trend (Hu *et al.*, 1996). Other sources of information are provided in the references for this chapter.

A brief overview of the general historical development of codes is considered next. Further sections elaborate on the response spectrum part of loading.

A worked example following Eurocode 8 (EN 1998-1:2004) and IBC (2000) is included in Sections 8.4.1. and 8.5.

8.1.1 Historical Development

Chronologically the development of earthquake engineering may be divided into three stages:

- a focus on the concept of lateral force, which covered the first half of the 20th century;
- the response spectrum theory, which originated in the 1940s;
- modern dynamic concepts developed from the 1960s onwards.

1755–1926

Early development of the codes was experience based. For example, the Lisbon earthquake of 1755 showed inadequacies of certain types of construction and building regulations were enacted to require specific construction details to be followed. In the USA, poor performance of unreinforced brick masonry bearing wall construction during that period was noticed following the Hayward earthquake in 1868. Based on this and similar observations in the California region, some local building codes began to enforce construction detailing of unreinforced masonry buildings. The details were prescriptive, regarding placement of out-of-plane bond anchors for preventing out-of-plane failure. Similarly, iron anchor bonds were introduced to prevent in-plane failure of unreinforced masonry walls.

1927

The first edition of the Uniform Building Code was published in the USA and this is believed to be the first modern code. As one would expect, the 1927 edition incorporated the lessons learnt in the intervening years since the 1868 earthquake in the San Francisco Bay area. The 1927 edition also included provisions based on another important observation. The primary cause of damage was observed to be the lateral shaking motion and also damage was more extensive for structures founded on soft soil deposits. The code required that such structures be designed for a lateral force equal to 10 % of the total weight coming on to a floor; the lateral force to be applied simultaneously to each roof and floor level. In the case of buildings on firm soil, design could be based on lateral force equal to only 3 % of the total weight coming on to the floor; an implicit recognition of reduced intensity of ground shaking on firm soil.

1933 Field Act & Riley Act

The Long Beach (1933) earthquake which had an estimated magnitude of 6.3 again caused serious damage to unreinforced masonry buildings, in addition to which several public schools were severely damaged or destroyed. Collapse of public schools caused panic and led to a sea change in the theory of how buildings should be designed. Immediately after the Long Beach earthquake, the State of California introduced two important pieces of legislation, the Riley Act and the Field Act (1933). The Field Act was introduced as an emergency measure.

Photographs illustrating the damage to public school buildings in the 1933 Long Beach Earthquake may be seen in Figures 8.1–8.3.

Figure 8.1 Long Beach, California, Earthquake, 10 March 1933. Damage to Compton Union High School. Photographed by W.W. Bradley, reproduced courtesy of the US Geological Survey.

Figure 8.2 Long Beach, California, Earthquake, 10 March 1933. Damage to John Muir School. Anonymous photograph, reproduced courtesy of the U.S. Geological Survey.

Figure 8.3 Long Beach, California, Earthquake, 10 March 1933. Damage to the Shop Building at Compton Junior High School, Compton, Ca. Photographed by H.M. Engle, reproduced courtesy of the US Geological Survey.

The following extract, about the introduction of the Field Act following the 1933 earthquake, is from the California Department of General Services website at http://www.excellence.dgs.ca.gov/StudentSafety/S7_7-1.htm 01/08/08 and readers may find it interesting.

The Field Act

Introduction

Parents expect a safe and secure learning environment for their children. Public school safety encompasses many aspects of security such as protecting children from the abusive behavior of other classmates or the dangers of toxic chemicals. Prevention of injury or death due to natural forces such as wind or ground motion depends upon the structural integrity of public school buildings and the bracing and anchoring of furniture and other objects within the classrooms to prevent them from falling on the children. Parents should be able to assume that once their child enters a public school that the building will shelter the pupils from the powerful forces of nature. These are the reasons the Field Act was written and enacted.

History of the Field Act

In 1933 the lateral force resistant design of public schools, as well as other buildings throughout the state, was based on estimated wind loads. Engineers assumed that buildings designed to withstand wind forces would also be able to withstand earthquake forces. The magnitude 6.3 Long Beach Earthquake of March 10, 1933 destroyed 70 schools and another 120 schools suffered major structural damage. Luckily, the earthquake occurred when the buildings were unoccupied. Hundreds of children might have perished if the earthquake had occurred only a few hours earlier.

Although there had been previous earthquakes in urban areas of California (San Francisco in 1906 and Santa Barbara in 1925) the significance of the Long Beach Earthquake was that it drove home to engineers, public officials and the general public that the need to develop measures to resist the effects of earthquakes is a necessary part of public policy. The level of damage and the extent of the deaths and injuries in Long Beach were perceived as unacceptable outcomes for an event, which could be repeated at any time. The great number of collapsed public schools led to a public outcry for a remedy to the situation of housing public school students in structures that were unsafe in earthquakes. On April 10, 1933, the Field Act was enacted. The Field Act requires that the building designs be based on high level building standards adopted by the state and plans and specifications be prepared by competent designers qualified by state registration. The quality of construction was to be enforced through independent plan review and independent inspection. Finally, the design professionals, independent inspector and the contractor had to verify under penalty of perjury that the building was constructed according to the approved plans.

The Field Act as adopted applied only to new construction; not to existing pre-1933 school buildings. Legislation to cover the criteria for continued use or abandonment of these pre-1933 school buildings was enacted under the Garrison Act of 1939. These pre-Field Act buildings were not retrofitted to conform to current codes until funding was made available in the 1970s. It is no coincidence that the state provided funding shortly after the magnitude 6.4 San Fernando Earthquake of 1971. California is now grappling with the problem of evaluating and retrofitting thousands of school buildings constructed before 1976. School Districts contemplating projects on such a campus are well advised to contact their DSA Regional Office in advance of project submission.

The Riley Act

The Riley Act made it mandatory that all buildings should be provided with lateral strength equivalent to 3 % of the weight of the structure. It was later adopted by the UBC and included in the NEHRP documents. However, the form and content has changed over the years.

Stringent rules for building plan review, supervision of construction details and high standards for structural design were introduced through the Field Act. It was introduced as an emergency measure but provisions from it have been retained. These important provisions were later introduced as part of the Uniform Building Code and remain in a modified form. In the following years, performance of buildings designed to Code were observed after each earthquake and modifications introduced for improved performance. This had an impact on the direction of earthquake engineering research initiated around that time and made an important contribution to understanding of building response. As yet there were no records of actual ground motion. The first accelerometer was installed in 1933 and the first acceleration record was obtained during the Long Beach (1933) earthquake. Theories of structural dynamics were evolving. The El Centro earthquake (1940) record became the more famous record and has been widely used by the engineering community for subsequent research work.

The idea of lateral force 'P' was introduced through the simple equation

$$P = \frac{W}{g} a_{\max} = kW,$$

where W is the total weight of the structure, $a_{\max}$ is the earthquake acceleration and $k = a_{\max}/g$ is the seismic coefficient. As we have noted, the static method and the value of seismic coefficient k both originated from field experience and were included in the UBC.

1946

UBC introduced seismic zones within the USA to take care of regional seismicity.

The Californian belt is a zone of high seismicity. Other regions of lower seismicity in the US were identified.

1952

The response spectrum theory was first developed and proposed by Biot (1943). It was Professor G.W. Housner and his colleagues at the California Institute of Technology, who championed the theory's acceptance as a design procedure. In 1952 a joint committee was formed by the American Society of Civil Engineers to develop recommendations for procedures and concepts for incorporation into codes. The recommendations relating lateral forces on buildings to their structural periods based on a design response spectrum were incorporated by SEAOC in the first edition of its Recommended Lateral Force Requirements and Commentary in 1952.

1958

Spectral concepts were adopted by the Uniform Building Code in its 1958 edition. The spectral theory combines the spectral properties of ground motion and the dynamic properties of the structure.

The total lateral force (identified as base shear) was given by the formula:

$$V = ZKCW$$

where

W is the total weight of the structure
Z is the zone coefficient relating design force to regional seismicity

This information could be obtained from the seismic zone map published in the 1958 UBC code. Four seismic zones were identified, zones 0 to 3. No seismic design was necessary for zone 0. Zones 1–3 were assigned values. Zone 3 was most severe and the factor was 1. Zones 1 and 2 were assigned values less than 1.

K is a structural system coefficient. Four types of construction were identified:

- Light timber frame construction; $K = 1.0$
- Building frame system. Lateral forces resisted by diagonal bracing or shear walls; $K = 1.0$
- Box type construction system. This type included the unreinforced masonry structure which had suffered severe damage in previous earthquakes as mentioned earlier; $K = 1.33$
- Moment resisting frame. Beams and columns rigidly connected to provide lateral stability. $K = 0.67$

C is the spectral amplification factor depending on the fundamental period of the structure; $C = \dfrac{0.05}{\sqrt{T}}$; where T is the fundamental period of the structure. An empirical formula was given in the code to assess the fundamental period.

The empirical formula above is approximate. More exact methods like those based on the Rayleigh method could also be used. Additional clauses were introduced which would restrict the value of C so that the total base shear for structures with K value equal to 1, was not more than 0.1 times the weight of the building. This view was based on earlier observations and in conformity with provisions contained in earlier codes. The lower bound of the base shear value was still maintained at 0.03 times the total weight, which was established in 1933 with the introduction of the Riley Act.

Once the base shear was obtained from the above equation, the lateral forces were distributed at each level according to the weight supported at that level. The analysis was for gravity loads and lateral loads. The combined forces in individual members were checked according to allowable stress theory. If seismic loads were considered the code permitted an increase in the allowable stress.

The introduction of response spectrum method in the codes was a great leap forward in terms of earthquake design.

1958 onwards

Design up to this point was performed on the basis of linear elastic theory. There were further earthquake engineering-related developments during the 1960s. These include:

- availability of computing facilities and wide application of computer techniques for non-linear analysis and new experimental techniques, to predict non-linear behaviour during strong motion earthquakes helped in the understanding of failure mechanisms of structures;
- many earthquake motion records accumulated for analyses;
- field observations of strong motion earthquakes provided new knowledge of behaviour of tall buildings, dams, foundations and other structures which were severely damaged.

These provided a large impetus to the development of dynamic theory and non-linear analysis which could not be undertaken earlier.

Designs in the State of California were carried out as laid down in the SEAOC Blue Book. It was observed that the forces experienced by the buildings during a major earthquake were far in excess of what was predicted or designed for. These actual observations pointed towards the thinking that structures subjected to much larger imposed loading could survive earthquake shaking with damage but not collapse as long as they were provided with tough, continuous lateral force resisting members which had the capability of deforming. The philosophy of earthquake design had to be revisited.

The SEAOC Blue Book published its recommendations in the 1960s. It adopted the following three-tiered performance goals:

- Resist minor earthquake shaking without damage.
- Resist moderate earthquake shaking without structural damage but possibly with some damage to nonstructural members.
- Resist major earthquake shaking with both structural and nonstructural damage but without endangering the lives of occupants.

If we ignore temporarily the second goal, we arrive at the essence of the dual design philosophy discussed in Chapter 7. The final intent was clearly laid down, that is to protect human lives first, even at the cost of some damage to capital investment. Thus life-safety emerged as the most dominant criteria. Protection of real estate or capital investments was not the major end-goal.

Although the last performance criteria have been refined and revisited over the years, the basic philosophy laid down by SEAOC in the 1960s remains the same.

Research in the 1960s

The last criteria laid down in the 'Blue Book' could only be achieved if structures deformed inelastically and dissipated energy. In the 1960s Newmark and his team, at the University of Illinois, initiated extensive research on inelastic performance of steel beam-columns leading to deformations several times the deformation at first yield but without collapse. The inelastic displacement leading to a useful limit of displacement was studied. The extent to which excursions in to the inelastic regime was allowed was defined by a new factor μ introduced by Newmark. This new factor, called the ductility factor, was defined as $\mu = \frac{u_m}{u_y}$ (see Chapter 6, Section 6.4).

u_m – defined as the useful limit of displacement.
u_y – displacement at yield.

Spectral concepts were introduced in 1958. Newmark introduced the factor μ, modified the spectral values of acceleration, velocity and displacement to produce the inelastic response spectrum (see discussions in Section 6.4). Other researchers also began to investigate post-yield behaviour of typical framing elements. It became clear at this early stage, that the way in which structural elements and their connections are detailed is important in realizing the ductility potential of the structural system. Controlling the ductility was the key to achieving the third goal – that of producing a design which would resist a major earthquake with both structural and non-structural damage but not collapse, thus ensuring that lives of occupants were not endangered. Simultaneous attention was focused on connections. It was noticed that reinforced concrete frames exhibited brittle failure unless they had proper reinforcing bars to take care of reverse loading.

Provisions for detailing of reinforced concrete frames were included in the 1967 edition of UBC. Refinement of detailing, supported by research, continued over the years and provisions for steel, timber and masonry followed later.

The Niigata (1964) earthquake in Japan produced failures due to liquefaction. The Alaska earthquake in the same year also caused liquefaction failure. This initiated research efforts to determine the causes of liquefaction and how to assess the potential for liquefaction at a site. Seed and Idriss (1971) introduced the first empirical assessment of liquefaction potential.

This led to important advances in seismic engineering concepts. The consideration of site conditions in terms of spectral shapes was deemed important. It was realized that spectral components would be different for different site conditions. Seed *et al.* (1976) published spectral curves for different soil conditions (see also Chapter 5).

1971

San Fernando Earthquake

The San Fernando region in California was well instrumented. The large magnitude earthquake, which struck at the rim of the San Fernando Valley in 1971, was significant in the sense that it made available a large amount of data to research workers and code developers working on refining the attenuation relationships for the region. The earthquake, which occurred near the city of Los Angeles, damaged modern buildings that were designed according to the current code, including several low-rise buildings, and exposed the weaknesses of the SEAOC code of the time. Following this major earthquake and its resultant damage, the ATC (Applied Technology Council) was formed with the objective of producing fresh recommendations to meet the SEAOC three-tiered philosophy of life safety.

1978

ATC publishes its ATC-3.06 report

The ATC-3.06 report is considered a milestone in terms of modernizing the code and bringing it in line with the state-of-the-art and reflects a major shift in thinking.

Amongst other recommendations it introduced dynamic analysis for earthquake design and adopted response spectra analysis as the preferred method of analysis. It also simplified the lateral force procedure, while maintaining clarity regarding its use. It laid down, in an orderly step-by-step method, the procedure for estimating the design loading parameters. Another feature was the introduction of the risk-based zoning map.

The provisions introduced for the first time a contemplated seismic force resisting system with redundant characteristics wherein the overall strength above the level of significant yield was to be obtained by plastification at other locations in the structure prior to formation of a complete plastic hinge mechanism. The definition of the term 'significant yield' in ATC-3.06 (1978) is quoted below:

'The term "significant yield" specifically is not the point where first yield occurs in any member, but is defined as the level causing complete plastification of at least the most critical region of the structure, such as formation of the plastic hinge of the structure.'

This is basically in line with the original thinking contained in the SEAOC policy document of the 1960s:

'Resist major earthquake shaking with both structural and nonstructural damage but without endangering the lives of occupants.'

The over-strength capacity represented by this continuous inelastic behaviour provides the necessary reserve strength that will be called upon while resisting extreme motions of the actual seismic forces.

The ATC-3.06 provisions are further developed in the '2000 NEHRP Recommended Provisions for Seismic Regulation for Buildings and Other Structures' and are discussed later in this section.

After the publication of the ATC report the Building Seismic Safety Council (BSSC) was formed. Its specific task was to develop the ATC-3.06 report into a set of usable BSSC guidelines which were later published under the auspices of NEHRP (Recommended Provisions for Seismic Regulation for Buildings, 1985) and have since then been revised every three years.

Major earthquakes 1971–1988

From 1971 to 1988 major earthquakes happened at Imperial Valley California, in 1979 and in Mexico City in 1985. Some important, notable observations from these earthquakes included:

- The introduction of a site factor to reflect the effect of soil conditions at the site;
- The frequency content and amplitude of ground shaking are affected by the soil;
- Introduction of a factor to account for the importance of the building;
- A more conservative design for important buildings, for example a hospital should be more conservatively designed than a shed;
- Introduction of inter-storey drift limits.

1988

Uniform Building Code adopts the ATC-3.06 recommendations.

1993

1991 NEHRP provisions adopted by ASCE-7.

1994

UBC response spectra based on site classification introduced. The UBC Response Spectra are shown in Figure 8.4.

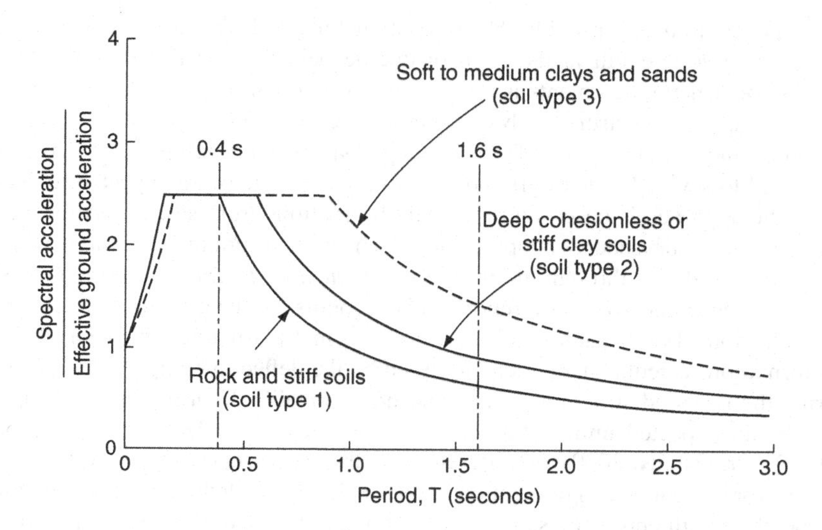

Figure 8.4 Response spectrum UBC 1994. 1994 Uniform Building Code © 1994 Washington DC; International Code Council. Reproduced with permission, all rights reserved.

2000, 2003

NEHRP Recommended Provisions for Seismic Regulation for Buildings and Other Structures published.

The NEHRP Provisions (2000) was a landmark publication. It was a comprehensive document representing the latest thinking which would form the basis of much of seismic design in the USA. The 2003 version introduced further refinements. This document has also influenced the code development in other parts of the world. Since then many new provisions have been introduced in various parts of the code. However, here only the ductile-behaviour based inelastic design philosophy will be discussed briefly.

In line with the original thinking postulated in the 1960s SEAOC document, the 2000 NEHRP Provisions assume a significant amount of non-linear behaviour. The extent of non-linear behaviour sustainable depends on the material used and the amount of ductility allowed, the structural system and the configuration. Chapter 4 of the Provisions deals with Structural Design Criteria and Chapter 5 deals with Structural Analysis Procedures.

As part of the general requirements of the NEHRP Provisions (2000), the structural design for acceptable seismic resistance includes:

- the selection of gravity and seismic-force-resisting systems that is appropriate to the anticipated intensity of ground shaking;
- the layout of these systems such that they provide a continuous, regular, and redundant load path capable of ensuring that the structures act as integral units in responding to ground shaking;
- proportioning the various members and connections such that adequate lateral and vertical strength and stiffness are present to limit damage in a design earthquake to acceptable levels.

These general requirements provide the foundation for good design practice which would have enough reserve strength in the form of ductile behaviour in the inelastic range. The proportioning of structural elements is on the basis of internal forces obtained from detailed linear elastic response spectrum analysis, which have been substantially reduced from the anticipated design ground motions. These concepts have been developed in Chapter 6. Thus when subjected to severe levels of ground shaking, many regions are expected to deform inelastically. The approach is based on historical observations from severe earthquakes in the past. Therefore, these procedures adopt the approach of proportioning structures based on the reduced forces introduced through the *response modification factor* 'R_d' in the provisions as originally intended in the ATC-3.06 report. This accounts for the ductility effects during the post yield behaviour. The actual inelastic deformations in the structure will be larger. Thus, the elastic deformations calculated are then amplified by the deflection amplification factor 'C_d'. Considering the intended structural behaviour and acceptable deformation levels, the storey drift limits for the expected amplified deformations are prescribed. The storey drift procedures in the NEHRP Provisions (2000) are different from those prescribed in earlier codes.

The values for the response modification factor, 'R_d', and the deflection amplification factor 'C_d' are specified in the provisions. The 'R_d' factor essentially represents the ratio of forces, which would develop under the design forces assuming entirely linearly elastic behaviour to the prescribed design forces at significant yield. In establishing 'R_d' values, consideration has also been given to the performance of different materials and systems in past earthquakes. An explanatory text on *the ductility reduction factor* 'R_d' may be found in the Commentary on Chapter 4 of the Provisions.

Controlling Storey Drift

Storey drift and lateral displacement are defined in the Provisions as follows:

- Storey drift is the maximum lateral displacement within a storey (i.e. deflection of one floor relative to the floor below due to the effects of seismic loads)
- Lateral displacement or deflection due to design forces is the absolute displacement of any point in the structure relative to the base. This is not 'storey drift' and is not to be used for drift control or stability considerations since it may create a false impression of the effects in critical storeys. Storey drift is important when considering other seismic requirements.

One of the main reasons for controlling storey drift is to control inelastic strain. The Provisions recognize that drift limitations are an imprecise and highly variable way of controlling strain but have been introduced in line with the current state of the knowledge of what the strain should be.

From the stability point of view it is important that flexibility is controlled. Stability of vertical members, as has been discussed earlier in Chapter 8, is a function of axial load and bending of the members. Controlling storey drift is a direct way of controlling '$P - \Delta$' effects (secondary moment from axial load and deflection) in vertical load carrying members. This is extremely important, as instability in a column could be disastrous. That is why in many practical designs a strong-column and weak-beam concept is preferred. The drift limits contained in the Provisions indirectly provide upper bounds to limit these effects.

Response Spectrum Procedure

The Provisions include the response spectrum method being applicable for linear response of complex, multi-degree-freedom structures and are based on the fact that the response is obtained from the superposition of responses from individual modes (the procedure is discussed in Chapter 5 of this book and a worked example included). The commentary in Chapter 5 of the Provisions, 'Structural Analysis Procedures', notes the following:

> The response of the structure, therefore, can be modelled by the response of a number of single-degree-of-freedom oscillators with properties chosen to be representative of the mode and the degree to which the mode is excited by the earthquake motion. For certain types of damping, this representation is mathematically exact, and for structures, numerous full scale tests and analyses of earthquake response of structures have shown that the use of modal analysis, with viscously damped single-degree-of-freedom oscillators describing the response of the structural modes, is an accurate approximation for analysis of linear response. (NEHRP, 2003)

Alternative Simplified Procedure

In recent years, practising engineers and building regulation authorities have become increasingly concerned about the Provisions being too complex to follow and consequently to implement. The earthquake is by its very nature a complex process; it is difficult to understand its mechanism and to implement earthquake resistant structures in practice. The motivation of the Code Drafting Committee has been to reflect the state-of-the art in design, and this has led to Provisions becoming complex with improvements in knowledge and understanding. In response to the growing concerns about complexity, a simplified set of rules was thus formulated for smaller and simpler structures. It is presented as a stand-alone document. The rules are prescriptive to a great extent.

Some of the important features are:

- The simplified procedure would apply to structures up to three storeys high, in Seismic Design Categories B, C, D and E, but would not be allowed for systems for which the design is typically controlled by drift.
- The table of basic seismic-force-resisting systems simplified to include only allowable systems, and deflection amplification factors are not used;
- design and detailing requirements consolidated into a single set of requirements applicable to all Seismic Design Categories.
- The procedure is limited to site classes A to D.

Reference documents

The following reference documents have been used:

American Concrete Institute (2008) *ACI 318-08 Building Code Requirements for Structural Concrete and Commentary*. Farmington Hills, MI: ACI.

American Institute of Steel Construction (2006) *AISC Steel Construction Manual – 13th Edition*. Chicago, IL: AISC.

American Institute of Steel Construction (2005) *AISC Seismic Provisions for Structural Steel Buildings*, March 9, 2005. Chicago, IL: AISC.

American Iron and Steel Institute (2007) *AISI Specification for the Design of Cold-formed Steel Structural Members*. Washington, DC. AISI.

American Society of Civil Engineers (2006) *ASCE 7/SEI 7-05 Minimum Design Loads for Buildings and Other Structures*. Reston, VA: ASCE.

2000

The three codes operating in the US were merged in to a single document, the International Building Code (IBC) in 2000.

Current Philosophy of Codes of Practice

General Discussion

Designing a structure where all its members are to remain elastic during a strong motion earthquake is economically prohibitive. Thus, the SEAOC committee initiated the three-tiered philosophy of design, referred to earlier, to resist forces due to such large magnitude earthquakes, way back in 1968. Eurocode 8, being recent in origin, follows a similar train of thought, established over the years.

Ductility Demands

Structural design has traditionally been force based. The structural members are sized according to internal forces obtained from linear elastic analysis. The analysis is carried out on the basis of the design acceleration spectrum, which is usually obtained by dividing the elastic acceleration spectrum by the 'reduction factor' denoted by 'R' in SEAOC, UBC/IBC codes or behaviour factor 'q' in Eurocode 8. The detailing of the members and the connections should be such as to develop the intended inelastic deformations associated with the reduction factor.

Structural Simplicity

The modelling of the structure should ensure that a clear and direct path for transmission of the seismic forces is available. Uniformity in plan and elevations as the preferred concept has been emphasized in other codes and is emphasized again in Eurocode 8.

We have to bear in mind that even for structures where simplicity and uniformity have been maintained and which are well designed, a large magnitude earthquake would be an extreme event and could well push the structure to its limits and expose all its unknown weaknesses (Northridge, 1994). Maintaining simplicity has the advantage that modelling, analysis and detailing and finally construction at site are subjected to much less uncertainty and consequently structures can be expected to perform better and cope better during such an extreme event.

Analysis Methods in Eurocode 8 (EN 1998-1)

Section 4 of EN 1998-1 provides the following analysis options to designers:

- Linear static analysis (usually referred to as the lateral force method in EN 1998-1);
- Modal response spectrum analysis (this is termed as linear-dynamic analysis in practice);
- Non-linear static analysis (known as 'push-over analysis');
- Non-linear dynamic analysis (step-by-step time history analysis considering all known non-linearities).

The modal analysis procedure is applicable for design of buildings without any limitations. In the linear method of analysis the design response spectrum, is the elastic response spectrum based on 5 % damping divided by the behaviour factor 'q'. The members are designed on the basis of linear elastic analysis and the design is consistent with the ductility demand. As discussed in Chapter 6, the displacements due to the seismic action are taken as equal to that obtained from the linear elastic analysis, multiplied by the behaviour factor 'q'. The behaviour factor 'q' is implicit in the non-linear analysis.

Readers should note that use of a linear method of analysis, does not imply that that seismic response will be linear elastic; it is simply a device for simplification of practical design within the framework of current force-based seismic design with elastic spectrum divided by the behaviour factor 'q'. Carrying out non-linear dynamic analysis is complicated and time consuming.

Linear Elastic Methods

The Eurocode 8 procedure for lateral force analysis is similar to those in use in the US codes discussed earlier. Linear static analysis is performed under a set of lateral forces in the two orthogonal directions 'X' and 'Y'. The intent is to simulate, through the application of these static forces, the peak inertial forces induced by the horizontal component of the seismic action in the two directions. The method is popular amongst engineers across the world because of the familiarity with linear elastic analysis. The version of the method in Eurocode 8 has been tuned to give similar results for storey shears to those obtained from the modal analysis, which in Eurocode 8 is referred to as reference method, at least for the class of structures where lateral force method is permitted to be used.

The modal response analysis is an elastic, linear dynamic analysis and Eurocode 8 points towards its use where both the lateral force method and the modal response method are applicable. Within Eurocode 8 the use of modal response spectrum analysis is applicable without any limitations. The overall inelastic performance is expected to be better when the members are sized according to internal forces obtained from a modal analysis.

8.2 Static Force Procedure

8.2.1 Base Shear Method

Loading provisions – recent codes

SEAOC 1999

The SEAOC 1999 document provides the relevant UBC 1997 clauses within brackets. The same is shown here.

SEAOC Clause 105 (UBC § 1630)

Cl. 105.2 Static Force Procedure

The following symbols are referred to:

T – Fundamental period of the structure
I – Importance Factor
C_v – Seismic response coefficient
C_a – Seismic response coefficient
W – Weight of the structure
R – Seismic reduction factor
R_o – Factor to allow for over-strength portion of R
R_d – Factor to allow for reduction in seismic response due to inelastic action

The total design base shear V shall be determined from the following:

$$V = \frac{C_v}{(R/I)T}W \tag{105.4}$$

where

$$R = R_d R_o \tag{105.5}$$

The total base shear need not exceed the following:

$$V = \frac{2.5C_a}{(R/I)}W \tag{105.6}$$

where

$$R = R_d R_0$$

Extract from SEAOC Commentary

The design base shear equations (105.4) and (105.6) provide the level of seismic design forces for a structural system. As an insight into the equations the SEAOC Commentary on the section states the following:

> Equation (105.4) represents the constant velocity portion of the spectrum and equation (105.6) represents the constant acceleration portion. The resulting force level is based on the assumption that the structure will undergo several cycles of inelastic deformation during major earthquake ground motion, and, therefore, the level is related to the type of structural system and its estimated ability to sustain these deformations and dissipate energy without collapse. The force level determined by the spectrum is used not only for the static lateral force procedure, but also as the lower bound for the dynamic lateral force procedure under section 106 (SEAOC, 1999).

The physical relationship between the base shear and the spectral representation of major earthquake ground motion is given in the Commentary:

> For a given fundamental period '*T*', the base shear equations (105.4) and (105.6) provide values that are equal to the fundamental ordinate of the acceleration response spectrum (C_V/T and $2.5C_a$ respectively) shown in Figure 8.5 times the total structure weight, divided by *R*. While only the fundamental mode period is employed, the additional response due to higher modes of vibration is represented and approximated by the use of the total weight of the structure *W*. (SEAOC, 1999).

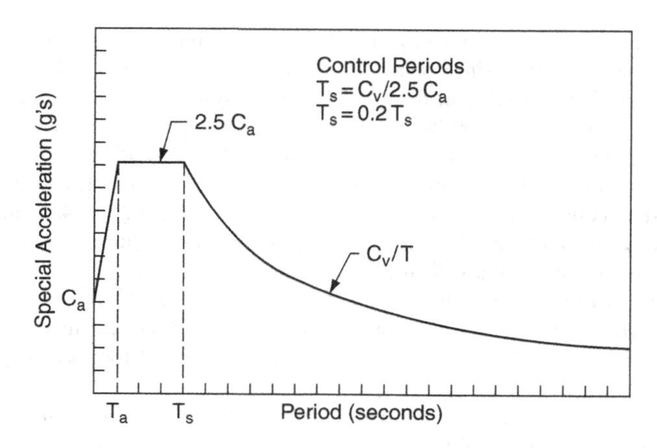

Figure 8.5 Response spectrum shape. Reproduced from SEAOC (1999) by permission of the Structural Engineers Association of California.

Two parameters, C_a and C_v introduced in Equations (105.4) and (105.6), are seismic response coefficients.

In order to determine C_a and C_v, the seismic zone factor 'Z' needs to be established.

The seismic zone factor 'Z', may be obtained from the seismic zone map in Figure 104-1 (SEAOC, 1999) or from Figure 16-2 (UBC, 1997).

Seismic Zone 4

In California there are two zones, zones 3 and 4. In previous editions of SEAOC, the seismic zone factor was determined from the relevant table and used throughout the zone.

Near Source Factor

In the 1999 edition, the near source factors N_a and N_v were introduced in Zone 4. The reasons are easy to follow intuitively and were finally incorporated in the requirements. It is now well documented that if the fault is near the site, it is capable of significantly more damage than if it was much further away. Tables (104-9) and (104-10) specify seismic coefficients, C_a and C_v, in terms of near source factors N_a and N_v, for each of the soil profile types and are applicable to Zone 4.

Commentary on 'R' (SEAOC, 1999)

> R is the reduction factor used to reduce the dynamic force level that would exist if the structure remained elastic to a design force level that includes the effect of ductility and over-strength. This force level is now at the strength level, rather than the working stress level as it was in previous editions of the SEAOC. The reduction factor is compatible with the R used in the 1997 NEHRP Provisions. (SEAOC, 1999)

The R factor is expressed in terms of two separate components R_o and R_d. This is done with the expressed intention of better representing the R factor for future systems. The SEAOC Commentary notes:

> R_o represents the over-strength portion of factor R. This over-strength is partially material-dependent and partially system-dependent. Since design levels are based on the first yield of the highest stressed element of the system, the maximum force level that the system can resist after the formation of successive hinges, bracing yield, or shear wall yield or cracking will be somewhat higher than the initial yield value. Designs are also based on minimum expected yield or strength values, whereas the average strength of a material could be significantly higher. Preliminary investigations of various material/system configurations indicated that R_o can vary from 2.25 to 4.5. However, R_o values selected were between 2.0 to 2.5. In future, it may be possible by pushover analysis and other methods to uniquely determine R_o for a structure.
>
> R_d is the portion of the R factor that represents the reduction in seismic response due to inelastic action of the structure. This factor is proportional to the overall ductility of a system and its materials, and varies from 1.2 for systems with low ductility to 3.4 for special systems with high ductility.
>
> The I factor is used in the base shear equations to emphasize its role in raising the yield level for important structures. (SEAOC, 1999)

The vertical distribution of force is outlined in Section 105.5.

Clause 105.5

Vertical Distribution of Force

This section outlines the vertical distribution of the force. The code stipulates that the total force is to be distributed over the height of the structure in the following manner:

$$V = F_t + \sum_{i=1}^{n} F_i \tag{105.14}$$

The concentrated force 'F_t' at the top, which is in addition to F_n, shall be determined from the formula:

$$F_t = 0.071TV \tag{105.15}$$

The value of T used for the purposes of calculating F_t shall be the period that corresponds with design base shear as computed using Equation (105.4). F_t need not exceed $0.25V$ and may be considered as zero where T is 0.7 seconds or less. The remaining portion of base shear (105.14) shall be distributed over the height of the structure, including Level n, according to the formula:

$$F_x = \frac{(V - F_t)w_x h_x}{\sum_{i=1}^{n} w_i h_i} \tag{105.16}$$

[Note: A corresponding analytical expression can be obtained from Equation (6.17) (see Chapter 6).]

$$\{F_i\} = \{I\}^T [M] \cdot \{\varphi_i\} \frac{\{\varphi_i\}^T [M] \{I\}}{\{\varphi\}^T [M] \{\varphi_i\}} \omega S_v$$

Equation (6.17) is in matrix form. Hence, for each mode component at each mass we have equivalent horizontal design force and total horizontal force.

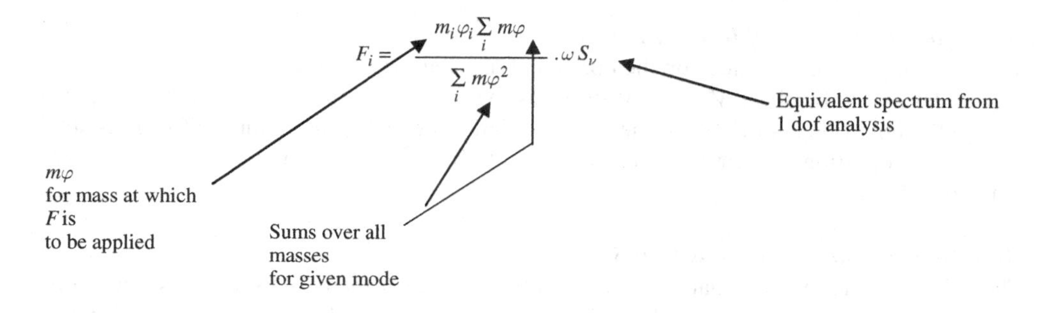

$$V = \sum_i F = \frac{(\sum m\phi)^2}{\sum m\varphi^2} .\omega S_\nu$$

From Equation (6.17) the lateral force at node 1 for mode 1 has been worked out and it follows that the lateral force at level i for mode m may be calculated as

$$F_{im} = \frac{m_i \varphi_{im}}{\sum m_i \varphi_{im}} \cdot V_m$$

It is common practice to approximate the first mode as

$$\varphi_1 = \frac{h}{H} \varphi_H$$

where

H is the height of the building
h is the height above ground
φ_H is the mode shape value at the top

With a linear representation of the first mode shape, the above equation for distribution of storey shear reduces to

$$\frac{w \cdot h}{\sum_i wh} \cdot V$$

where V is the total base shear, which is equivalent to Equation (105.16).

This form has been adopted in the code because observations on the vibrations of a large number of tall buildings demonstrate that the first mode shape generally is quite close to a straight line.

As a general check at this stage, the reader is referred to the numerical example in Section 8.5.

[Also note that maximum stress in structure is *not* the sum of maximum of modal components as the maxima will not occur simultaneously.]

At each level designated as x, the force F_x shall be applied over the area of the building in accordance with the mass distribution at that level. Structural displacements and design seismic forces shall be calculated as the effect of forces F_x and F_t applied at the appropriate levels above the base.

Horizontal distribution of force (§ 105.6)

This section provides guidance on the horizontal distribution of force within a storey.

The design storey shear V_x, in any storey is the sum of the forces F_x and F_t above that storey. V_x should be distributed to the various elements of the vertical lateral force resisting system in proportion to their rigidities. The rigidity of the diaphragm also needs to be taken in to consideration.

Horizontal Torsional Moments (§ 105.7)

The SEAOC code requires that provisions shall be made for increased shears resulting from horizontal torsion in situations where diaphragms are not flexible. Additional guidance is contained in this section.

The requirements for Simplified Design Base Shear procedure are contained in Section 105.2.3 of SEAOC.

8.3 IBC 2006

8.3.1 Introducing Mapped Spectral Accelerations

The IBC 2003 and 2006 editions introduced a series of national seismic hazard maps for the USA and its territories. The figures in IBC 2006 will be referred to in this section. The maps were developed by the US Geological Survey and are in form of contours of spectral acceleration. The contours are presented in terms of % g (acceleration due to gravity) and the spacing is 0.02g. The contours are for 2 % probability of exceedance in 50 years.

A sample plot (Figure 1613.5 (1), IBC 2006) of the contour maps provided in the code is shown in Figure 8.6. The maps also provide enlarged plots of the 'boxed' regions 1, 2, 3 and 4 designated on the maps. In the highly seismic prone areas of the Western California belt the plots are very close to each other and difficult to read. The discussions (in small print in the 'box' in Figure 8.6) also refer to software being available to allow determination of site class B map values by longitude and latitude.

8.3.2 Dynamic Analysis Procedures

Response Spectrum Method

For the purpose of constructing the response spectrum, two controlling points, namely 0.2 seconds for short period response and at 1.0 seconds for long period, are specified.

The contour plots have been developed for 0.2 and 1.0 seconds (see Figure 8.6). The plots, as may be seen, are for site class B. Maps for both the 0.2 seconds spectral acceleration and 1.0 seconds have been provided in the code. Since the plots are for site class B (see Table 8.1 for site classification) the coefficients are introduced to adjust values of S_s and S_1 for other site classes. The code provisions are now introduced.

Site coefficients and adjusted maximum considered earthquake spectral response acceleration parameters (§ 1613.5.3).

The code specifications are discussed in the following paragraphs:

The mapped values for site class B are adjusted through coefficients F_a and F_v. Coefficient F_a is to account for site response effects on short period ground shaking. Coefficient F_v is to account for site response effects on ground shaking of longer periods. Thus

$$S_{MS} = F_a S_s \qquad \text{(Equation 16-37 – IBC 2006)}$$

$$S_{M1} = F_v S_1 \qquad \text{(Equation 16-38 – IBC 2006)}$$

where

F_a – site coefficient defined in Table 8.2,

F_v – site coefficient defined in Table 8.3,

S_s – the mapped spectral accelerations for short periods (see Figure 1613.5 (1) through to 1613.5 (14), IBC 2006),

S_1 – the mapped spectral accelerations for 1-second period (see Figure 1613.5 (1) through to 1613.5 (14), IBC 2006).

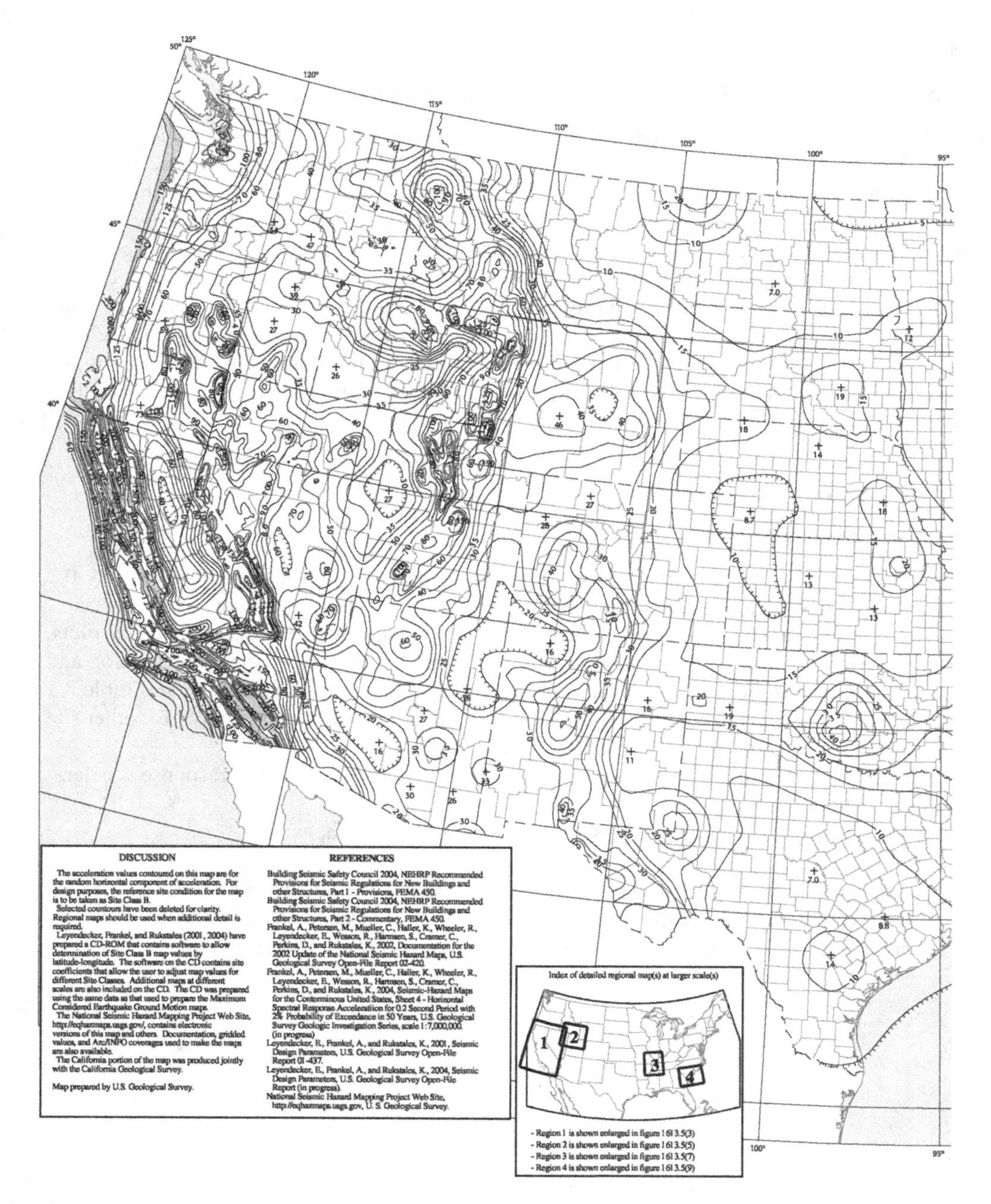

Figure 8.6 Maximum considered earthquake ground motion (Figure 1613.5(1)). 2006 International Building Code © 2006 Washington DC; International Code Council. Reproduced with permission, all rights reserved.

Table 8.1 Site class definitions.

SITE CLASS	SOIL PROFILE NAME	AVERAGE PROPERTIES IN TOP 100 feet, AS PER SECTION 1615.1.5		
		Soil shear wave velocity, v_s, (ft/s)	Standard penetration resistance, N	Soil undrained shear strength, s_u, (psf)
A	Hard Rock	$v_s > 5{,}000$	N/A	N/A
B	Rock	$2{,}500 < v_s \leq 5{,}000$	N/A	N/A
C	Very dense soil and soft rock	$1{,}200 < v_s \leq 2{,}500$	$N > 50$	$s_u \geq 2{,}000$
D	Stiff soil profile	$600 \leq v_s \leq 1{,}200$	$15 \leq N \leq 50$	$1{,}000 \leq s_u \leq 2{,}000$
E	Soft soil profile	$v_s < 600$	$N < 15$	$s_u < 1{,}000$
E	–	Any profile with more than 10 feet of soil having one or more of the following characteristics: 1. Plasticity index PI > 20, 2. Moisture content w ≥ 40 %, and 3. Undrained shear strength s_u < 500 psf		
F	–	Any profile containing soils having one or more of the following characteristics: 1. Soils vulnerable to potential failure or collapse under seismic loading such as liquefiable soils, quick and highly sensitive clays, collapsible weakly cemented soils. 2. Peats and or highly organic clays (H > 10 feet of peat and/or highly organic clay where H = thickness of soil). 3. Very high plasticity clays (H > 25 feet with plasticity index PI > 75) 4. Very thick soft medium stiff clays (H > 120 feet).		

Source: 2006 International Building Code © 2006 Washington DC; International Code Council. Reproduced with permission, all rights reserved.

For SI: foot = 304.8 mm, 1 square foot = 0.0929 m^2, 1 pound per square foot – 0.0479 kPa. N/A = Not applicable.

Table 8.2 Values of site coefficent F_a as a function of site class and mapped spectral response acceleration at short periods $(S_s)^a$.

SITE CLASS	MAPPED SPECTRAL RESPONSE ACCELERATION AT SHORT PERIODS				
	$S_s \leq 0.25$	$S_s = 0.05$	$S_s = 0.75$	$S_s = 1.00$	$S_s \geq 1.25$
A	0.8	0.8	0.8	0.8	0.8
B	1	1	1	10	1
C	1.2	1.2	1.1	1	1
D	1.6	1.4	1.2	1.1	1
E	2.5	1.7	1.2	0.9	0.9
F	Note b	Note b	Note b	Note b	Note b

Source: 2006 International Building Code © 2006 Washington DC; International Code Council. Reproduced with permission, all rights reserved.

Table 8.3 Values of site coefficent F_v as a function of site class and mapped spectral response acceleration at 1-second period $(S_1)^a$.

SITE CLASS	MAPPED SPECTRAL RESPONSE ACCELERATION AT SHORT PERIODS				
	$S_s \leq 0.25$	$S_s = 0.05$	$S_s = 0.75$	$S_s = 1.00$	$S_s \geq 1.25$
A	0.8	0.8	0.8	0.8	0.8
B	1	1	1	10	1
C	1.2	1.2	1.1	1	1
D	1.6	1.4	1.2	1.1	1
E	2.5	1.7	1.2	0.9	0.9
F	Note b	Note b	Note b	Note b	Note b

Source: 2006 International Building Code © 2006 Washington DC; International Code Council. Reproduced with permission, all rights reserved.

Notes

a. Use straight line interpolation for intermediate values of mapped spectral response acceleration at short period, S_s.

b. Site specific geotechnical investigation and dynamic site response analyses shall be performed to determine appropriate values, except that for structures with periods of vibration equal to or less than 0.5 seconds, values of F_a for liquefiable soils are permitted to be taken equal to the values for the site class determined without regard to liquefaction in Section 1615.1.5.1.

Design spectral response acceleration parameters are required to be determined in order to construct the response acceleration parameters. Five per cent damped design spectral response acceleration at short periods, S_{DS}, and at 1-second period, S_{D1}, shall be determined from Equations 16-39 and 16-40, respectively:

$$S_{DS} = \frac{2}{3} S_{MS} \qquad \text{(Equation 16-39 – IBC 2006)}$$

$$S_{D1} = \frac{2}{3} S_{M1} \qquad \text{(Equation 16-40 – IBC 2006)}$$

where:

S_{MS} = the maximum considered earthquake spectral response accelerations for short period as determined in 1613.5.1 (see Figures 1613.5 (1) through to 1613.5 (14), IBC 2006).
S_{M1} = the maximum considered earthquake spectral response accelerations for a 1-second period as determined in 1613.5.1 (see Figures 1613.5 (1) through to 1613.5 (14), IBC 2006).

Site classification for seismic design
Site classification for site class C, D or E shall be determined from Table 1613.5.5 (IBC 2006).

Construction of design response acceleration spectrum (IBC 2006)
This section of IBC 2006 refers to ASCE 7 (2006)
The design response spectrum may now be constructed using the steps are indicated below:

1. For periods less than or equal to T_0, the design response acceleration, S_a to be determined from

$$S_a = 0.6 \frac{S_{DS}}{T_0} T + 0.4 S_{DS} \qquad \text{(Equation 11-4-5, ASCE 7, 2006).}$$

2. For periods greater than or equal to T_0 and less than or equal to T_s the design response acceleration, S_a shall be taken equal to S_{DS}.
3. For periods greater than T_s and less than or equal to T_L, the design spectral response acceleration, S_a shall be determined from

$$S_a = \frac{S_{DI}}{T} \qquad \text{(Equation 11-4-6 – ASCE 7, 2006).}$$

4. For periods greater than T_L, S_a shall be taken as given by

$$S_a = \frac{S_{DI} T_L}{T^2} \qquad \text{(Equation 11-4-7, ASCE 7, 2006)}$$

where

S_{DS} = the design spectral response acceleration at short periods and determined as above (Equation 16-37 – IBC 2006)
S_{DI} = the design spectral response acceleration at 1-second period and determined as above (Equation 16-38 – IBC 2006)
T = fundamental period (in seconds) of the structure (sec. 12 – ASCE 7, 2006)
$T_0 = 0.2 S_{DI}/S_{DS}$
$T_s = S_{DI}/S_{DS}$
T_L = long-period transition period(s) (Section 11.4.5 – ASCE 7, 2006).

The simplified design procedure is dealt with in ASCE 7 (2006) and the general form of the design response spectrum is shown in Figure 11.4.1 of the same standard.

8.4 Eurocode 8

Introduction

Eurocode 8, Design of Structures for Earthquake Resistance, covers the earthquake-resistant design and construction of buildings and other civil engineering works. Its stated purpose is no different to those stated in the opening paragraphs of this chapter, 'to protect the human lives and the buildings that are being designed; ensure that safe design and construction practices are being followed; and ensure that, structures that are important for civilian protection remain operational.'

Eurocode 8 has six parts as shown in Table 8.4.

EN 1999-1 covers in separate sections the design and detailing rules for buildings constructed with the main structural materials:

- concrete
- steel
- composite (steel-concrete)
- timber
- masonry

It also covers design of buildings using base isolation.

Table 8.4 Eurocode 8 parts and key dates (achievement or expectation, as of January 2005).

Eurocode 8 part	Title	Approval by SC8 for formal voting	Availability from CEN of approved EN in English French and German to CEN members
EN 1998-1	General Rules, Seismic Actions, Rules for Buildings	Jul. 02	Dec. 04
EN 1998-2	Bridges	Sep. 03	Oct. 05
EN 1998-3	Assessment and Retrofitting of Buildings	Sep. 03	June 05
EN-1998-4	Silos, Tanks, Pipelines	Mar. 05	Jun. 06
EN 1998-5	Foundations, Retaining Structures, Geotechnical Aspects	Jul. 02	Nov. 04
EN 1998-6	Towers, Masts, Chimneys	Jul. 04	Jun. 05

Source: Reproduced from Fardis *et al.* (2005) by permission of Thomas Telford.

Use of Eurocode 8

Eurocode 8 is not a standalone code. It is written in such a way that it is to be applied with other relevant parts of Eurocode, as part of the Eurocode packages detailed in Table 8.4. Each package refers to a specific type of civil engineering structure and construction material.

Elastic Response Spectrum

Seismic Zones

It is to be noted that Eurocode 8 has been written to be applicable in various parts of Europe, and for that reason the document refers to various parameters, like the peak ground acceleration a_{gR} and the reference return period T_{CNR} being defined by the National Authorities. Extracts from EN 1998 will help to clarify:

> For the purpose of EN 1998, national territories shall be subdivided by the National Authorities into seismic zones, depending on the local hazard. By definition, the hazard within each zone is assumed to be constant. (Clause 3.2.1, (1)P)
>
> For most applications of EN 1998, the hazard is described in terms of a single parameter, i.e. the value of the reference peak ground acceleration on type A ground, a_{gR}. Additional parameters required for specific types of structure are given in the relevant parts of EN 1998. (Clause 3.2.1, (2)P)

Note: The reference peak ground acceleration on type A ground, a_{gR}, for use in a country or parts of the country, may be derived from zonation maps found in its National Annex.

> The reference peak ground acceleration, chosen by the National Authorities for each seismic zone, corresponds to the reference return period T_{NCR} of the seismic action for the no-collapse requirement (or equivalently the reference probability of exceedance in 50 years, P_{NCR}) chosen by the National Authorities. An importance factor λ_1 equal to 1,0 is assigned to this reference return period. For return periods other than the reference return period, the design ground acceleration on type A ground a_{gR} times the importance factor λ_1 ($a_g = \lambda_1 \cdot a_{gR}$). (Clause 3.2.1, (3)P)

Identification of Ground Types

Five types of ground profiles A, B, C, D and E are identified to account for the influence of local soil conditions (Cl. 3.1.2 (1), EN 1998-1).

The site should be identified according to the average shear wave velocity. If the site consists of several layers then the average shear wave velocity may be computed from the expression provided (Cl. 3.1.2(3), EN 1998-1).

Design Spectrum

Eurocode 8 also takes into account the capacity of the structure to dissipate energy through ductile behaviour. This is achieved by performing an elastic analysis with a response spectrum, which has been scaled down with respect to the elastic one. The reduction is introduced through the behaviour factor 'q'. As mentioned earlier, this avoids the explicit inelastic non-linear structural analysis.

The relevant periods of the design spectrum and the corresponding spectral ordinate $S_d(T)$, may be obtained from Clause 3.2.2.2 and Table 3.2 (EN 1998–1)

8.4.1 A Worked Example

The dimensions of a storage shed with a gable frame are shown in Figure 8.7.

The frames are at an interval of 6 m. The shed is located in a seismic zone subjected to a moderate to strong earthquake with M_s (surface wave magnitude) ranging from 5.6 to 6.1. From the National Annex it is found that the seismic zone corresponds to a peak ground acceleration of 0.06 g. It is required to assess the design lateral forces.

Clause 3.2.1

Assume reference return period, hence $\lambda_1 = 1$ and $a_g = 0.06$.

The ground is type 'D' obtained from soil investigation at a nearby site. So assume ground type 'D' for this location as well.

Clause 3.2.2.2 (EN 1998-1:2004) as $M_s \geq 5.5$ adopt response spectrum parameters from Table 3.2 (EN 1998-1:2004).

Ground Type	*S*	$T_{B(sec)}$	$T_{C(sec)}$	$T_{D(sec)}$
D	1.35	0.20	0.8	2.0

Clause 4.2.5, Table 4.3 (EN 1998-1:2004)
Importance Class: II (specified for this project)
Note:

Clause 4.2.5(5)P (EN 1998-1:2004)
The value of $\lambda_1 = 1$ for importance class II shall be by definition equal to 1.0.

Seismicity
Check for low seismicity

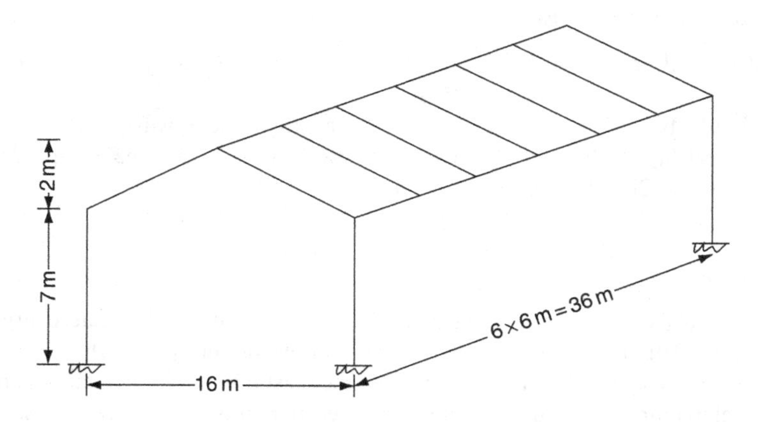

Figure 8.7 Storage shed (isometric view).

Clause 3.2.1(4) $a_g \cdot S = 0.06 \times 1.35\text{g} = 0.081\text{g} < 1.0\text{g}$ (low seismicity)
[In case of low seismicity, reduced or simplified design procedures for certain types or categories of structures may be used.]

Clause 4.3.3.2 (EN 1998-1:2004)
The Lateral Force Method of Analysis may be adopted as conditions of Clause 4.3.3.2 are met.

Lateral Force Method
Clause 4.3.3.2.2 (3) (EN 1998-1:2004)
Fundamental period $T_1 = C_t \cdot H^{3/4}$ ($H < 40\text{m}$)
$C_t = 0.05$,
$H = 8$ m (see Figure 8.7)

$$T_1 = C_t \cdot H^{3/4} = 0.05.8^{3/4} = 0.24\ (\text{sec}) < 4.T_C = 4 \times .8 = 3.2\ (\text{sec})$$
$$< 2.0\ (\text{sec})$$

(Table 3.2, Clause 3.2.2.2, EN-1998-1:2004)
Hence Clause 4.3.3.2.1a (EN 1998-1:2004) applies.

Vertical elastic response spectrum
Type I spectra – Clause 3.2.2.3, Table 3.4

Spectrum	a_{vg}/a_g	$T_{B(sec)}$	$T_{C(sec)}$	$T_{D(sec)}$
Type I	0.90	0.05	0.15	1.0

$a_{vg} = 0.90 \times 0.06 \times 1.0\ (\lambda_1 = 1) = 0.054\text{g}$.
Clause 4.3.3.5.2(1) (EN 1998-1:2004) $a_{vg} < 0.25\text{g}$
Vertical action is neglected.

Design spectrum for elastic analysis
Clause 3.2.2.5 (4)P, Equation 3.14 (EN 1998-1:2004)

$$T_B \leq T \leq T_C:$$
$$S_d(T) = a_g \cdot S \cdot \frac{2.5}{q} = 0.06g \times 1.35 \times \frac{2.5}{1.50} = 0.135g.$$

Behaviour factor 'q'= 1.5
(Low dissipative behaviour)

Criteria for structural regularity
The building is designed as 'regular'. Clause 4.2.3.1(P).

Clause 4.2.3.(3), Table. 4.1 (EN 1998-1:2004)

Regularity		Allowed Simplification	Behaviour Factor
Plan	Elevation	Modal Linear Elastic Analysis	(for linear analysis)
YES	YES	Lateral Force* *Conditions 4.3.3 (2a) is satisfied as above	Reference Value** ** Table 6.1 Clause 6.1.2, (EN 1998-1:2004)

Conditions of clause 4.3.3.2.1(2) are met (EN 1998-1:2004).
Clause 4.2.3 (4) – Conditions of Clause 4.2.3.2 and 4.2.3.3 are satisfied.
Clause 4.2.3 (5) – Regularity of building structure is not impaired by other characteristics.
Clause 4.2.3 (6) – $q = 1.5$ as above.
Clause 4.2.3 (7) – Building is regular in elevation.
Clause 4.2.3.2 (EN 1998-1:2004) – Criteria for regularity in plan
Clause 4.2.3.2 (1) to (4) are satisfied.
Clause 4.2.3.2 (5) – $\lambda = L_{\max}/L_{\min} = 36/16 = 2.25 < 4.0$
Clause 4.2.3.2 (6) – $e_{ox} = 0.0$.
Clause 4.2.3.3 (EN 1998-1:2004) – Criteria for regularity in elevation are all satisfied.

Structural Analysis
Clause 4.3.2 (EN 1998-1:2004) Accidental torsional effects

$$e_{ai} = \pm 0.05 \cdot L_i = 0.05 \times 36 = 1.8$$

where

e_{ai} – accidental eccentricity of storey mass i from its nominal location, applied in the same direction at all floors.
L_i is the floor dimension perpendicular to the direction of seismic action.

Clause 4.3.3.1 (EN 1998-1:2004) – General
Clause 4.3.3.1(1) – Basis of 'linear elastic behaviour' adopted.
Clause 4.3.3.1(3)a – Lateral force method of analysis is adopted.
Clause 4.3.3.1 (8) and (9) are satisfied – Increase seismic load by a factor of 1.25.

Base Shear
Clause 4.3.3.2.2 (EN 1998-1:2004)

$$F_b = S_d(T_1) \cdot m \cdot \lambda$$

Where

$S_d(T_1)$ – is the ordinate of the design spectrum at period T_1.
T_1 is the fundamental period of vibration of the building for the lateral motion in the direction considered.

m is the total mass of the building above the top of rigid basement.
λ is the correction factor=1.0; [Cl. 4.3.3.2.2 (1)P].
$F_b = 0.135\text{g} \times 1.0 \times m = 0.135 \times m$ g.

Torsional moment: $F_b \times e_{ai} = 0.135.m.e_{ai}\text{g}$.
These loads are applied at the eaves level of the structure (Figure 8.7).

Clause 4.3.2 (1)P
$e_{ai} = \pm 0.05L_i = 0.05 \times 36 = 1.8$
Clause 4.3.3.1(8) & (9)
Design base shear $F_{bd} = 0.135 \times 1.8 \times$ m.g. $= 0.243$ m.g.
Next we need to consider load combination.

Combination of Seismic Action with Other Loads
Clause 3.2.4
Design base shear (with load combination) $F_{bd} = 0.243\{\sum G_{k,j} + \sum \psi_{E,i} \cdot Q_{k,i}\} \cdot g$
Where

$G_{k,j}$ – summation of permanent action
$Q_{k,i}$ – summation of variable action

Clause 4.2.4
$\psi_{E,i} = \varphi \cdot \psi_{2i}$
Referring to Table 4.2 (Values of φ for calculating $\psi_{E,i}$ – EN 1998-1:2004)
For Category A-C; Roof $\phi = 1.0$ (Categories as defined in EN 1991-1-1:2002)
Referring to Clause 4.2.4(1) and Table A.1.1 (EN 1990:2002)
$\psi_{2,i} = 0.2$ (for variable action due to snow)
$\psi_{E,i} = 1.0 \times 0.2 = 0.2$

Design base shear (with load combination) $F_{bd} = 0.243\{\Sigma G_{k,j} + \Sigma 0.2..Q_{k,i}\} \cdot g$

(Note: only variable action due to snow has been considered; if there are other variable actions, they need to be included as well.)

8.5 A Worked Example (IBC 2000)

The worked example included in this section reproduces sections from Seismic Design Using Structural Dynamics *(reproduced from Ghosh (2004) by permission of the International Code Council, Inc, Washington, DC. All rights reserved).*

8.5.1 General

A 20-storey reinforced concrete building is designed following the requirements of the International Building Code (IBC) 2000 edition. The building is located on IBC Site Class D. Both dynamic and static lateral-force procedures are used as the basis of design.

The building is symmetrical about both principal plan axes. Along each axis a dual system (concrete shear walls with special moment-resisting frames of SMRF) is utilized for resistance to lateral forces.

A dual system is defined as a structural system with the following features: (IBC 1617.6.1)

1617.6.1 Dual Systems. For a dual system, the moment frames shall be capable of resisting at least 25 % of the design forces. The total seismic force resistance is to be provided by the combination of the moment frame and the shear walls or braced frames in proportion to their stiffness.

8.5.2 Design Criteria

A typical plan and elevation of the building are shown in Figures 8.8 and 8.9 respectively. The member sizes for the structure are chosen as follows:

Spandrel beams (width = 34 in.)	34 × 24 in.
Interior beams	34 × 24 in.
Columns	34 × 34 in.
Shear walls: Grade to 9th floor 10th floor to 16th floor 17th floor to roof	 16 in. thick 14 in. thick 12 in. thick
Shear wall boundary elements	34 × 34 in.

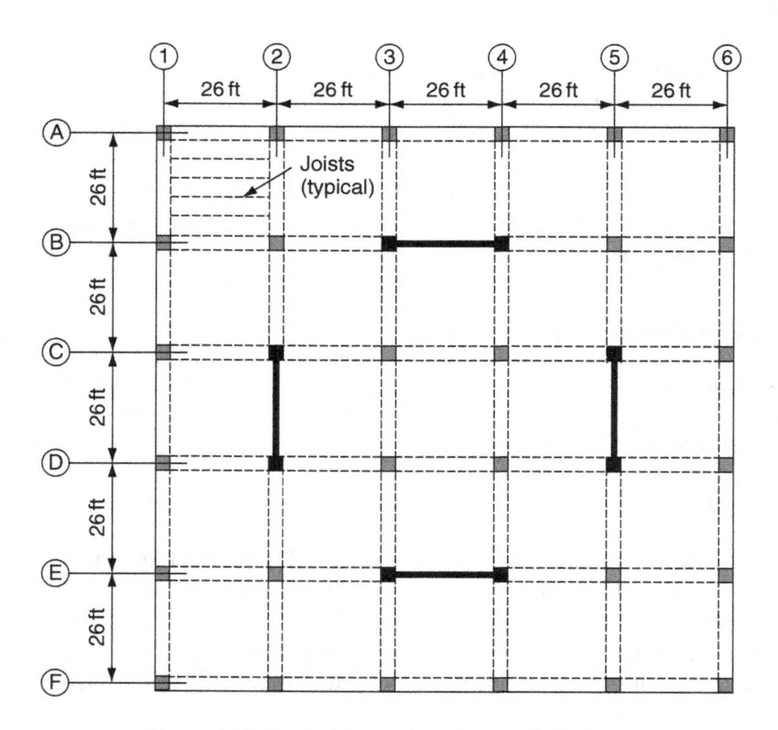

Figure 8.8 Typical floor plan of example building.

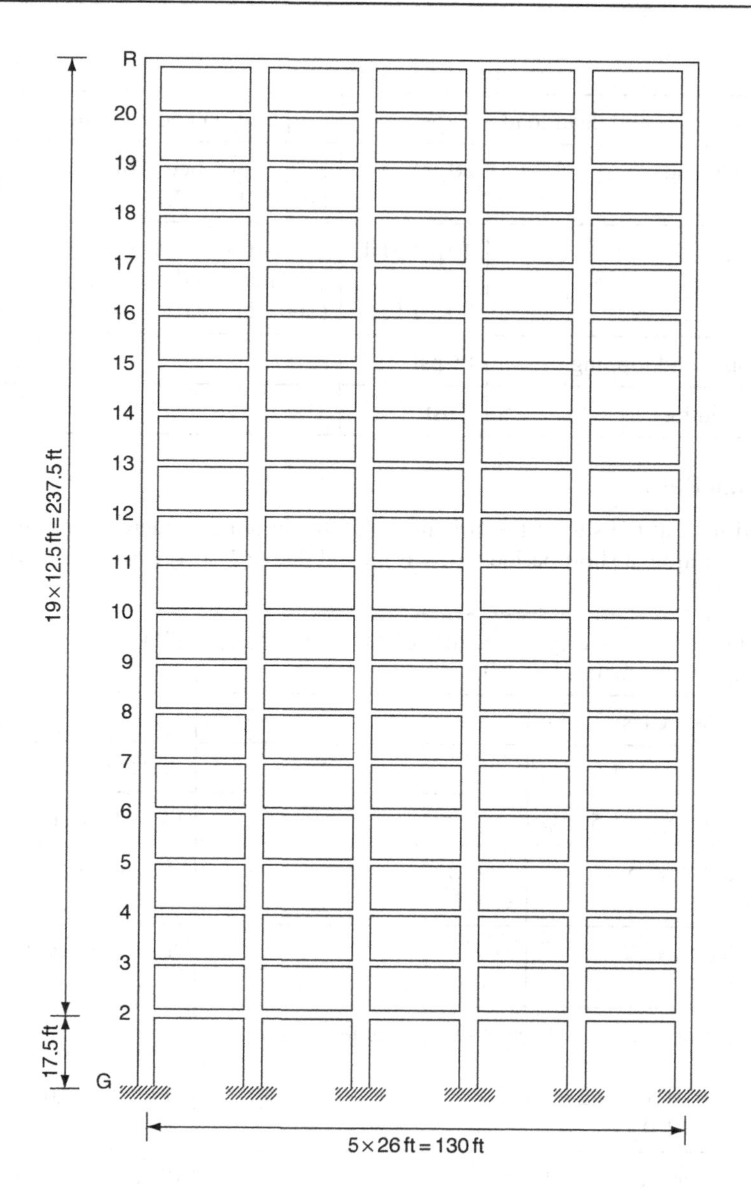

Figure 8.9 Elevation of example building.

Other relevant design data are as follows:

Material Properties

Concrete	$f'_c = 4{,}000$ psi (all members)

All members are constructed of normal weight concrete ($w_c = 145$ pcf)

Reinforcement	$f_y = 60{,}000$ psi

Service Loads

Superimposed dead load	20 psf	... partition and equipment
Live load	80 psf	... per practice, Minimum 50 psf (T 1607.1)
Loads on roof	10 psf SDL 20 psf LL	... roofing +200 kips for penthouse and equipment (1607.11.2.1)
Joists and topping	86 psf	
Cladding	8 psf	

Seismic Design Data

It is assumed that, at the site of the structure, the maximum considered earthquake spectral response acceleration at short periods, $S_s = 1.5g$, and that at 1-second period, $S_1 = 0.6g$.

Assume standard occupancy or Seismic Use Group = I ... seismic importance factor, I = 1.0		(T 1604.5)
Use Default Site Class ... D		
Site coefficient $F_a = 1.0$		[T 1615.1.2(1)]
Site coefficient $F_v = 1.5$		[T 1615.1.2(2)]
Soil-modified $S_s = S_{MS}$ $= 1.0 \times 1.5$ g	$= F_a S_s$ $= 1.5$ g	(Equation 16-16)
Soil-modified $S_1 = S_{M1}$ $= 1.5 \times 0.6$ g	$= F_v S_1$ $= 0.9$ g	(Equation 16-17)
Design Spectral Response Acceleration Parameters (at 5 % damping): At short periods: $S_{DS} = 2/3 S_{MS}$ $= 2/3 \times 1.5 = 1.0$ g		(Equation 16-18)
At 1-second period: $S_{D1} = 2/3 S_{M1}$ $= 2/3 \times 0.9 = 0.6$ g		(Equation 16-19)
Dual system (RC shear walls with SMRF) $R = 8$; $C_d = 6.5$		(Table 1617.6)
Where: R and C_d are the response modification factor and deflection amplification factor, respectively.		

Seismic Design Category: Based on both S_{DS} [Table 1616.3(1)] and S_{D1} [Table 1616.3(2)], the seismic design category (SDC) for the example building is D.

8.5.3 *Design Basis*

Calculation of the design basis shear and distribution of that shear along the height of the building using the equivalent lateral force procedure (which is used in a majority of designs) is not appropriate and is not allowed by the International Building Code for buildings exceeding 240 feet in height in SDC D (T 1616.6.3, Item 3). In these cases, the dynamic lateral force procedure (1618) must be used. In this example, the dynamic procedure will be used as the height of the building is 255 feet in SDC D (more than 240 feet). However, for comparison purposes, the equivalent lateral force procedure (1617.4) will also be illustrated.

8.5.4 *Gravity Loads and Load Combinations*

Weights at each Floor Level

Table 8.5 shows the weights (self weight $+SDL$) at each floor level. The weights are calculated as follows:

Table 8.5 Lateral forces by equivalent force procedure using approximate method.

Floor Level x	Weight w_x, kips	Height h_x, ft	$w_x h_x^k$, ft-kips	Lateral Force F_x, kips	Story Shear V_x, kips
1	2	3	4	5	6
21	2987	255.0	6,611,897	392	392
20	3338	242.5	6,890,261	409	801
19	3338	230.0	6,401,592	380	1181
18	3338	217.5	5,923,177	351	1532
17	3352	205.0	5,478,250	325	1857
16	3366	192.5	5,040,491	299	2157
15	3366	180.0	4, 591,376	272	2429
14	3366	167.5	4,154,270	246	2675
13	3366	155.0	3,729,711	221	2897
12	3366	142.5	3,318,309	197	3094
11	3366	130.0	2,920,757	173	3267
10	3380	117.5	2,548,410	151	3418
9	3394	105.0	2,188,593	130	3548
8	3394	92.5	1,835,054	109	3657
7	3394	80.0	1,499,709	89	3746
6	3394	67.5	1,184,252	70	3816
5	3394	55.0	890,873	53	3869
4	3394	42.5	622,547	37	3906
3	3394	30.0	383,628	23	3929
2	3559	17.5	190,174	11	3940
Σ	67,246	Σ	66403331	3940	

$V = 3940$ kips, T (sec.) $= 1.28$
$k = 1.39$

$w_i =$	$(86 + 20\{10 \text{ psf for roof}\}) \times 130^2$ {+200 kips for roof}	... *SDL* + Joists
	$+8 \times 12.5^* \times 130 \times 4$	... Cladding
	$+150 \times 12.5^* \times (34/12)^2 \times 36$	... Column selfweight
	$+150 \times 23.1 \times 34/12 \times 17/12 \times 56$	... Beam selfweight
	$+150 \times 12.5^* \times 23.1 \times (h/12) \times 4$	... Shear wall selfweight
	*{15.0 for 2nd floor & 6.25 for roof} (h = wall thickness)	
	=3,559 kips for floor 2	
	=3,394 kips for floors 3 to 9	
	=3,380 kips for floor 10	
	=3,366 kips for floors 11 to 16	
	=3,352 kips for floor 17	
	=3,338 kips for floors 18 to 20	
	=2,987 kips for roof (floor 21)	
Total weight of the building: $W = \sum_{i=1}^{20} w_i = 67{,}246$ kips		

Table 8.6a Service-level axial forces due to DL and LL in an interior column.

Floor Level	DL psf	DL kips	Cum.DL kips	LL psf	Supported Area - psf	RLL psf	RLL kips	Cum.LL kips
1	2	3	4	5	6	7	8	9
21	177	119.7	120	20	676	12	8.1	8.1
20	198	133.8	253	80	1352	32	21.6	29.7
19	198	133.8	387	80	2028	32	21.6	51.3
18	198	133.8	521	80	2704	32	21.6	72.9
17	198	133.8	655	80	3380	32	21.6	94.5
16	199	134.5	790	80	4056	32	21.6	116.1
15	199	134.5	924	80	4732	32	21.6	137.7
14	199	134.5	1059	80	5408	32	21.6	159.3
13	199	134.5	1193	80	6084	32	21.6	180.9
12	199	134.5	1328	80	6760	32	21.6	202.5
11	199	134.5	1462	80	7436	32	21.6	224.1
10	200	135.2	1597	80	8112	32	21.6	245.7

Table 8.6a *(continued)*

Floor Level	DL psf	DL kips	Cum.DL kips	LL psf	Supported Area – psf	RLL psf	RLL kips	Cum.LL kips
1	2	3	4	5	6	7	8	9
9	201	135.9	1733	80	8788	32	21.6	267.3
8	201	135.9	1869	80	9464	32	21.6	288.9
7	201	135.9	2005	80	10140	32	21.6	310.5
6	201	135.9	2141	80	10816	32	21.6	332.1
5	201	135.9	2277	80	11492	32	21.6	353.7
4	201	135.9	2413	80	12168	32	21.6	375.3
3	201	135.9	**2549**	80	12844	32	21.6	**396.9**
2	210	142.0	**2690**	80	13520	32	21.6	**418.5**

DL = Dead Load, LL = Live Load, RLL = Reduced Live Load

Table 8.6b Service-level axial forces due to DL and LL in an edge column.

Floor Level	DL psf	DL kips	Cum.DL kips	LL psf	Supported Area – psf	RLL psf	RLL kips	Cum.LL kips
1	2	3	4	5	6	7	8	9
21	177	66.4	66	20	375	16.5[a]	6.2	6.2
20	198	74.3	141	80	750	32	15.6	21.8
19	198	74.3	215	80	1125	32	12	33.8
18	198	74.3	289	80	1500	32	12	45.8
17	198	74.3	363	80	1875	32	12	57.8
16	199	74.6	438	80	2250	32	12	69.8
15	199	74.6	513	80	2625	32	12	81.8
14	199	74.6	587	80	3000	32	12	93.8
13	199	74.6	662	80	3375	32	12	105.8
12	199	74.6	737	80	3750	32	12	117.8
11	199	74.6	811	80	4125	32	12	129.8
10	200	75.0	886	80	4500	32	12	141.8
9	201	75.4	962	80	4875	32	12	153.8
8	201	75.4	1037	80	5250	32	12	165.8
7	201	75.4	1112	80	5625	32	12	177.8
6	201	75.4	1188	80	6000	32	12	189.8
5	201	75.4	1263	80	6375	32	12	201.8
4	201	75.4	1338	80	6750	32	12	213.8
3	201	75.4	**1414**	80	7125	32	12	**225.8**
2	210	78.8	**1493**	80	7500	32	12	**237.8**

DL = Dead Load, LL = Live Load, RLL = Reduced Live Load

[a] Based on the expression $R = 1.2 - 0.001\, A_t$ (1607.11.2.1)

Table 8.6c Service-level axial forces due to DL and LL in a shear wall.

Floor Level	DL psf	DL kips	Cum.DL kips	LL psf	Supported Area – psf	RLL psf	RLL kips	Cum.LL kips
1	2	3	4	5	6	7	8	9
21	177	239.3	239	20	1352	12	16.2	16
20	198	267.7	507	80	2704	32	43.2	59
19	198	267.7	775	80	4056	32	43.2	103
18	198	267.7	1042	80	5408	32	43.2	146
17	198	267.7	1310	80	6760	32	43.2	189
16	199	269.0	1579	80	8112	32	43.2	232
15	199	269.0	1848	80	9464	32	43.2	275
14	199	269.0	2117	80	10816	32	43.2	319
13	199	269.0	2386	80	12168	32	43.2	362
12	199	269.0	2655	80	13520	32	43.2	405
11	199	269.0	2924	80	14872	32	43.2	448
10	200	270.4	3195	80	16224	32	43.2	491
9	201	271.8	3467	80	17576	32	43.2	535
8	201	271.8	3738	80	18928	32	43.2	578
7	201	271.8	4010	80	20280	32	43.2	621
6	201	271.8	4282	80	21632	32	43.2	664
5	201	271.8	4554	80	22984	32	43.2	707
4	201	271.8	4825	80	24336	32	43.2	751
3	201	271.8	5097	80	25688	32	43.2	794
2	210	283.9	**5381**	80	27040	32	43.2	**837**

DL = Dead Load, LL = Live Load, RLL = Reduced Live Load

8.5.5 Gravity Load Analysis

Service-level axial forces due to dead and live loads for shear wall, edge column, and interior column at different floor levels are given in Table 8.6. Live load reduction factors were used as follows:

(1607.9.2)

$R = r\,(A - 150)$		
≤ 0.6		
	$\leq 23.1(1 + D/L)$ for floors other than the roof $= 0.6\,(1607.11.2)$ for flat roof with tributary area $A_t \geq 600\,\text{ft}^2$ $= 1.2 - 0.001A_t$ for flat roof with tributary area $200\,\text{ft}^2 < A_t < 600\,\text{ft}^2$	
Where: R	$= 0.08$ for floors other than the roof	
A	$= 676\ (= 26 \times 26)\ \text{ft}^2$ for roof	(interior column)
	$= 2 \times 676\,\text{ft}^2$ for floor 20 & so on	(interior column)
	$= 375\ (= 26 \times 14.42)\ \text{ft}^2$ for roof	(edge column)

	$= 750\ \text{ft}^2$ for floor 20 & so on	(edge column)
	$= 1352\ (= 26 \times 52)\ \text{ft}^2$ for roof	(shear wall)
	$= 2 \times 1352\ \text{ft}^2$ for floor 20 & so on	(shear wall)

In Table 8.6, the reduced live load (RLL) is calculated as $RLL = L\,(1 - R)$.

8.5.6 Load Combinations For Design

The following load combinations are used in the strength design method for concrete:

(1) $U = 1.4D + 1.7L$	(ACI 318-99 Equation 9-1)
(2) $U = 1.2D + f_1L + 1.0E$	(Equation 16-5)
(3) $U = 0.9D \pm 1.0E$	(Equation 16-6)
Where D = dead load effect	
L = live load effect	
$f_1 = 0.5$	(1605.2)
$E = \rho Q_E + 0.2S_{DS}D$	When the effects of gravity and seismic loads are additive (Equation 16-28)
$E = \rho Q_E - 0.2S_{DS}D$	When the effects of gravity and seismic loads are counteractive. (Equation 16-29)
Q_E	The effect of horizontal seismic forces
ρ	A reliability factor based on system redundancy

8.5.7 Equivalent Lateral Force Procedure (1617.4)

Design Base Shear (1617.4.1)

$V = C_S W$	(Equation 16-34)
Where $C_S = \frac{S_{D1}I_E}{RT}$	(Equation 16-36)
$\leq \frac{S_{DS}I_E}{R}$	(Equation 16-35)
$\geq 0.044S_{DS}I_E$	(Equation 16-37)
$\geq \frac{0.5S_1I_E}{R}$	{for SDC E and F or where $S_1 \geq 0.6g$} (Equation 16-38)

For the example building considered

$S_{DS} = 1.0\text{g}$
$S_{D1} = 0.6\text{g}$
$S_1 = 0.6\text{g}$
$R = 8$
$I_E = 1.0$

Approximate fundamental period T_a (1617.4.2.1)

$$T_a = C_T(h_n)^{3/4} \quad (16\text{-}39)$$

$C_T = 0.02$ for dual systems
$h_n =$ total height $= 255$ ft.
$T_a = 0.02 \times (255)^{3/4} = 1.28$ sec

$$\frac{S_{D1}I_E}{RT}W = \frac{0.6 \times 1 \times 67{,}246}{8 \times 1.28} = 3{,}940 \text{ kips governs}$$

$$\frac{S_{DS}I_E}{R}W = \frac{1.0 \times 1 \times 67{,}246}{8} = 8{,}406 \text{ kips}$$

$$0.044 S_{DS} I_E W = 0.044 \times 1.0 \times 1 \times 67{,}246 = 2{,}959 \text{ kips}$$

Since $S_1 = 0.6g$, Equation 16-38 is applicable for the example building in SDC D.

$$\frac{0.5S_1I_E}{R}W = \frac{(0.5)(0.6)(1.0)}{8}(67{,}246) = 2{,}522 \text{ kips}$$

Use $V = 3{,}940$ kips

8.5.8 Vertical Distribution of Base Shear (1617.4.3)

Distribute base shear as follows:

$$F_x = C_{vx}V \quad (16\text{-}41)$$

$$C_{vx} = \frac{w_x h_x^k}{\sum_{i=1}^{20} w_i^k h_i^k} \quad (16\text{-}42)$$

$$T = 1.28 \text{ sec}$$

$$k = 1 \leq 1 + 0.5(T - 0.5) \leq 2 = 1.39 \quad (1617.4.3)$$

Distribution of the design base shear along the height of the building is shown in Table 8.5.

8.5.9 Lateral Analysis

A three-dimensional analysis of the structure was performed under the lateral forces shown in Table 8.5, using the SAP 2000 computer program. To account for accidental torsion, the mass at each level was assumed to be displaced from the centre of mass by a distance equal to 5 % of the building dimension perpendicular to the direction of force (1617.4.4.4). In the model,

Table 8.7 Calculation of period by rational method (equivalent lateral procedure).

Floor Level x	Weight W_x, kips	Lateral Force F_x, kips	Displacement δ_x, in	$W_x\delta_x^2$, kip-in.^{2}	$F_x\delta_x$, kip-in.
1	2	3	4	5	6
21	2987	392	10.20	310601	4001
20	3338	409	9.75	317097	3985
19	3338	380	9.27	286794	3521
18	3338	351	8.77	256932	3083
17	3352	325	8.26	228441	2683
16	3366	299	7.72	200388	2308
15	3366	272	7.16	172326	1949
14	3366	246	6.58	145520	1621
13	3366	221	5.98	120307	1323
12	3366	197	5.37	97028	1057
11	3366	173	4.75	75993	823
10	3380	151	4.13	57715	625
9	3394	130	3.52	42055	457
8	3394	109	2.92	28955	318
7	3394	89	2.34	18636	209
6	3394	70	1.80	10969	126
5	3394	53	1.30	5701	69
4	3394	37	0.85	2462	31
3	3394	23	0.48	782	11
2	3559	11	0.20	140	2
Σ	67,246		Σ	2,378,842	28,202

rigid diaphragms were assigned at each level, and rigid end offsets were defined at each end of each member so that the results were automatically generated at the face of each support.

According to 1618.1 the mathematical model must consider cracked section properties. The stiffnesses of members used in the analyses were as follows:

For columns and shear walls, $I_{eff} = I_g$

For beams, $I_{eff} = 0.5I_g$ (considering slab contribution)

$P - \Delta$ effects are considered in the lateral analysis. It may be noted, that this effect is allowed to be neglected in many situations as explained later (1617.4.6.2).

Lateral displacements of the example building, computed elastically under the distributed lateral forces of Table 8.5, are shown in Table 8.7.

8.5.10 Modification of Approximate Period

The use of period by the approximate method (1617.4.2.1) often results in a conservative design. It is appropriate to use a more rational method for computation of period to reduce the design forces.

However, the modified period must not exceed the approximate period by a factor (referred to as coefficient C_u) shown in Table 1617.4.2. The Rayleigh-Ritz procedure, given by the following equation, is used as a rational method.

$$T = 2\pi\sqrt{\sum w_i\delta_i^2/g\sum F_i\delta_i}$$

where the values of F_i represent any lateral force distributed approximately in accordance with the principles of Equations 16-41 and 16-42 or any other rational distribution. The elastic deflections, δ_i, shall be calculated using applied lateral forces, F_i.

Table 8.7 shows values of F_i and the corresponding δ_i based on the approximate period. The modified period can be found as follows:

$$T = 2\pi\sqrt{\sum w_i\delta_i^2/g\sum F_i\delta_i}$$

$$= 2\pi\sqrt{(2{,}378{,}842/386 \times 28{,}202} \text{ (see Table 8.7)}$$

$$= 2.937 \text{ sec}$$

$$\leq 1.2 \times T \text{ from approximate method} \qquad \text{(T 1617.4.2 for } S_{D1} > 0.4\text{)}$$

$$\leq 1.2 \times 1.28 = 1.536 \text{ sec}$$

governs.

8.5.11 Revised Design Base Shear

Using the modified period of $T = 1.536$ seconds, the design base shear is recalculated as:

$$V = \frac{S_{D1}I_E}{RT}W = \frac{0.6 \times 1 \times 67{,}246}{8 \times 1.536} = 3{,}284 \text{ kips}$$

governs.

$$\leq \frac{S_{DS}I_E}{R}W = \frac{1.0 \times 1 \times 67{,}246}{8} = 8{,}406 \text{ kips}$$

$$\geq 0.044S_{DS}I_E W = 0.044 \times 1.0 \times 1 \times 67{,}246 = 2{,}959 \text{ kips}$$

Since $S_1 = 0.6g$, Equation 16-38 is applicable for the example building in SDC D.

$$V \geq \frac{0.5S_1I_E}{R}W = \frac{(0.5)(0.6)(1.0)}{8}(67{,}246) = 2{,}522 \text{ kips}$$

Use $V = 3{,}284$ kips.
Distribute the base shear as follows:

$$F_x = C_{vx}V \qquad (16\text{-}41)$$

$$C_{vx} = \frac{w_x h_x^k}{\sum_{i=1}^{20} w_i h_i^k} \qquad (16\text{-}42)$$

$$T = 1.536 \text{ sec}$$

$$k = 1 \leq 1 + 0.5(T - 0.5) \leq 2 = 1.52 \qquad (1617.4.3)$$

The distribution of the design base shear along the height of the building is shown in Table 8.8.

Table 8.8 Lateral force by equivalent lateral force procedure using period from rational analysis.

Floor Level x	Weight w_x, kips	Height h_x, ft	$w_x h_x{}^k$, ft-kips	Lateral Force F_x, kips	Story Shear V_x, kips
1	2	3	4	5	6
21	2987	255.0	13,438,884	344	344
20	3338	242.5	13,914,857	356	700
19	3338	230.0	12,840,714	328	1028
18	3338	217.5	11,796,400	302	1330
17	3352	205.0	10,827,953	277	1607
16	3366	192.5	9,882,799	253	1859
15	3366	180.0	8,925,196	228	2088
14	3366	167.5	8,001,448	205	2292
13	3366	155.0	7,112,751	182	2474
12	3366	142.5	6,260,443	160	2634
11	3366	130.0	5,446,031	139	2774
10	3380	117.5	4,690,662	120	2894
9	3394	105.0	3,970,793	102	2995
8	3394	92.5	3,275,782	84	3079
7	3394	80.0	2,627,861	67	3146
6	3394	67.5	2,030,462	52	3198
5	3394	55.0	1,487,930	38	3236
4	3394	42.5	1,006,019	26	3262
3	3394	30.0	592,901	15	3277
2	3559	17.5	274,321	7	3284
Σ	67,246		128,404,210	3284	

V = 3284 kips, T (sec) = 1.536
k = 1.52

8.5.12 Results of Analysis

The maximum shear force and bending moment at the base of each shear wall (between ground and second floor) were found to be 1,571 kips and 118,596 ft-kips respectively (Table 8.11).

Because of the location of the shear wall within the plan of the building, the earthquake-induced axial force in each shear wall is equal to zero.

The lateral displacements at every floor level (δ_{xe}) are shown in Table 8.9. The maximum inelastic response displacements (δ_x) and storey drifts are computed and shown in Table 8.9.

δ_x is calculated as per 1617.4.6.1

$$\delta_x = \frac{C_d \delta_{xe}}{I_E} \qquad (16\text{-}46)$$

8.5.13 Storey Drift Limitation

According to 1617.3, the calculated storey drift, Δ, as shown in Table 8.9, shall not exceed 0.020 times the storey height (Table 1617.3 for Seismic Use Group 1 and all other buildings).

Table 8.9 Lateral displacements and drifts (with revised T) of example building by equivalent lateral force procedure (in.)

(along outer frame line F)

Floor Level	δ_{xe}	C_d	δ_x	Drift, Δ
1	2	3	4	5
21	6.47	6.50	42.06	1.76
20	6.20	6.50	40.30	1.89
19	5.91	6.50	38.42	1.95
18	5.61	6.50	36.47	2.15
17	5.28	6.50	34.32	2.21
16	4.94	6.50	32.11	2.28
15	4.59	6.50	29.84	2.41
14	4.22	6.50	27.43	2.47
13	3.84	6.50	24.96	2.54
12	3.45	6.50	22.43	**2.60**
11	3.05	6.50	19.83	2.60
10	2.65	6.50	17.23	2.60
9	2.25	6.50	14.63	2.47
8	1.87	6.50	12.16	2.41
7	1.50	6.50	9.75	2.28
6	1.15	6.50	7.48	2.15
5	0.82	6.50	5.33	1.82
4	0.54	6.50	3.51	1.56
3	0.30	6.50	1.95	1.11
2	0.13	6.50	0.85	0.85
1	0.00	0.00	0.00	0.00

$\delta_x = C_d\delta_{xe}/I \quad \Delta = \delta_{x,i} - \delta_{x,i-1}$

(along shear wall line E)

Floor Level	δ_{xe}	C_d	δ_x	Drift, Δ
1	2	3	4	5
21	6.22	6.50	40.43	1.69
20	5.96	6.50	38.74	1.82
19	5.68	6.50	36.92	1.95
18	5.38	6.50	34.97	2.02
17	5.07	6.50	32.96	2.08
16	4.75	6.50	30.88	2.21
15	4.41	6.50	28.67	2.34
14	4.05	6.50	26.33	2.41
13	3.68	6.50	23.92	2.41
12	3.31	6.50	21.52	**2.54**
11	2.92	6.50	18.98	2.47
10	2.54	6.50	16.51	2.47
9	2.16	6.50	14.04	2.41
8	1.79	6.50	11.64	2.34

Table 8.9 (*continued*)

Floor Level	δ_{xe}	C_d	δ_x	Drift, Δ
1	2	3	4	5
7	1.43	6.50	9.30	2.15
6	1.10	6.50	7.15	2.02
5	0.79	6.50	5.14	1.76
4	0.52	6.50	3.38	1.50
3	0.29	6.50	1.89	1.11
2	0.12	6.50	0.78	0.78
1	0.00	0.00	0.00	0.00

$\delta_x = C_d\delta_{xe}/1 \quad \Delta = \delta_{x,i} - \delta_{x,i-1}$

Floor	*Maximum allowable drift*	*Largest drift*	*Table 8.9*
1st	$0.02 \times 17.5\,\text{ft} = 4.2\,\text{in}$	> 0.85 in	o.k
Others	$0.02 \times 12.5\,\text{ft} = 3.0\,\text{in}$	> 2.60 in	o.k

8.5.14 $P - \Delta$ Effects

According to 1617.4.6.2, $P - \Delta$ effects on storey shears and moments, the resulting member forces and moments, and storey drifts induced by these effects need not be considered when the stability coefficient, θ, as determined by the following formula, is equal to or less than 0.1.

$$\theta = \frac{P_x\Delta}{V_x h_{sx} C_d} \tag{16-47}$$

where

P_x = the total unfactored vertical force
Δ = the design storey drift
V_x = the seismic shear force acting between level x and $x - 1$
h_{xx} = the storey height below x
C_d = the deflection amplification factor

In the lateral analysis performed using the SAP 2000 computer program, the $P - \Delta$ effects are included. However, for illustration purposes, the stability coefficient is calculated as shown in Table 8.10. As the maximum stability coefficient θ (= 0.044) is less than 0.1, the $P - \Delta$ effect could have been neglected.

8.5.15 Redundancy Factor, ρ, (1617.2)

Typically, in a dual system, the shear walls carry the largest proportion of the storey shear at the base of the structure. The redundancy factor is expressed as follows:

$$1 \leq \rho = \rho_1 = 2 - \frac{20}{r_{\max_1}\sqrt{A_1}} \leq 1.5 \tag{16-32}$$

Table 8.10 Calculation of stability coefficient.

Story Level	DL psf	LL psf	Area sq.ft	P_x kips	V_x kips	h_{sx} ft	Drift, Δ in.	θ
1	2	3	4	5	6	7	8	9
21	177	12	16900	3194	344	12.5	1.76	0.017
20	198	32	33800	7774	700	12.5	1.89	0.021
19	198	32	50700	11661	1028	12.5	1.95	0.023
18	198	32	67600	15548	1330	12.5	2.15	0.026
17	198	32	84500	19435	1607	12.5	2.21	0.027
16	199	32	101400	23423	1859	12.5	2.28	0.029
15	199	32	118300	27327	2088	12.5	2.41	0.032
14	199	32	135200	31231	2292	12.5	2.47	0.035
13	199	32	152100	35135	2474	12.5	2.54	0.037
12	199	32	169000	39039	2634	12.5	2.60	0.040
11	199	32	185900	42943	2774	12.5	2.60	0.041
10	200	32	202800	47050	2894	12.5	2.60	0.043
9	201	32	219700	51190	2995	12.5	2.47	0.043
8	201	32	236600	55128	3079	12.5	2.41	0.044
7	201	32	253500	59066	3146	12.5	2.28	0.044
6	201	32	270400	63003	3198	12.5	2.15	0.043
5	201	32	287300	66941	3236	12.5	1.82	0.039
4	201	32	304200	70879	3262	12.5	1.56	0.035
3	201	32	321100	74816	3277	12.5	1.11	0.026
2	210	32	354900	85886	3284	17.5	0.85	0.016

where

A_1 = floor area of diaphragm immediately above the first storey, ft^2

$$r_{\max_1} = \frac{\text{maximum design shear in any of the walls at base}}{\text{total design base shear}} \times \frac{10}{l_w}$$

$$= \frac{1{,}571}{3{,}284} \times \frac{10}{28.83} = 0.166$$

$$A_1 = 132.83 \times 132.83\text{ft}^2$$

$$\rho = 2 - \frac{20}{0.166 \times \sqrt{132.83 \times 132.83}} = 1.09$$

For dual systems, the value of ρ need not exceed 80 % of the value calculated above. (1617.2.2)

$$80\% \text{ of above} = 0.8 \times 1.09 = 0.87 < 1$$

Use $\rho = 1$

8.5.16 *Dynamic Analysis Procedure (response spectrum analysis)*

As explained earlier, a dynamic analysis procedure is required for this example building (having height > 240 feet in Seismic Design Category D). The response spectrum analysis method (1618.1) was used, utilizing the SAP 2000 computer program.

The following items are worth mentioning in conjunction with the analyses carried out.

Self weight is automatically considered by SAP 2000. The superimposed dead load (SDL) needs to be computed and assigned to relevant joints as masses.

SDL on each floor:

$= (86 + 20\{10 \text{ for roof}\}) \times 130^2$ [Joists + SDL]
$+8 \times 12.5$ (15 for 2nd floor and 6.25 for roof) $\times 130 \times 4$ [Cladding]
$-150 \times (26\text{–}34/12) \times 34/12 \times 7^*/12 \times 56$ [Equivalent self weight for joists]
$= 1,521$ kips [1,532 for 2nd floor and 1,526 for roof including 200 kips]
* [86 psf for joists gives 7 inches of equivalent concrete slab thickness]

Masses magnitude 0.475, 0.95, and 1.90 kip-sec²/ft are assigned to each corner, edge, and interior joint, respectively, based on the above loads. These values are obtained as follows:

On each floor, there are 4 corner joints, each with a tributary area $X = 13 \times 13$ ft (mass assigned to each $= m$), 16 edge joints, each with tributary area of 2X (mass assigned to each $= 2m$), and 16 interior joints, each with a tributary area of 4X (mass assigned to each $= 4m$). Thus the total mass on each floor becomes

$$(m \times 4) + (2m \times 16) + (4m \times 16) = [\text{SDL(kips)}]/32 \text{ ft/sec}^2$$

$$\text{or, } m = \text{SDL}/3{,}200 \text{ (kip} - \text{sec}^2\text{/ft)}$$

Considering the magnitude of SDL = 1,521 kips on each floor (approximately),

$$m = 0.475 \text{ kip-sec}^2\text{/ft}$$

The magnitude of mass to be assigned to each corner joint is equal to m or 0.475 kip-sec²/ft. Similarly, the masses assigned to each edge and interior joint would be 0.95 kip-sec²/ft ($2m$) and 1.90 kip-sec²/ft ($4m$), respectively.

8.5.17 *Mode Shapes*

The 3-D analysis by SAP 2000 yielded the following periods for the first four modes:

Mode	*Period (sec)*	*Participating mass (%)*
1	2.485	71.2
2	0.659	14.8
3	0.300	6.1
4	0.178	3.1

As seen from the above, consideration of modes 1 (period = 2.485 sec), 2 (period = 0.659 sec), and 3 (period = 0.3 sec) should be adequate for lateral load analysis, as they account for 92.1 % (more than 90 %) of the participating mass (1618.2). The periods and mode shapes of these three modes are given in Table 8.12.

Table 8.11 Summary of design axial force shear force and bending moment for shear wall between grade and level 2.

Loads	Symbol	Axial Force (kips)	Shear Force (kips)	Bending Moment (ft-kips)
Dead Load	D	5381	0	0
Live Load	L	837	0	0
Lateral Load	EQ	0	1571	118,596
Load Combinations				
1	$1.4D + 1.7L$	8956	0	0
2	$1.2D + (\rho E_h + 0.2DS_{DS}) + 0.5L$	7952	1571	118,596
3	$0.9D - (\rho E_h + 0.2DS_{DS})$	3767	−1571	−118,596

Max. axial force with lateral force 7952

Table 8.12 Comparison of periods from SAP 2000 and STAAD analysis.

Case	Period (sec.)	SAP2000 3D	STAAD 3D	STAAD 2D	STAAD-3D/ STAAD-2D	STAAD-3D/ SAP2000	STAAD-2D/ SAP2000
1	T_1	2.031	2.130	2.292	0.93	1.05	1.13
	T_2	0.578	0.593	0.628	0.94	1.03	1.09
	T_3	0.280	0.283	0.309	0.92	1.01	1.10
2	T_1	2.485	2.583	2.768	0.93	1.04	1.11
	T_2	0.659	0.671	0.704	0.95	1.02	1.07
	T_3	0.300	0.303	0.330	0.92	1.01	1.10
3	T_1	2.827	2.960	3.162	0.94	1.05	1.12
	T_2	0.742	0.756	0.788	0.96	1.02	1.06
	T_3	0.338	0.340	0.366	0.93	1.01	1.08
4	T_1	2.928	3.104	3.318	0.94	1.06	1.13
	T_2	0.805	0.835	0.871	0.96	1.04	1.08
	T_3	0.386	0.397	0.420	0.95	1.03	1.09
5	T_1	2.617	2.786	2.980	0.93	1.06	1.14
	T_2	0.729	0.757	0.793	0.95	1.04	1.09
	T_3	0.365	0.359	0.383	0.94	0.98	1.05

Note 1: Definition of different Cases

Case	Column	Beam	Wall
1	$1.0/_g$	$1.0/_g$	$1.0/_g$
2	$1.0/_g$	$0.5/_g$	$0.5/_g$
3	$0.7/_g$	$0.35/_g$	$0.35/_g$
4	$0.7/_g$	$0.35/_g$	$0.35/_g$
5	$0.5/_g$	$0.5/_g$	$0.5/_g$

Note 2: Only the first three modes (not necessarily in sequence) contributing more than 90% mass participation are taken into consideration.

The three modes considered in modal analysis have periods 2.485, 0.659, and 0.300 seconds.

8.5.18 *Verification of Results from SAP 2000*

To check the accuracy of results obtained from the SAP 2000 computer program, the example building was analysed using STAAD-III. Both three-dimensional and equivalent two-dimensional analyses were performed to compute the modal periods. Five cases were considered, each with different stiffnesses assigned to beams, columns, and shear walls. Table 8.12 shows the comparison between the results obtained from both the computer programs. It shows a good comparison between the results from SAP 2000 and STAAD–III. In addition, the table shows that considering an equivalent two-dimensional model is quite reasonable. It may be noted that the command for rigid diaphragms was not available in the version of STAAD-III used and it was necessary to assign rigid diagonal truss elements at each floor level.

8.5.19 *L_m and M_m for each mode shape*

According to 1618-4, the portion of base shear contributed by the mth mode, V_m, shall be determined from the following equations:

$$V_m = C_{sm}\bar{W}_m \qquad (16\text{-}51)$$

$$\bar{W} = L_m^2/M_m \qquad (16\text{-}21)$$

$$L_m = \sum_{i=1}^{n} w_i\phi_{im}$$

$$M_m = \sum_{i=1}^{n} w_i\phi_{im}^2$$

where:

C_m = the modal seismic response coefficient determined in Equation 16-53.
$\bar{W}_m$ = the effective modal gravity load.
w_i = the portion of total gravity load, W, of the building at level i.
ϕ_{im} = the displacement amplitude at the ith level of the building when vibrating in its mth mode.

From Table 8.13:

$L_1 = 33{,}440$ kips/g $\quad L_2 = -14{,}622$ kips/g $\quad L_3 = 10{,}202$ kips/g
$M_1 = 23{,}347$ kips/g $\quad M_2 = 21{,}480$ kips/g $\quad M_3 = 24{,}445$ kips/g

8.5.20 *Modal Seismic Design Coefficients, C_{sm}*

$$C_{sm} = \frac{S_{am}}{(R/I_E)} \qquad (16\text{-}53)$$

where: S_{am} = the modal design spectral response acceleration at period T_m determined from either the general design response spectrum of 1615.1 or the site-specific response spectrum per 1615.2.

In the example considered here, the general procedure of 1615.1 will be followed. Under this procedure, the spectral response acceleration, S_a, can be expressed by the following equations (Figure 1615.1.4):

For $T > T_S$ $\quad S_a = S_{D1}/T$
For $T_0 \leq T \leq T_S$ $\quad S_a = S_{DS}$
For $T \leq T_0$ $\quad S_a = 0.6S_{DS}T/T_0 + 0.4S_{DS}$
where $T_S = S_{D1}/S_{DS}$, and $T_0 = 0.2T_S$

According to 1618.4 (Exception), when the general response spectrum of 1615.1 is used for buildings on Site Class D, E or F sites, the modal seismic design coefficients for modes other than the fundamental mode that have periods less than 0.3 seconds are permitted to be determined by the following equation:

$$C_{sm} = \frac{0.4S_{DS}}{(R/I_E)}(1.0 + 5.0T_m) \qquad (16\text{-}54)$$

For the example building considered, the periods from the second and third modes are greater than or equal to 0.3 seconds. Equation 16-53 is therefore used for the following calculations;

For the example building, $T_s = 0.6/1.0 = 0.6\,\text{sec}$
$T_0 = 0.6/5 = 0.12\,\text{sec}$

Mode 1: $T_1 = 2.485\,\text{sec} > 0.60\,\text{sec}$ $\quad C_{s!} = \dfrac{0.6\,g}{2.485 \times (8/1)} = 0.0302\,g$

Mode 2: $T_2 = 0.659\,\text{sec} > 0.60\,\text{sec}$ $\quad C_{s2} = \dfrac{0.6\,g}{0.659 \times (8/1)} = 0.1138\,g$

Mode 3: $T_3 = 0.300\,\text{sec} > 0.12\,\text{sec}$ $\quad C_{s3} = \dfrac{1.0\,g}{(8/1)} = 0.1250\,g$
$< 0.60\,\text{sec}$

8.5.21 *Base Shear using Modal Analysis*

$$V_m = C_{sm}\bar{W}_m = \frac{L_m^2}{M_m}C_{sm}$$

Mode 1 $\quad V_1 = \dfrac{33,440^2}{23,347\,g} \times 0.0302\,g = 1,446\,\text{kips}$

Mode 2 $\quad V_2 = \dfrac{(-14,622)^2}{21,480\,g} \times 0.1138\,g = 1,133\,\text{kips}$

Mode 3 $\quad V_3 = \dfrac{10,202^2}{24,445\,g} \times 0.1250\,g = 532\,\text{kips}$

The modal base shears are combined by the SRSS method to give the resultant base shear:

$$V = \left[1{,}446^2 + 1{,}133^2 + 532^2\right]^{1/2} = 1{,}912 \text{ kips}$$

The participating mass (*PM*) for each of the above three modes is determined as:

$$\text{Mode 1:} \quad PM_1 = \frac{33,440^2}{23,347 \times 67,246} = 0.712$$

$$\text{Mode 2:} \quad PM_2 = \frac{(-14,622)^2}{21,480 \times 67,246} = 0.148$$

$$\text{Mode 3:} \quad PM_3 = \frac{10,202^2}{24,445 \times 67,246} = 0.063$$

Please note that these participating masses are equal to or very close to the values from SAP 2000 (Section 8.5.17)

$$\sum PM = 0.712 + 0.148 + 0.063 = 0.923 > 0.90$$

Therefore, consideration of the above three modes (1,2 and 3) is sufficient as per 1618.2.

8.5.22 Design Base Shear using Static Procedure

The design base shear using the static lateral force procedure was computed in the previous section using a fundamental period of 1.536 seconds (i.e., $= C_u \; x \; T_a$) and was found to be 3,284 kips.

8.5.23 Scaling of Elastic Response Parameters for Design

Section 1618.7 stipulates that the base shear using modal analysis must be scaled up when the base shear calculated using the equivalent lateral force procedure is greater. However, it is permitted to use a fundamental period of $T = 1.2C_uT_a$ instead of C_uT_a in the calculation of base shear by the equivalent lateral force procedure.

Based on the new period of $T = 1.2 \times 1.536 = 1.843$ seconds, the design base shear is re-calculated as follows:

$$\begin{aligned} V &= \frac{S_{D1}I_E}{RT}W = \frac{0.6 \times 1 \times 67{,}246}{8 \times 1.843} = 2{,}736 \text{ kips} \\ &\leq \frac{S_{DS}I_E}{R}W = \frac{1.0 \times 1 \times 67{,}246}{8} = 8{,}406 \text{ kips} \\ &\geq 0.044 S_{DS}I_E W = 0.044 \times 1.0 \times 67{,}246 = 2{,}959 \text{ kips } (\textit{governs}). \end{aligned}$$

Since $S_1 = 0.6g$, Equation 16-38 is applicable for the example building in SDC D.

$$V \geq \frac{0.5 S_1 I_E}{R} W = \frac{(0.5) \times (0.6) \times (1.0)}{8} (67{,}246) = 2{,}522 \text{ kips}$$

Use $V = 2{,}959$ kips

2,959 kips (equivalent base shear) > 1,912 kips (modal base shear)

Therefore, the modal forces must be scaled up per Equation 16-59.

Scale factor = 2,959/1,912 = 1,548

The modified modal base shears are as follows:
$V_1 = 1.548 \times 1{,}446 = 2{,}238$ kips
$V_2 = 1.548 \times 1{,}133 = 1{,}754$ kips
$V_3 = 1.548 \times 532 = 824$ kips

$$V = \left[2{,}238^2 + 1{,}754^2 + 824^2\right]^{1/2} = 2{,}960 \text{ kips}$$

8.5.24 Distribution of Base Shear

Lateral force at level i (1 to 20) for mode m (1 to 3) is to be calculated as (see Equations 16-55 and 16-56):

$$F_{im} = \frac{w_i \phi_{im}}{\sum w_i \phi_{im}} V_m$$

The distribution of the modal base shear for each mode is shown in Table 8.13.

8.5.25 Lateral Analysis

Three dimensional analysis of the structure was performed for each set of modal forces using the SAP 2000 computer program. To account for accidental torsion, the mass at each level was assumed to be displaced from the centre of mass by a distance equal to 5 % the building dimension perpendicular to the direction of force (1617.4.4.4). In the model, rigid diaphragms were assigned at each level, and rigid-end offsets were defined at the ends of each member so that results were automatically obtained at the face of each support.

The stiffnesses of members used in the analyses were as follows:

For columns and shear walls, $I_{eff} = I_g$
For beams, $I_{eff} = 0.5 I_g$

Table 8.14 shows the shear force and bending moment at each floor level of each shear wall (the four shear walls are identical in every respect and are subject to the same forces) due to each considered mode and the resultant load effects. Because of the location of the shear walls

Table 8.13 Calculation of L_m and M_m and distribution of modal base shear for example building.

Mode 1, $T_1 = 2.485$ sec
$V_1 = 2238$ kips

Floor Level i	w_i kips	ϕ_{i1}	$w_i\phi_{i1}$ kips	$w_i\phi_{i1}^2$ kips	$w_i\phi_{i1}/\Sigma w_i\phi_{i1}$	F_{i1} $V_1 \times$ col. 6 kips
1	2	3	4	5	6	7
21	2987	1	2987.0	2987.0	0.0893	200
20	3338	0.9585	3199.6	3066.9	0.0957	214
19	3338	0.9140	3051.0	2788.7	0.0912	204
18	3338	0.8676	2895.9	2512.4	0.0866	194
17	3352	0.8184	2743.1	2244.9	0.0820	184
16	3366	0.7664	2579.6	1977.0	0.0771	173
15	3366	0.7122	2397.2	1707.3	0.0717	160
14	3366	0.6558	2207.4	1447.6	0.0660	148
13	3366	0.5972	2010.1	1200.4	0.0601	135
12	3366	0.5372	1808.2	971.3	0.0541	121
11	3366	0.4758	1601.6	762.0	0.0479	107
10	3380	0.4142	1399.8	579.8	0.0419	94
9	3394	0.3531	1198.3	423.1	0.0358	80
8	3394	0.2928	993.7	290.9	0.0297	67
7	3394	0.2347	796.7	187.0	0.0238	53
6	3394	0.1800	610.9	109.9	0.0183	41
5	3394	0.1294	439.1	56.8	0.0131	29
4	3394	0.0849	288.1	24.5	0.0086	19
3	3394	0.0478	162.3	7.8	0.0049	11
2	3559	0.0196	69.9	1.4	0.0021	5
Σ	67,246		33,440	23,347	1.0000	2238
	$W =$		$L_1 \times g =$	$M_1 \times g =$		$V_1 =$

Mode 2, $T_2 = 0.659$ sec
$V_2 = 1754$ kips

Floor Level i	w_i kips	ϕ_{i2}	$w_i\phi_{i2}$ kips	$w_i\phi_{i2}^2$ kips	$W_i\phi_{i2}/\Sigma W_i\phi_{i2}$	F_{i2} $V_2 \times$ col. 13 kips
8	9	10	11	12	13	14
21	2987	1	2987.0	2987.0	−0.2043	−358
20	3338	0.8065	2692.0	2171.1	−0.1841	−323
19	3338	0.5981	1996.4	1194.0	−0.1365	−239
18	3338	0.3836	1280.4	491.1	−0.0876	−154
17	3352	0.1683	564.0	94.9	−0.0386	−68
16	3366	−0.0381	−128.1	4.9	0.0088	15
15	3366	−0.2308	−777.0	179.4	0.0531	93

(continued overleaf)

Table 8.13 (*continued*)

Floor Level i	W_i kips	ϕ_{i2}	$W_i\phi_{i2}$ kips	$W_i\phi_{i2}^2$ kips	$W_i\phi_{i2}/\Sigma W_i\phi_{i2}$	F_{i2} $V_2 \times$ col. 13 kips
8	9	10	11	12	13	14
14	3366	−0.4033	−1357.7	547.6	0.0928	163
13	3366	−0.5488	−1847.1	1013.6	0.1263	222
12	3366	−0.6614	−2226.4	1472.6	0.1523	267
11	3366	−0.7372	−2481.3	1829.1	0.1697	298
10	3380	−0.7736	−2614.8	2022.8	0.1788	314
9	3394	−0.7706	−2615.5	2015.6	0.1789	314
8	3394	−0.7302	−2478.3	1809.6	0.1695	297
7	3394	−0.6566	−2228.7	1463.4	0.1524	267
6	3394	−0.5565	−1888.7	1051.0	0.1292	227
5	3394	−0.4382	−1487.3	651.8	0.1017	178
4	3394	−0.3123	−1059.8	330.9	0.0725	127
3	3394	−0.1905	−646.5	123.1	0.0442	78
2	3559	−0.0857	−305.1	26.2	0.0209	37
Σ	67,246		−14,622	21,480	1.0000	1754
	$W =$		$L_2 \times g =$	$M_2 \times g =$		$V_2 =$

Mode 3, $T_3 = 0.300$ sec
$V_3 = 824$ kips

Floor Level i	w_i kips	ϕ_{i3}	$w_i\phi_{i3}$ kips	$w_i\phi_{i3}^2$ kips	$w_i\phi_{i3}/\Sigma w_i\phi_{i3}$	F_{i3} $V_3 \times$ col.20 kips
15	16	17	18	19	20	21
21	2987	1.0000	2987.0	2987.0	0.2928	241
20	3338	0.6386	2131.6	1361.3	0.2089	172
19	3338	0.2501	834.9	208.8	0.0818	67
18	3338	−0.1231	−410.8	50.6	−0.0403	−33
17	3352	−0.4388	−1470.9	645.5	−0.1442	−119
16	3366	−0.6562	−2208.9	1449.6	−0.2165	−178
15	3366	−0.7654	−2576.3	1971.8	−0.2525	−208
14	3366	−0.7571	−2548.5	1929.6	−0.2498	−206
13	3366	−0.6355	−2139.1	1359.4	−0.2097	−173
12	3366	−0.4186	−1409.2	589.9	−0.1381	−114
11	3366	−0.1381	−464.9	64.2	−0.0456	−38
10	3380	0.1668	563.9	94.1	0.0553	46
9	3394	0.4543	1541.8	700.4	0.1511	125
8	3394	0.6854	2326.2	1594.3	0.2280	188
7	3394	0.8300	2816.9	2337.9	0.2761	228
6	3394	0.8711	2956.5	2575.3	0.2898	239

(*continued overleaf*)

Table 8.13 (*continued*)

Floor Level i	w_i kips	ϕ_{i3}	$w_i\phi_{i3}$ kips	$w_i\phi_{i3}^2$ kips	$w_i\phi_{i3}/\Sigma w_i\phi_{i3}$	F_{i3} $V_3 \times$ col.20 kips
15	16	17	18	19	20	21
5	3394	0.8078	2741.6	2214.7	0.2687	221
4	3394	0.6558	2225.8	1459.6	0.2182	180
3	3394	0.4465	1515.4	676.6	0.1485	122
2	3559	0.2217	789.0	174.9	0.0773	64
Σ	67,246		10,202	24,445	1.0000	824
	$W =$		$L_3 \times g =$	$M_3 \times g =$		$V_3 =$

within the plan of the building, the earthquake-induced axial force in each shear wall is equal to zero.

The above values at the base level from the dynamic procedure can be compared with the corresponding values from the static procedure as follows: (subscripts d and s represent results

Table 8.14 Internal forces in a shear wall due to lateral forces given in Table 8.13.

	Shear Force (kips)			
Storey Level	Mode 1	Mode 2	Mode 3	Resultant
1	2	3	4	5
20–21	−348	93	54	**364**
19–20	−98	−152	158	**240**
18–19	−33	−258	193	**324**
17–18	47	−334	183	**384**
16–17	120	−387	140	**429**
15–16	188	−388	61	**435**
14–15	250	−362	−29	**441**
13–14	308	−306	−119	**450**
12–13	363	−226	−194	**469**
11–12	415	−126	−244	**498**
10–11	460	−16	−257	**527**
9–10	521	107	−244	**585**
8–9	565	234	−184	**639**
7–8	617	356	−100	**719**
6–7	670	469	4	**817**
5–6	726	569	113	**930**
4–5	789	656	216	**1048**
3–4	859	728	301	**1166**
2–3	942	795	368	**1287**
1–2	1070	843	396	**1419**

(*continued overleaf*)

Table 8.14 (*continued*)

		Bending Moment (ft-kips)			
Storey Level 1	Section 2	Mode 1 3	Mode 2 4	Mode 3 5	Resultant 6
20–21	top	−1208	751	−196	**1436**
	bottom	5556	−1911	−476	**5895**
19–20	top	−6863	2724	265	**7389**
	bottom	8094	−823	−2235	**8437**
18–19	top	−9449	1651	2030	**9805**
	bottom	9865	1572	−4440	**10932**
17–18	top	−11298	−734	4256	**12096**
	bottom	10712	4909	−6541	**13477**
16–17	top	−12231	−4086	6398	**14395**
	bottom	10730	8927	−8151	**16163**
15–16	top	−12332	−8145	8060	**16834**
	bottom	9982	12993	−8817	**18606**
14–15	top	−11665	−12280	8787	**19081**
	bottom	8540	16803	−8423	**20645**
13–14	top	−10298	−16187	8455	**20966**
	bottom	6446	20010	−6967	**22147**
12–13	top	−8269	−19516	7055	**22338**
	bottom	3737	22338	−4625	**23116**
11–12	top	−5609	−21984	4755	**23181**
	bottom	421	23559	−1700	**23624**
10–11	top	−2323	−23360	1854	**23548**
	bottom	−3426	23558	1365	**23845**
9–10	top	1518	−23518	−1211	**23598**
	bottom	−8030	22175	4258	**23966**
8–9	top	6142	−22282	−4125	**23478**
	bottom	−13206	19351	6425	**24293**
7–8	top	11367	−19593	−6329	**23519**
	bottom	−19076	15147	7573	**25508**
6–7	top	17319	−15500	−7526	**24430**
	bottom	−25689	9643	7476	**28440**
5–6	top	24054	−10078	−7482	**27132**
	bottom	−33131	2964	6069	**33813**
4–5	top	31660	−3442	−6123	**32430**
	bottom	−41522	−4755	3428	**41934**
3–4	top	40265	4279	−3518	**40644**
	bottom	−51007	−13375	−240	**52732**
2–3	top	50019	12954	137	**51670**
	bottom	−61793	−22897	−4735	**66069**
1–2	top	61141	22586	4642	**65344**
	bottom	−79869	−37335	−11567	**88920**

Table 8.15 (along shear wall line E).

Floor Level	Mode 1 $\delta_{xe\ 1}$	Mode 2 $\delta_{xe\ 2}$	Mode 3 $\delta_{xe\ 3}$	Resultant δ_{xe}	C_d	δ_x	Drift Δ
21	4.07	−0.52	0.08	4.11	6.50	26.70	1.16
20	3.91	−0.42	0.05	3.93	6.50	25.54	1.22
19	3.73	−0.31	0.03	3.74	6.50	24.32	1.27
18	3.54	−0.20	0.00	3.55	6.50	23.05	1.32
17	3.34	−0.09	−0.03	3.34	6.50	21.73	1.37
16	3.13	0.02	−0.04	3.13	6.50	20.36	1.41
15	2.91	0.12	−0.05	2.92	6.50	18.95	1.46
14	2.68	0.21	−0.05	2.69	6.50	17.49	1.50
13	2.44	0.29	−0.04	2.46	6.50	16.00	1.53
12	2.20	0.35	−0.03	2.23	6.50	14.47	1.56
11	1.95	0.39	−0.01	1.99	6.50	12.91	1.58
10	1.70	0.41	0.02	1.74	6.50	11.33	1.58
9	1.44	0.40	0.04	1.50	6.50	9.75	1.57
8	1.20	0.38	0.05	1.26	6.50	8.18	1.54
7	0.96	0.34	0.06	1.02	6.50	6.65	1.48
6	0.74	0.29	0.07	0.80	6.50	5.17	1.39
5	0.53	0.23	0.06	0.58	6.50	3.78	1.26
4	0.35	0.16	0.05	0.39	6.50	2.52	1.08
3	0.20	0.10	0.03	0.22	6.50	1.45	0.84
2	0.08	0.04	0.02	0.09	6.50	0.61	0.61

from dynamic and static procedures, respectively).

$$\frac{V_d}{V_s} = \frac{1{,}479}{1{,}571} = 0.90$$

$$\frac{M_d}{M_s} = \frac{88{,}920}{118{,}596} = 0.75$$

The lower ratio of dynamic-to-static moments reflects the different distribution of lateral forces along the height of the building obtained from dynamic analysis. This also shows the possible advantage of doing dynamic analysis.

Resultant lateral displacements (square root of the sum of the squares of modal displacements) at every floor level, δ_{xe}, are shown in Table 8.15. The maximum inelastic response displacements, δ_x, and storey drifts are computed and shown in Table 8.15.

8.5.26 Storey Drift Limitation

According to Section 1617.3 the calculated storey drifts, Δ, as shown in Table 8.15 shall not exceed 0.020 times storey height (Table 1617.3 for Seismic Use Group I and all other buildings).

Floor	*Maximum allowable drift*	*Largest drift*	*(Table 8.15)*
1st	$0.02 \times 17.5\,\text{ft} = 4.2\,\text{in.}$	$> 0.61\,\text{in}$	o.k.
Others	$0.02 \times 12.5\,\text{ft} = 3.0\,\text{in}$	$> 1.58\,\text{in}$	o.k.

8.5.27 $P - \Delta$ Effects

According to Section 1617.4.6.2, $P - \Delta$ effects on storey shears and moments, the resulting member forces and moments, and storey drifts induced by these effects need not be considered when the stability coefficient, θ, as determined by the following formula is equal to less than 0.1.

$$\theta = \frac{P_x \Delta}{V_x h_{sx} C_d} \qquad (16\text{-}47)$$

where

P_x = the total unfactored vertical force
Δ = the design storey drift
V_x = the seismic shear force acting between level x and $x - 1$
h_{sx} = the storey height below level x
C_d = the deflection amplification factor

In the lateral analysis performed using SAP 2000, $P - \Delta$ effects are included.

8.5.28 Redundancy Factor, ρ

At the base, $r_{max} = (1{,}419/2{,}960) \times 10/29.17 = 0.164$ (assuming 38×38-inch boundary elements):

$\rho = \mathbf{1}$ (as computed under equivalent lateral force procedure).

8.6 References

American Concrete Institute (2008) *ACI 318-08 Building Code Requirements for Structural Concrete and Commentary*. Farmington Hills, MI: ACI.

American Institute of Steel Construction (2005) *AISC Seismic Provisions for Structural Steel Buildings*, March 9, 2005. Chicago, IL: AISC.

American Institute of Steel Construction (2006) *AISC Steel Construction Manual*. Chicago, IL: AISC.

American Iron and Steel Institute (2007) *AISI Specification for the Design of Cold-formed Steel Structural Members*. Washington, DC. AISI.

American Society of Civil Engineers (2006) *ASCE 7/SEI 7-05 Minimum Design Loads for Buildings and Other Structures*. Reston, VA: ASCE.

Applied Technology Council (1978) *Tentative Provisions for the Development of Seismic Regulations for Buildings*. Report no. ATC-3.06. ATC, Washington, DC.

Biot, M.A. (1943) 'Analytical and experimental methods in engineering seismology'. *Transactions of the Am. Soc. Civil Engineers*, ASCE, **180**: 365–75.

British Standards Institution (2005) BS EN 1998-1: 2005 Eurocode 8: *Design of Structures for Earthquake Resistance Part 1: General Rules, Seismic Actions and Rules for Buildings*. Milton Keynes: BSI.

California Department of General Services Excellence in Public Educational Facilities website (http://www.excellence.dgs.ca.gov/StudentSafety/S7_7-1.htm) 18/07/08.

California Field Act (1933).

California Riley Act (1933).

Fardis M.N. *et al.* (2005) *Designer's Guide to EN 1998-1 and EN 1998-5. Eurocode 8: Design of Structures for Earthquake Resistance. General Rules, Seismic Actions, Design Rules for Buildings, Foundations and Retaining Structures*. Thomas Telford, London.

Ghosh, S.K. (2003), 'Seismic Design Using Structural Dynamics (IBC 2000)', International Code Council, IL60478–5795, USA.
Hu, Y., Liu, S. and Dong, W. (1996) *Earthquake Engineering*. E and FN Spon, London.
International Code Council (1997) *Uniform Building Code (UBC) 1997*. Washington, DC. ICC.
International Code Council (2000) *International Building Code (IBC) 2000*. Washington, DC. ICC.
International Code Council (2003) *International Building Code (IBC) 2003*. Washington, DC. ICC.
International Code Council (2006) *International Building Code (IBC) 2006*. Washington, DC. ICC.
National Earthquake Hazard Reduction Program (NEHRP) (2000) *Recommended Provisions for New Buildings and Other Structures* (FEMA 368). NEHRP, Gaithersburg MD; FEMA, Washington DC.
National Earthquake Hazard Reduction Program (NEHRP) (2003) *Recommended Provisions and Commentary for Seismic Regulations for New Buildings and Other Structures* (FEMA 450). NEHRP, Gaithersburg MD; FEMA, Washington DC.
Seed, H.B. and Idriss, I.M. (1971) 'Simplified procedure for evaluating soil liquefaction potential'. *J. Soil Mechanics and Foundations Division*, ASCE, **97**, SM9, 1249–73.
Seed, H.B., Ugas, C. and Lysmer, J. (1976) 'Site dependent spectra for earthquake-resistance design'. *Bull. Seism. Soc. Am.* **66**(1): 221–43.
Structural Engineers Association of California (1999) *Recommended Lateral Force Requirements and Commentary (SEAOC Blue Book)*. Washington, D.C. International Code Council (ICC).

9

Inelastic Analysis and Design Concepts (with Particular Reference to H-Sections)

9.1 Introduction

It is a fact of life that earthquakes come in all sizes – small as well as large. The present state of the art does not permit an accurate prediction of the exact location of an earthquake or its size. Seismic hazard models based on probabilistic theories are in use. Informed decision-making helps with the selection of a scenario for large earthquakes. The engineer is still faced with the task of designing for small earthquakes and large earthquakes. Large earthquakes are rare, with a low probability. Buildings subject to large earthquakes are designed to accept large deformations but not to collapse. Large deformations/rotations of the joints in the building inevitably involve excursions into the inelastic regions of the material stress-strain curve. The material used should be ductile and able to deform without premature brittle fracture. This is where inelastic design concepts play a part. Ductility helps in dissipating some of the energy, which is one of the prime requirements during a large earthquake. However, it comes at a price. There is a quantum leap in the analytical complexity with the introduction of inelastic design. Structural load carrying members in a space frame are subjected to biaxial moments. This has been an active topic of research since the 1950s. Consequently, a large volume of literature exists on the subject containing both classical elastic solutions and solutions manifesting inelastic behaviour. An excellent treatise on beam column (usually referred to as a member in space subjected to axial load and bending moment) behaviour under *static* loading conditions is by Chen and Atsuta (1977). An extensive list of references is also given at the end of each chapter. Basic concepts of inelastic analysis are introduced in the next section.

The focus will primarily be on *inelastic* behaviour.

The aim of this chapter is to capture the essence of inelastic beam-column behaviour in the subsequent sections, rather than cover in detail the immense volume of literature that exists on the subject. Inelastic behaviour will be further exemplified with a brief summary of full-scale laboratory tests and analytical methods.

Fundamentals of Seismic Loading on Structures Tapan Sen

9.2 Behaviour of Beam Columns

In the past when computing facilities were scarce, three-dimensional space structures were idealized as consisting of a series of planar structures, i.e. two-dimensional structures with loading in the same plane. Designs were based on reasonable approximations coupled with good engineering judgement. The situation was not very satisfactory and called for refinements in design.

With respect to inelastic analysis, it is very important to note at this point that inelastic behaviour is *load-path* dependent. What is load path dependent? Consider the loading situation shown in Figure 9.1.

The loading conditions shown in (a) and (b) are statically equivalent. In (a) the axial load is applied first and then the end moments. In (b) the axial load and end moments are increased simultaneously. The inelastic response, however, in these two situations could be different as the load paths are different. For this reason the solution, unfortunately, has to rely on step-by-step incremental procedure that follows exactly the load path or the history of loading under static as well as dynamic conditions.

A beam column is the basic sub-assemblage of a frame. Hence, inelastic design concepts are best introduced through beam column behaviour.

The solution of beam-columns loaded about both the axes requires consideration of the geometrical shape of the cross section, compatibility with end conditions and equilibrium of internal and external forces for equilibrium. Before the differential equations governing

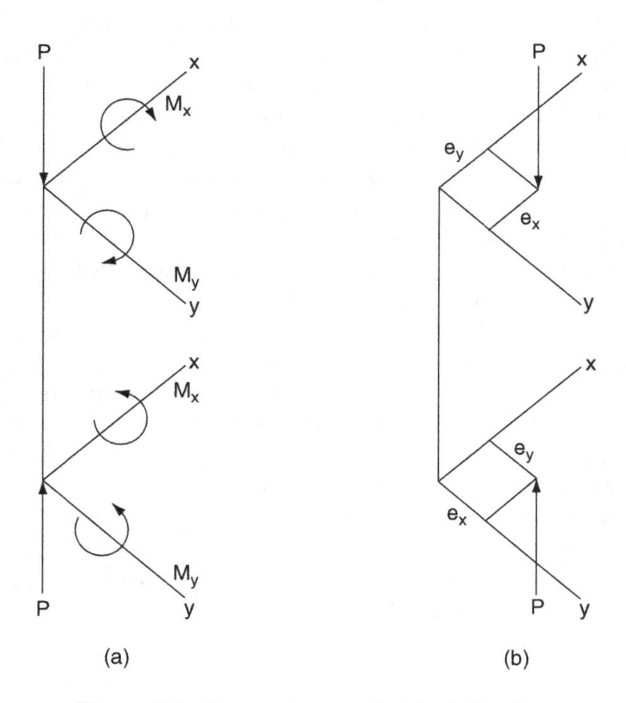

Figure 9.1 Beam-column under biaxial loading.

the behaviour of beam-columns are introduced, the problems encountered are discussed conceptually.

As mentioned earlier, behaviour of bi-axially loaded beam-columns has been studied extensively in the last fifty years or so both in USA and Europe and the research findings have slowly found their way into subsequent code revisions.

Consider the H-section column under biaxial loading shown in Figure 9.2.

There is always a small amount of unavoidable manufacturing imperfection that we need to take note of. Unfortunately, all columns will manifest them. Thus, for all practical purposes, even if the beam column was loaded about the major axis, minor axis moment would be introduced due to minor axis imperfections. This is further magnified by the presence of axial loads. If the major axis moment was gradually increased the column would gradually deflect about the major axis until instability sets in. The deformed position shows that the cross section, in addition to deflections about 'x' and 'y' axes, has also undergone a certain amount of twist. This is a feature of open cross sections like the one shown in Figure 9.2. Along with the twist of the section, there is another phenomenon known as 'warping'. We need to include this effect when formulating the equations of equilibrium of a beam column in space.

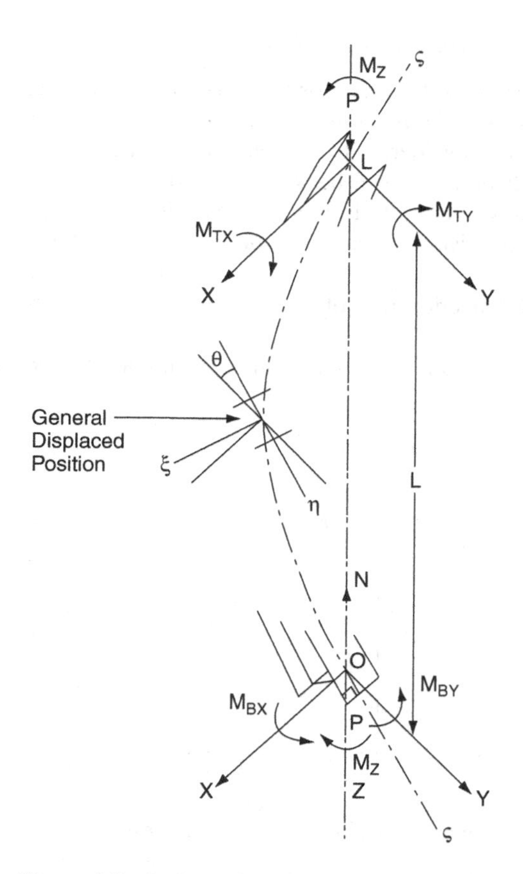

Figure 9.2 Deformed configuration of a column.

Beam columns have been categorized according to their ability to sustain the increasing load without out-of-plane instability. Thus beam columns have been categorized in to two main groups:

1. short (or stocky) columns
2. intermediate or long columns

9.2.1 Short or Stocky Beam Column

All beam columns would deflect under loading, but the use of the term 'short beam column' implies that the effect of lateral deflection does not affect its in plane behaviour. Or in other words, the effect of lateral deflection on the overall geometry may be ignored in the analysis. The ultimate load carrying capacity of the beam column would be limited by the full plastic yielding of the material of the cross section, provided local buckling of web/flange do not occur.

9.2.2 Long or Intermediate Length Beam Column

With reference to Figure 9.2, for a long or intermediate length beam column the instability of the member arising from the magnification of primary moments caused by the presence of the axial load must be considered. Long columns reach their out-of-plane elastic critical buckling load due to 'lateral torsional buckling' (LTB) before it reaches its plastic moment capacity. The buckling is referred to in literature as '*bifurcation*' type. Classical solution of the differential equations exists and has been reported in various texts (Galambos, 1968; Chen and Atsuta, 1977).

The schematic load deflection plot of a beam column in space (Figure 9.2) is shown in Figures 9.3a and 9.3b.

In the discussions which follow, the load path is axial load applied first followed by major axis moment to failure.

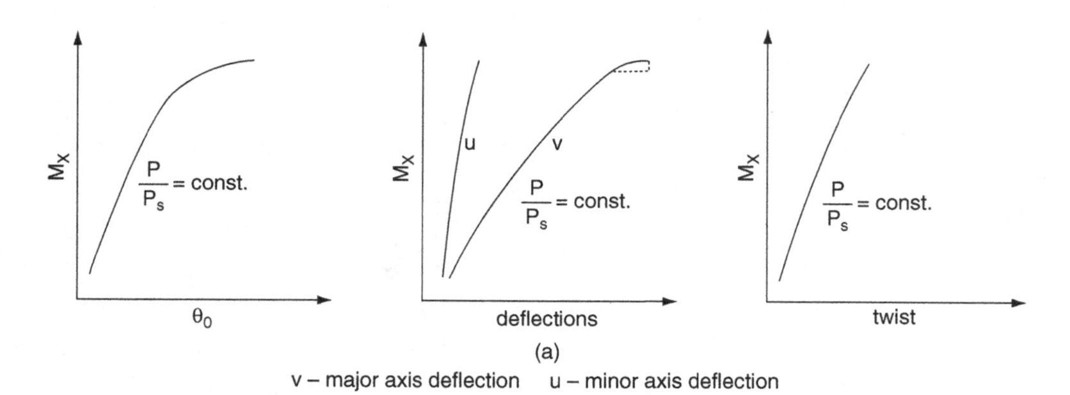

Figure 9.3a In-plane failure characteristics single curvature (S/C) bending – constant axial load.

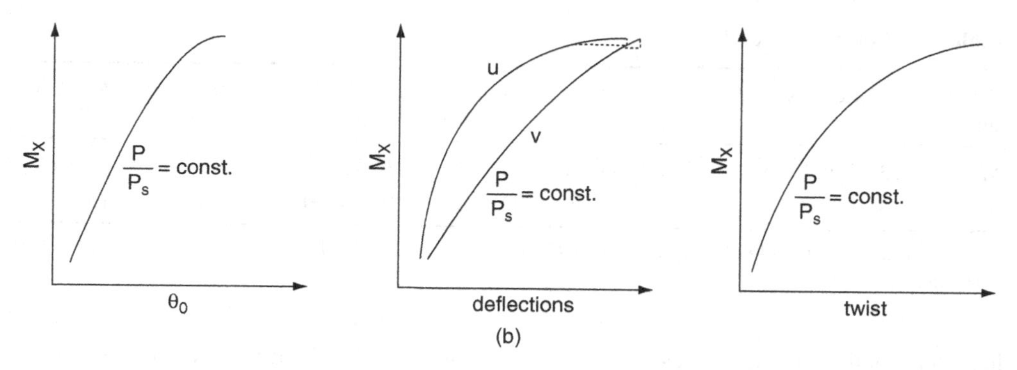

Figure 9.3b Lateral torsional buckling (S/C bending) – constant axial load.

In-Plane Characteristics

Referring to Figure 9.3a, it may be noted that:

1. During the application of major axis moment increments the growth of minor axis deflections is small, the minor axis deflection curve being nearly vertical.
2. The growth of twisting deformations is also small.

The failure is primarily in-plane.

Out-of-Plane Failure Characteristics

Referring Figure 9.3b, it may be noted that:

1. During increase of major axis moment, the increase of minor axis deflections is initially small, but they progressively increase at a faster rate with further increase of major axis moment. The slope of the minor axis deflection curve near failure is less than the slope of the major axis deflection curve. This indicates predominantly out-of-plane failure due to lateral torsional buckling (LTB).
2. The growth of the twisting deformations is excessive. The slope of the curve is nearly zero, an indication of very little torsional resistance.

9.3 Full Scale Laboratory Tests

A limited series of full scale tests were conducted at Imperial College laboratories with which the author was associated (Sen, 1976). Their purpose was to investigate the stability of H-columns in the high axial load range (0.6–0.9P_T). Eight full scale tests were conducted on 203mm × 203mm Universal column sections of varying lengths and weights. Results of only two columns are being reported here – one for 'short column – l/ry = 45 and another column of intermediate slenderness range – l/ry = 59. The columns were all tested to establish their

Table 9.1 Column test results.

Column number	UC sections	L/r_y	L (m)	P/P_T*	T value** (N/mm^2)	Axial loads (N)	Failure moment (N · m)	Mode of failure
A2	203/203@46 kg	45	2.286	0.81	506.95	174.3×10^4	3.468×10^4	In-plane
B1	203/203@71 kg	59	3.124	0.80	1162.3	229.1×10^4	13.66×10^4	LTB

* P_T – Ultimate axial load either at tangent modulus load or squash load.
** Horne, M.R. (1964a).

load deformation characteristics in the inelastic range. The details of the test programme are summarized in Table 9.1.

Photographs showing a general view of the entire loading rig are shown in Figure 9.4. The major axis moment arm arrangement is shown in Figure 9.5.

A special feature of the tests was the use of specially built load cells at each end to measure directly the axial loads and bending moments in the inelastic range. The instrumentations were placed on the load cells. The load cells were manufactured from high tensile Maraging steel with 18 % Nickel having Cobalt and Molybdenum as major alloying elements. The load cells were placed at the ends of the column (see Figure 9.5).

Figure 9.4 General view of the entire rig (viewed from the jack end).

Figure 9.5 Close-up view showing details of the major axis moment arm arrangement.

9.3.1 *Test Results*

The experimental research programme undertaken conveniently divides itself into two main groups according to slenderness ratios. The columns may be grouped as follows:

GROUP A Short columns – slenderness ratio $l/r_y \approx 45$
GROUP B Long columns – slenderness ratio $l/r_y \approx 60$

9.3.2 *Mode of Failure*

It was interesting to note that the modes of failure of the columns in Group A and B were different. All columns in Group A performed according to short column concepts, i.e. the failure was in the plane of the applied moment.

The failure of all columns under Group B followed a standard pattern. While no noticeable out-of-plane deformations were recorded in the early stages of loading, the final collapse was due to a combination of lateral bending and twisting. Only representative experimental plots are presented in Section 9.3.3.

9.3.3 Experimental Plots

Column A2

The experimental moment versus end rotation plot for column A2 is shown in Figure 9.6a. Although this was column was of the lighter rolling type (203/203 @46 kg) and therefore torsionally less stiff compared to the heavier rolling section (203/203 @71 kg) but being a short column it performed in-plane. Despite the experiment being conducted till a large joint rotation was reached, the failure was due to in-plane loading. It reflects broadly the moment-rotation characteristics at this high axial load ($P/P_T = 0.81$, Table 9.1). No noticeable out of plane deflections were recorded (Figure 9.6b). All the columns in Group A failed in the plane of the moment.

Column B1

This column was of the heavier rolling type (203/203 @71 kg) but performed differently. This column was tested at $P/P_T = 0.80$ (Table 9.1). The column was initially stable about the minor axis under axial load only and continued to remain so during the early stages of application of the major axis moment. Then the out-of-plane deflections progressively increased as the major axis moments were increased and eventually equalled the magnitude of the major axis deflections. The twisting deformations recorded also increased progressively. The final collapse was due to a combination of lateral bending and twisting. Figures 9.7a, 9.7b and 9.7c show the end moment vs. end rotation, end moment vs. major and minor axis deflections and end moment vs. twist values for column B1. All the columns in Group B ($l/r_y \approx 60$) failed out-of plane. Biaxial deterioration of stability had evidently taken place.

The conclusions drawn from the results of these tests led to design concepts.

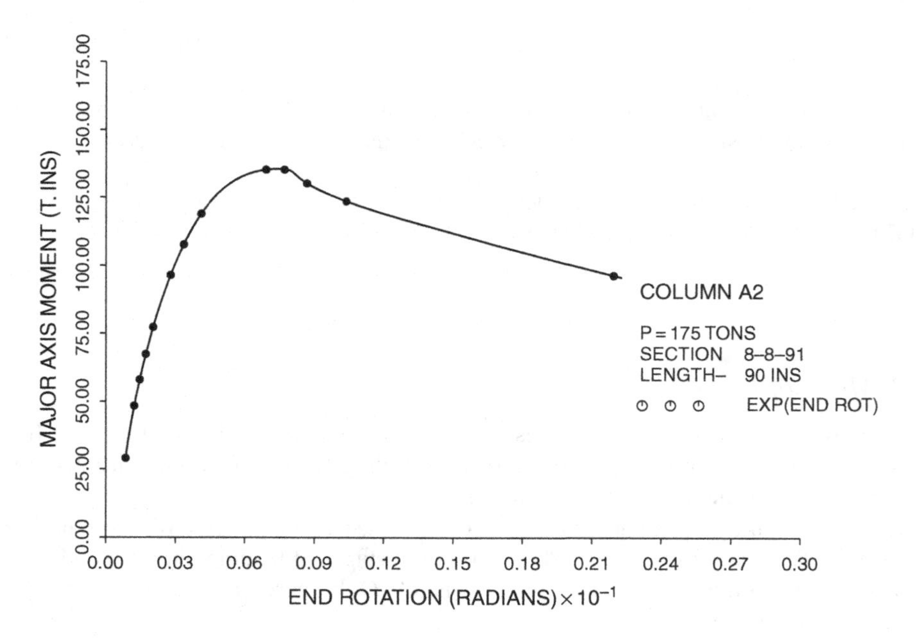

Figure 9.6a Experimental moment-end rotation plot.

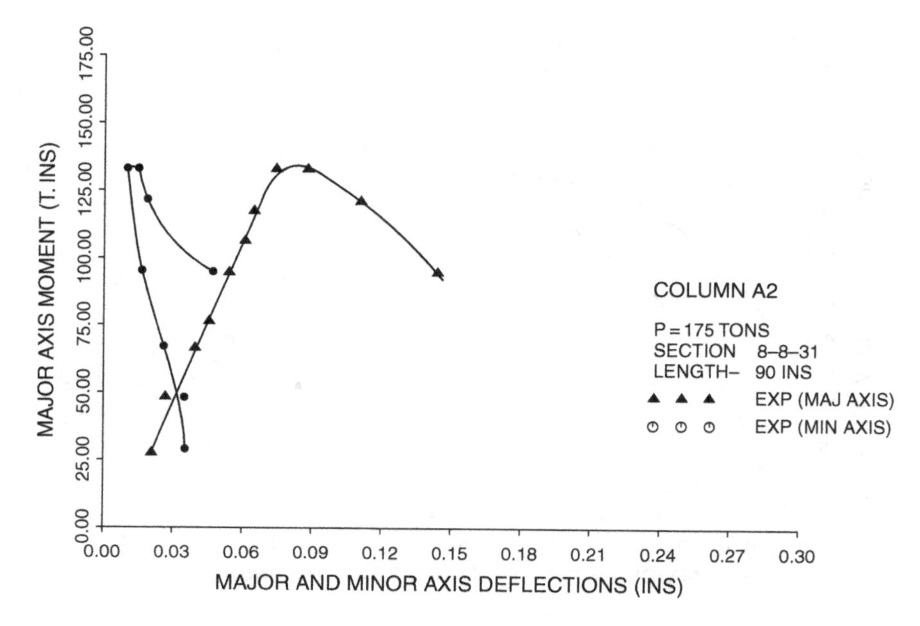

Figure 9.6b Experimental moment-mid column height deflection plot.

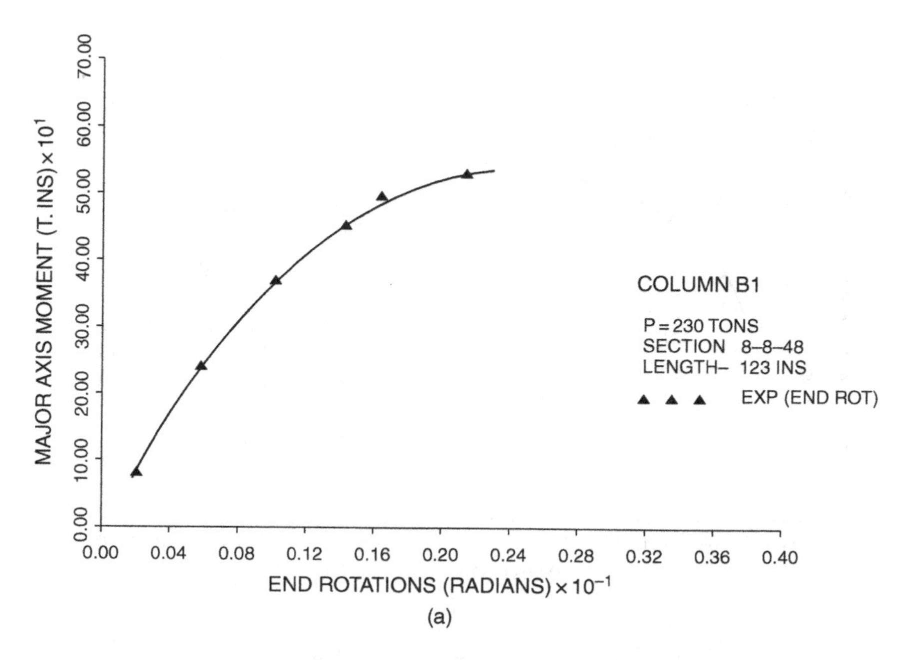

Figure 9.7a Experimental moment-end rotation plot.

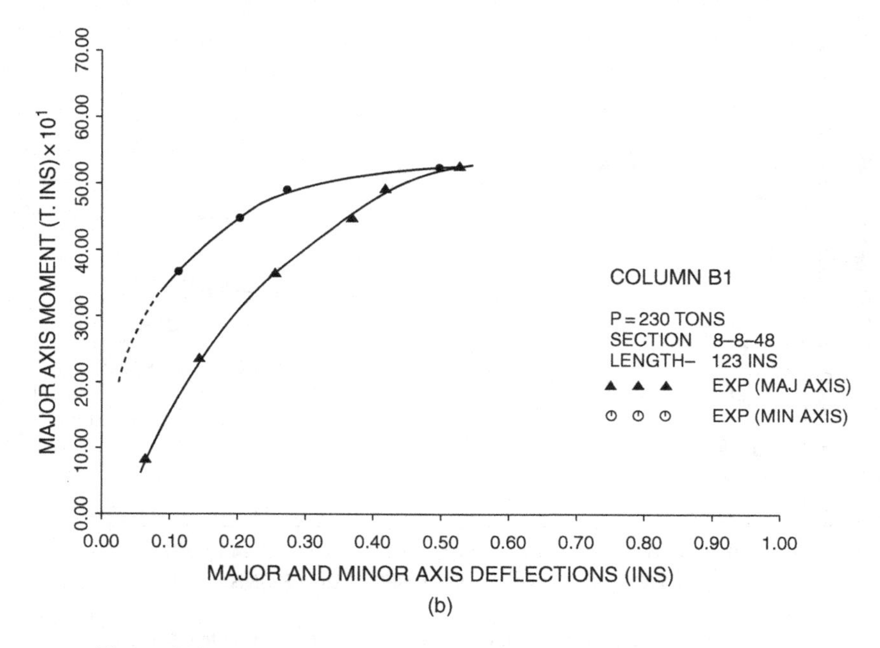

Figure 9.7b Experimental moment-mid column height deflection plot.

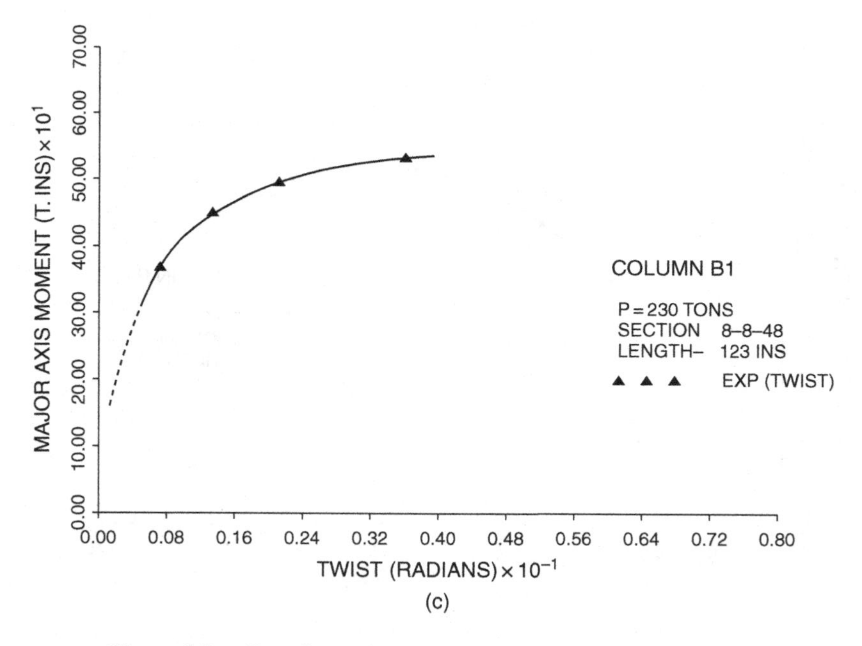

Figure 9.7c Experimental moment-mid column height twist plot.

Procedures for incremental analysis of inelastic beam-columns with particular reference to H-sections have been reported by Sen (1976) and have also been well covered in the literature (Chen & Atsuta, 1977).

Torsion affects the in-plane ultimate load carrying capacity of H-Columns. We have also seen through laboratory tests that at around slenderness ratio ($l/r_y \approx 59$) the columns at high axial loads failed out-of-plane. The ultimate in plane capacity of H-Columns including and excluding the effect of torsion have been investigated and reported by Virdi and Sen (1981).

9.4 Concepts and Issues: Frames Subjected to Seismic Loading

The experimental plots exemplify out-of-plane failure. In design, this type of behaviour must be avoided. The collapse of a column in a multi-storey frame in this mode could be catastrophic. Column design curves are recommended because there the slenderness ratio, at the desired axial load, is restricted to provide only in plane behaviour (Wood, 1958, 1974). There is also the requirement for good seismic performance which means that the moment–end rotation plot be 'flat topped' (Horne, 1964a, b), i.e. avoid local buckling and allow for sustained ductile behaviour.

The design aspects and behaviour of a multi-storey frame to resist seismic forces during a severe earthquake have been extensively discussed in the literature. The results of the research findings and laboratory tests have helped to formulate important sections of the code. The basic intent of the codes has been to provide provisions such that structures have the ability to withstand intense ground shaking without collapse, but with some significant structural damage. A structure is considered to exhibit ductile behaviour if it is capable of undergoing large inelastic deformations without significant degradation in strength and without development of instability and collapse. Compare the moment rotation curves for a beam column that exhibited in-plane behaviour introduced earlier in Section 9.2 and Figure 9.3a. This was referred to as 'the moment rotation curve being flat-topped' (Horne, 1964). The dual design philosophy referred to earlier (Chapter 7, Section 7.1) also aims to achieve the same goals.

The codes of practice have recognized that this form of behaviour has introduced new factors for design forces, which depend on the particular structural system involved. The design forces are related to the amount of ductility the structure is estimated to possess. Generally, structural systems with more ductility are designed for lower forces than less ductile systems. In this way, ductile systems are called upon to resist demands in terms of variability of seismic loading whose maximum values are significantly greater than the elastic strength limit of the individual members of the system. The caveat is that the system of connections, joining the beams to the columns should also perform in the same ductile manner as presumed above.

In highly seismic zones, the design intent is to allow beams and panel zones to have weaker member strength compared to the columns. This is to ensure inelastic deformations in the beams and panel zones first thus protecting the columns from plastic hinging. This is sometimes referred to as the 'strong column concept' and is currently implicit in the codes (UBC, 1997).

9.5 Proceeding with Dynamic Analysis (MDOF systems)

How do we proceed with dynamic analysis of multi-storey frames (MDOF systems) subjected to seismic loading? The static inelastic performance of beam columns at high axial loads was highlighted in Section 9.2. Inelastic analysis including the effects of lateral torsional buckling is complex and an approximate solution method could only be resorted to (Sen, 1976).

In the dynamic case under seismic loading the response determination is even more complex. An accurate determination of the inelastic response of a frame that includes all aspects of member lateral torsional buckling and structure '$P-\Delta$' effects would only be possible through a finite element analysis package which includes large displacement and the desired plasticity models. The modelling should be through proper elements and must be able to reflect ductile behaviour. Section sizes should be chosen so as to avoid local and flexural torsional buckling effects. If, however, lateral torsional buckling of beam elements are to be included in the analysis then the modelling itself becomes enormously complicated. Special beam elements to cater for torsion and warping effects are usually not suitable.

Hence, it is reasonable to assume that, if local and lateral torsional buckling effects *are avoided* then the dynamic formulation may proceed based on standard methods for multi degree freedom systems. The elastic continuum is divided into elements interconnected only at a finite number of nodes. The finite element (FE) method is best suited to enhance this spatial discretization. The equations of motion are established by assembling the element matrices (Bathe and Wilson, 1976).

The equation to be solved for the multi-degree freedom system (refer to Chapter 3) takes the form

$$[M]\{\ddot{r}\} + [C]\{\dot{r}\} + [K]\{r\} = \{R\} \tag{9.1}$$

The vector $\{r\}$, which is a function of time, contains the displacements of all unconstrained degrees of freedom at the nodes. The dot (•) on the top denotes derivative with respect to time. The matrices $[M], [C]$ and $[K]$ represent the mass, damping and static-stiffness matrix, and are constant for a linear elastic system. The solution method is incremental in the time domain. For non linear systems the stiffness damping matrices are upgraded at each time step (usually very small). The vector $\{R\}$ denotes the prescribed loads which are a function of time. Viscous damping is assumed in Equation (9.1). The radiation of energy due to propagation of waves through the soil medium, away from the region of interest (see Chapter 10) belongs to this type of damping.

A brief discussion on '$P-\Delta$' effects and lateral torsional buckling follows next.

'$P-\Delta$' Effect

'$P-\Delta$' effect is caused by gravity loads acting on the displaced configuration of the structure, thus accentuating the out-of-plane lateral torsional buckling tendencies.

It may be noted that the '$P-\Delta$' effect in a multi-storey frame depends on the properties of individual storeys. '$P-\Delta$' effects reduce the effective resistance of each storey by an amount approximately equal to $P_i\delta_i/h_i$, where P_i, δ_i and h_i are the vertical force, inter-storey deflection and height of storey 'i' respectively. It follows from this that large '$P-\Delta$' effects, may lead to effective negative storey stiffness due to large vertical forces (P_i) or large storey drifts (δ_i) or both. The level of deformation at which negative stiffness occurs depends strongly on local and lateral torsional buckling (FEMA 355C, 2000).

Assessing the severity of '$P-\Delta$' effect is a complex problem. It has been flagged as a potential seismic hazard that needs to be considered explicitly and more realistically than is done at present (FEMA 355C, 2000). The severity depends not only on the structural characteristics but also on the intensity of the seismic ground motions and their frequency characteristics. No simple procedures seem to be available at present to assess the collapse hazard due to '$P-\Delta$' effect. The severity of '$P-\Delta$' effect has been investigated by many authors including Gupta and Krawinkler (2000).

9.5.1 Lateral Torsional Buckling

Referring to our earlier discussion, beam columns are members subjected to a combination of axial load and bending moment. Columns bent about the major axis (in the plane of the web) will have a tendency to out-of-plane failure by a combination of bending and twisting depending on the axial load and slenderness ratio (refer to results of the reported laboratory experiments – Section 9.2). A very good summary of all relevant research on elastic and inelastic torsion is given by Trahair (1993). Torsional effects have also been investigated by Bertero (1998), De La Llera and Chopra (1995, 1996), and Hahn and Liu (1994).

We are faced with the situation where lateral torsional buckling in beam columns can inhibit in-plane behaviour. The out-of-plane stability of all members should be ensured under all loading conditions and not just for the critical load case. The minimum plastic resistance of frame members should be used in the finite element (FE) design. Where differential settlement of foundations is a design criterion, this should be taken into account in checking the out-of-plane stability (BS 5950: 2001).

Interaction equations for beam columns failing by lateral torsional buckling in several representative design specifications have been reported by Galambos (1998). The methods for determining the quantities in the interaction equations differ in all the specifications. Hence, comparisons can only be made for a specific member under a specified loading condition. For a general flavour of the differences, such a comparison for the W8x40 section is reproduced in Figure 9.8 (Galambos, 1998). It may be observed that substantial ductile column behaviour is only appropriate at relatively low values of P_u/P_y. The curves are also somewhat divergent when the axial force is low.

9.5.2 Column Strength Curves

In view of the fact that the column is the key member in resisting gravity loads in buildings, research efforts have been extensive in this field since the 1940s. An excellent account of research efforts to date is reported by Galambos, (1998). In practice, columns come with initial imperfections and certain patterns of residual stresses are always built into them during the manufacturing process. The SSRC column design strength curves, which account for all the effects that influence significantly their load carrying capacity, may be found in the above reference.

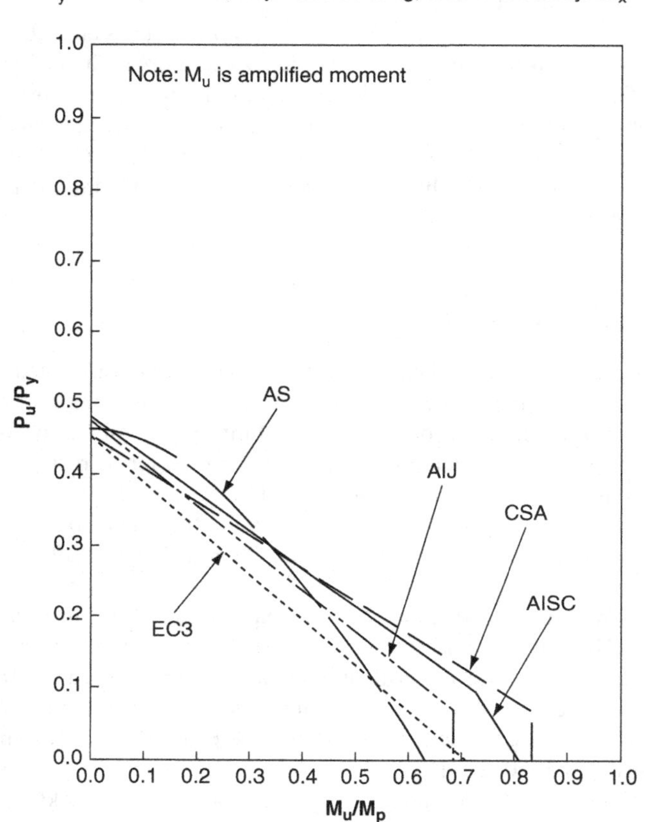

Figure 9.8 Lateral Torsional Buckling Interaction Curve.
Note: AISC, 1993 – American Institute of Steel Construction
CSA, 1994 – Canadian Standards Association
AS, 1990 – Australian Standard AS 4100-90
EC3, 1993 – Eurocode 3
AIJ, 1990 – Architectural Institute of Japan
[M_u – amplified moment; M_p – fully plastic moment capacity; P_u – ultimate axial load;
$P_y = A_s \cdot F_y$; A_s – area of steel]
Reproduced from Galambos (1998) by permission of John Wiley & Sons Inc.

9.5.3 Dynamic Analysis

In proceeding with dynamic analysis, the next step is to choose an appropriate design earthquake. This is discussed in Chapter 7. It is then necessary to establish/select a hysteresis model during seismic loading. The level of damping is also important and needs careful consideration.

This is where results from the deaggregation of the uniform hazard spectrum computations (see Chapter 6) may be used. A step-by-step direct integration is carried out next. The procedure for step-by-step analysis is outlined in Chapter 2. Efficient finite element packages with non-linear capability are available (for example ABAQUS).

The important results of the computation are the maximum values of base shear, overturning moment, lateral deflection, storey drift and storey ductility. These are then checked against the prescribed design criteria laid out in the codes. If the values are not acceptable then the design needs to be modified and checked again.

9.6 Behaviour of Steel Members under Cyclic Loading

Buildings are required to withstand cyclic loading. While the design intent under operating conditions has focused on members remaining elastic, the same is not practical under a severe earthquake, often referred to as the *ductility level earthquake*. As mentioned earlier in Chapter 7, socio-economic constraints do not justify an elastic design when the probability of such a severe earthquake is so low. Failure mode control under a 'ductility level earthquake', is the current focus of research activities. An important underlying concept associated with this method of failure mode control is the extent of energy absorption during cyclic loading when beams or columns are no longer elastic.

The behaviour of a steel member under cyclic bending and with a constant axial force (a beam column) is shown in Figure 9.9.

The behaviour of steel beams is similar to that of steel columns except that the axial load is not present in this case.

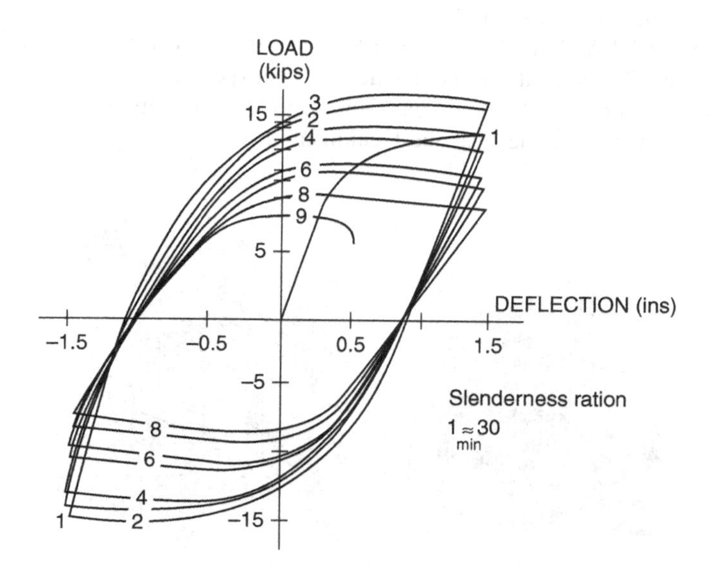

Figure 9.9 Hysteresis loops for a steel member under cyclic bending and a constant axial force of $N = 0.3N_y$ (after Vann *et al.*, 1973) [$N_y = A_s f_y$ where A_s is the sectional area of the member and f_y is the specified yield stress].

The failure of columns in tall buildings would be catastrophic; hence these buildings are designed to introduce plasticity in the beams first. However, some column hysteretic behaviour is most likely in the event of an unexpectedly strong earthquake. Lateral torsional buckling and lateral instability are major factors that prevent full plastic moment capacity of the columns and beams from being reached.

Energy absorption is a function of the area under the hysteresis loop. The material must be ductile. The degree to which ductility should be enhanced (debatable in many cases) depends on a variety of factors and also on the engineer concerned.

It should be borne in mind that the primary objectives before enhanced ductility is introduced into the design are to ensure that:

1. Plastic hinges develop in beams first; modern day finite element (FE) packages are capable of non-linear analysis and can incorporate material behaviour to reflect loading and unloading (hysteretic behaviour shown in Figure 9.9).
2. Prevention of lateral of lateral torsional buckling/instability in columns/beams (as discussed previously).
3. Prevention of local buckling in members.

9.6.1 FE Analysis

Analysis of a high rise building due to a strong motion earthquake is a complex task, especially when cyclic stress reversals are encountered. To appreciate the size and complexity of the analysis that may be encountered in practice, an example of a high rise building and its FE model is shown in Figure 9.10.

This is one of the buildings where analysis was undertaken at the Lawrence Livermore Laboratories in the USA, following the Northridge (1994) earthquake, when many of the buildings collapsed. The severity of the damage was unexpected. It may be pointed out that, for reasons mentioned earlier, in buildings of this kind, columns ought to be prevented from out-of-plane lateral torsional buckling and local instability.

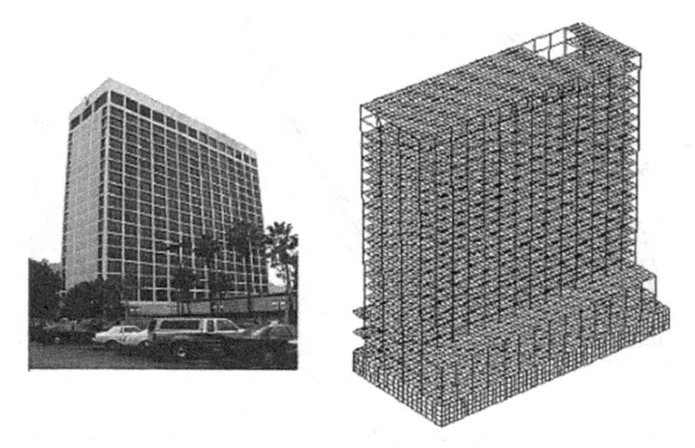

Figure 9.10 An example of a high rise building and its FE model. Reproduced by permission of the University of California Lawrence Livermore National Laboratory and the US Department of Energy.

The pay-off in terms of benefits of an accurate non-linear analysis is high. Numerical simulations of non-linear behaviour of soil and structures on this scale can provide valuable insight into how structures respond during a strong motion earthquake.

Some of the important findings include:

1. Simpler buildings (mass/stiffness distribution) perform better during earthquakes. According to Dowrick (1987), firstly it is easier to understand the overall earthquake behaviour of a simpler building rather than a more complex one. Secondly, it is easier to translate into drawings and develop simpler structural details than more complicated ones.
2. Symmetry and regularity in plan and elevation are to be favoured
3. Lack of symmetry in plan (mass distribution / stiffness) leads to torsional effects which are difficult to assess properly and could be destructive. Non-linear analyses are often intractable.

The Lawrence Livermore Laboratories web site (www.llnl.gov 18/07/08) has several examples of buildings with *asymmetrical* layout and the subsequent damage they encountered during the Northridge (1993) earthquake. Current research on seismology is also published.

9.6.2 A Note on Connections

Welded steel moment frames (WSMF) have been the preferred connection system within the engineering community for resisting seismic loading on the presumption that the connections fulfil their intended functions at these high demands. Extensive yielding and plastic deformation without of loss of strength are expected to occur. As we may anticipate, the intended plastic deformations consist of plastic rotations developing within the beams and at their connection to the columns. As the theory of inelastic deformation suggests, this should result in the benign dissipation of the earthquake energy being delivered to the building. This has been the primary design goal in dealing with large to moderate earthquakes. Damage is restricted to moderate yielding and localized buckling of the steel members without premature brittle fractures. Brittle fracture at the connections took place during the aftermath of the Northridge earthquake in 1994. This was unanticipated and led to significant economic losses in terms of direct and indirect costs. Hence, one of the primary duties at the design stage is to preclude brittle fractures at the joints. The behaviour of welded steel moment frame structures under seismic loading is discussed in an extensive state of the art report (FEMA 355C, 2000). Further research on connections has been reported by Uang *et al.* (2000).

9.6.3 A Note on the Factors Affecting the Strength of Columns

During an earthquake the column member is subjected to cyclic loading. The behaviour of the steel column is complicated due to a variety of effects. In the inelastic region, the deterioration in performance of steel structures comes primarily from local instabilities, i.e. local and lateral torsional buckling (FEMA 355C). Other effects discussed in FEMA 355C include 'strength deterioration', 'strength capping' and 'stiffness degradation'. All these effects would lead to a more rapid decay of the member's resistance. It is interesting to note that 'negative element stiffness' has been observed in all tests carried out under the SAC program (a partnership between SEAOC, ATC and California Universities for Research in Earthquake Engineering) and other steel assembly test programs. It may add considerably to the '$P - \Delta$' effect (FEMA 355C, 2000).

Another important effect during cyclic loading is low cycle fatigue. It is not unusual for failure to take place due to low cycle fatigue.

Current research efforts are under way to define the parameters more accurately leading to a safe design. Performance-based earthquake engineering is discussed in Chapter 12.

9.7 Energy Dissipating Devices

9.7.1 Introduction

The governing target of the dual design philosophy (see Chapter 8), is to design buildings that will allow some permanent deformation of the structure to take place during a strong motion earthquake without causing a total collapse, thereby preventing both loss of life and economic disruption. Permanent deformation of the structure that takes place during a tremor is accompanied by a dissipation of energy due to the hysteretic behaviour of the material (see Section 9.6 for hysteretic behaviour of steel).

The objective of adding energy damping devices to new and existing construction is to dissipate much of the earthquake-induced energy in these disposable elements in order to minimize the inelastic yielding in load-carrying elements. This implies that structural drift has to be controlled at the same time. Hence the energy dissipation must be achieved within the displacement limits imposed.

Being disposable, the devices may be replaced as necessary after an earthquake. Developed primarily since the 1970s, many such devices are now available.

Research carried out at the Earthquake Engineering Research Center (EERC), California (Aiken and Kelly, 1990) investigated the behaviour of two types of damper: a viscoelastic shear damper and a friction damper. The viscoelastic shear damper was designed using an energy approach. The friction device selected had almost perfectly rectangular hysteretic behaviour for which an iterative non-linear analysis was adopted. A nine storey moment-resisting frame, mounted on a shake (vibrating) table, represented the basic structure of the study and was tested with both types of energy absorbers. Extensive earthquake simulator testing was carried out. Both devices performed well and it was concluded, were capable of withstanding strong motion earthquakes.

Studies on friction devices have also been reported by Anagnostides *et al.* (1989).

Energy dissipation devices have also been used in retrofitting existing buildings. Islam (2001) reported the installation of fluid viscous dampers in an existing building. The building was located very close to the San Andreas fault in the earthquake-prone state of California. Subsequent analytical work on the retrofitted structure showed significant improvement in controlling storey drift.

9.7.2 Current Practice

In the last two decades a number of innovative techniques have been introduced to enable buildings and structures to better withstand strong motion earthquakes. Some of these recent techniques, where the general aim has been to *'channel'* the seismically induced forces to specially installed devices, are briefly covered.

An overview of the current state of innovative seismic design is presented in Figure 9.11.

Earthquake resistance techniques may be classified into two broad categories: 'passive' control and 'active' control.

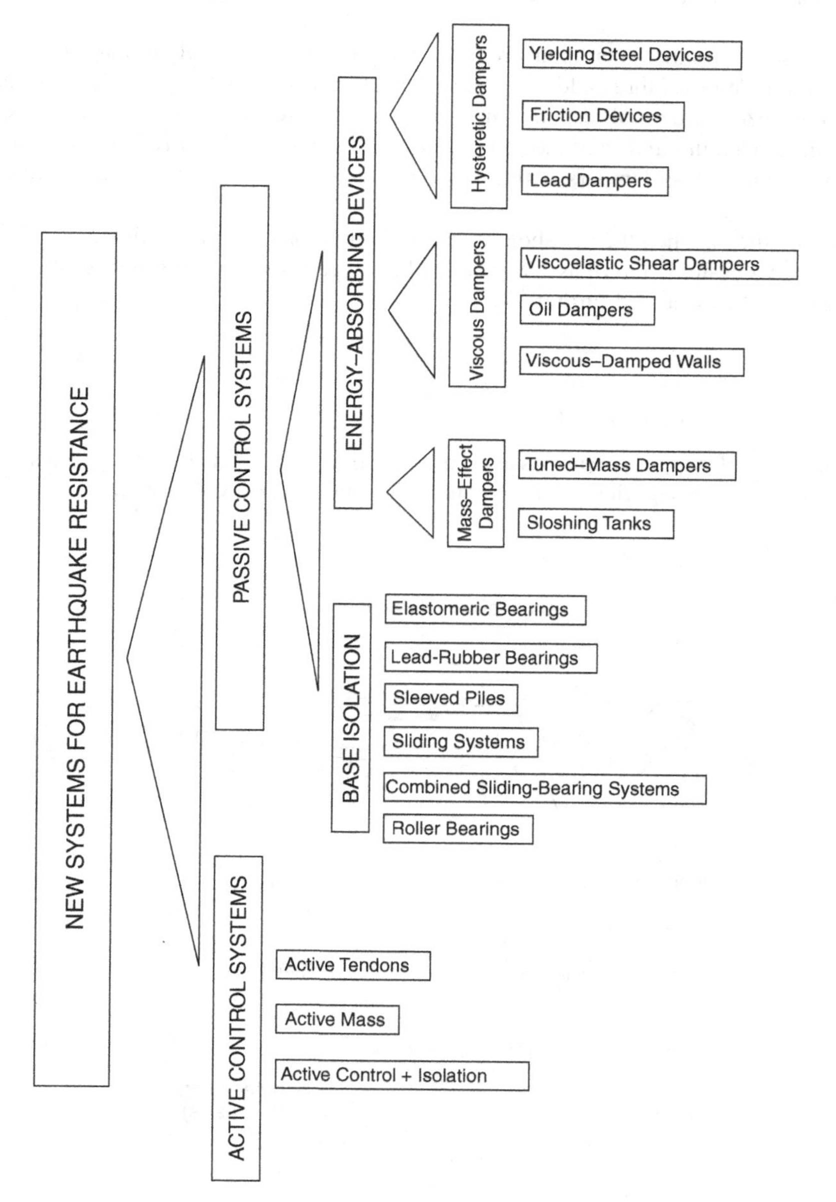

Figure 9.11 New systems being introduced for improved seismic resistance. Reproduced from Aiken and Kelly (1990) by permission of the Earthquake Engineering Research Center (EERC).

Passive systems require no active intervention or energy source. Seismic resistance is achieved by isolating the entire building from the majority of ground motion, or by damping the resonant oscillations of the frame.

Energy absorbing devices form part of the passive control system and are often referred to as supplemental damping devices. Some of the hardware in use is shown in the next section.

Active systems, on the other hand, are more complex. They typically monitor the incoming ground motion. When a threshold is reached the active system is triggered. The system then seeks to mobilize a structural response in such a manner as to counteract and reduce excessive displacements (Iemura and Pradono, 2003). Some of the methods currently under investigation include: tuned mass dampers, active tensioning of tendons, and reliance on liquid sloshing (Lou *et al.*, 1994).

In most methods, when the threshold is reached, monitoring is in real time.

Fewer active systems than passive systems have been installed in practice, therefore the former will not be discussed further.

9.7.3 Damping Devices in Use

Different types of supplemental passive damping devices are in use today. These may be classified into three categories: hysteretic, metallic, and viscous and viscoelastic.

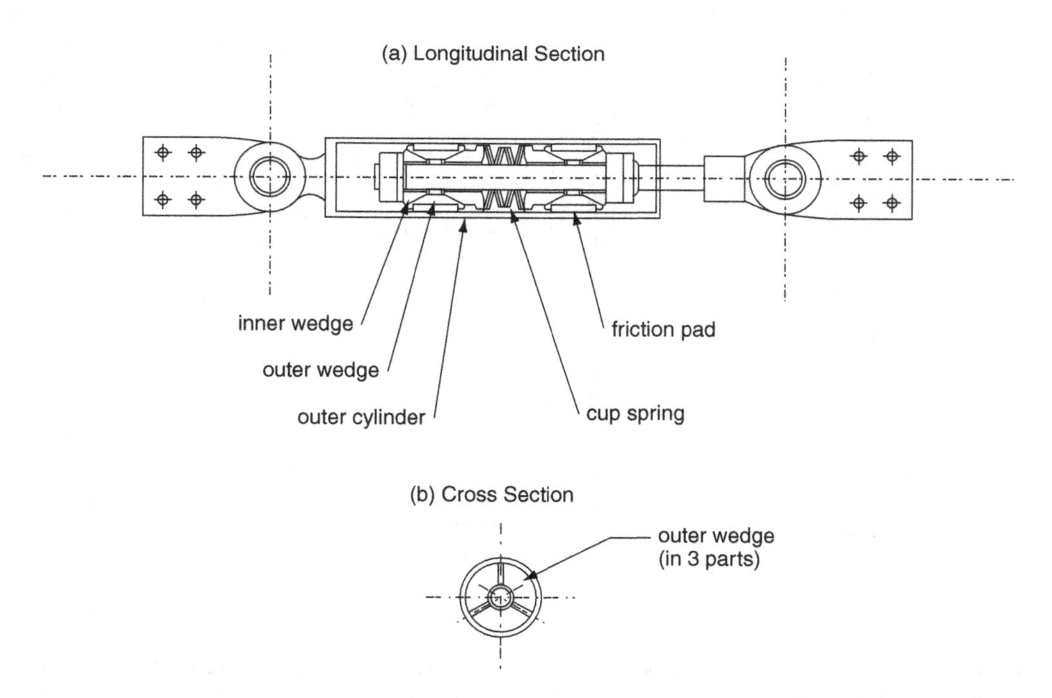

Figure 9.12 Cross section of a Sumitomo friction damper. (a) EERC test frame with friction dampers. Reproduced from Aiken and Kelly (1990) by permission of the Earthquake Engineering Research Center (EERC).

(a) Hysteretic

Hysteretic damping systems include devices based on friction and devices based on yielding.

An EERC study (Aiken and Kelly, 1990) investigated a friction device developed by Sumitomo Metal Industries in Japan. A cross-section of the Sumitomo friction damper is shown in Figure 9.12. These high performance friction devices were originally developed for shock absorption application on railways. The damper consists of a series of wedges under a normal compressive load by a spring. The wedges acting against each other during seismically induced motion produce frictional damping. The cross-section of a Sumitomo friction damper is shown in the EERC report and reproduced here. The dampers exhibit bilinear hysteretic behaviour, which is not difficult to simulate analytically with current software.

The test frame in the EERC study is shown in Figure 9.13a.

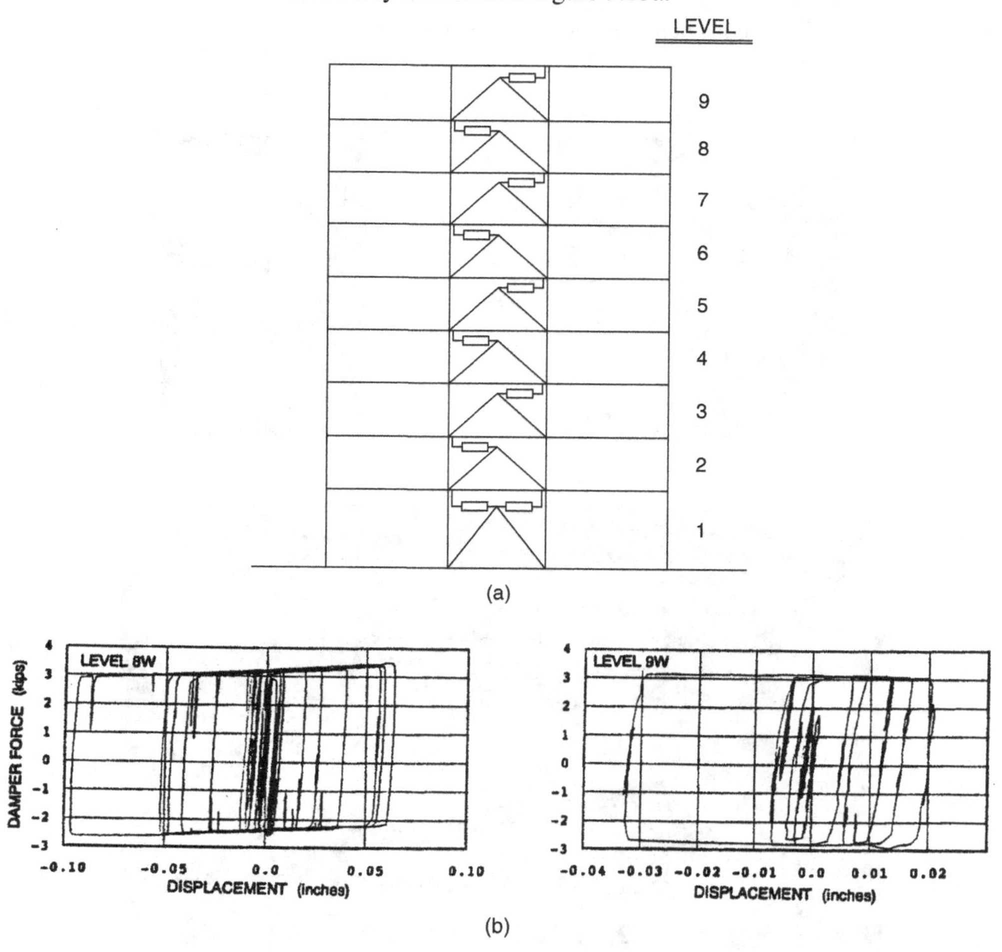

Figure 9.13 Hysteretic loop obtained for the friction damper (FD) device in the top two storeys in the EERC study (ec-400 test, PGA:0.753 g).
PGA: peak ground (table) acceleration.
Reproduced from Aiken and Kelly (1990) by permission of the Earthquake Engineering Research Center (EERC).

(b) Metallic Dampers

Metallic dampers dissipate energy through hysteretic behaviour of metals subjected to plastic deformation. A wide variety of such metallic yielding dampers has been tested in recent times. One of these in use, the triangular added damping and stiffness (TADAS) damper uses triangular steel plates as shown in Figure 9.14. This device increases both stiffness and damping. The hysteresis loop displays stable behaviour as shown. The device, installed in a building frame, may also be seen.

(c) Viscous and Viscoelastic

In these devices, the viscous damping element may be fluid or solid.

Figure 9.14 (a) TADAS device (b) device under load (c) force-displacement diagram of a device and (d) TADAS device in the Core Pacific City Building, Taipei, Taiwan. Reproduced from Chopra (2001). Courtesy K.C. Tsai.

Viscoelastic Damper

Use of copolymers to dissipate energy in shear deformations has been developed, mainly in the USA. In this device, layers of copolymers are bonded with steel plates as seen in Figure 9.15. The hysteresis loop is almost elliptical, as may be seen and stable. Energy is dissipated by relative motion between the steel plates on the outside and the centre plate of the device. The device in a diagonal bracing configuration in a test frame is shown in the figure.

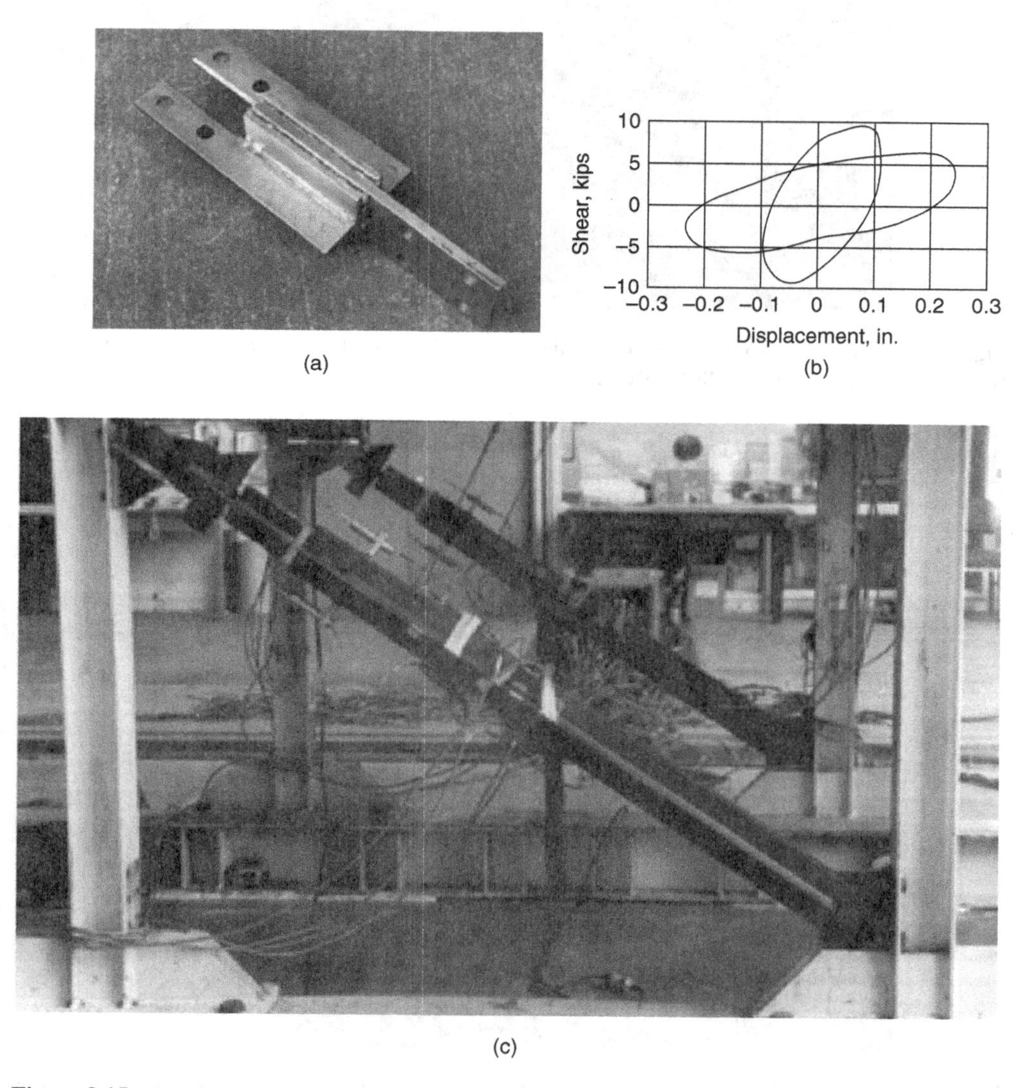

Figure 9.15 (a) Viscoelastic shear damper; (b) force-displacement diagram of a damper; and (c) diagonal bracing configuration with viscoelastic damper. Reproduced from Chopra (2001) by permission of I.D. Aiken.

Figure 9.16 A close-up view of a viscous damping device installed as part of a braced structure. Reproduced by permission of Taylor Devices Europe.

Figure 9.17 A viscous damping device in a building after completion. Reproduced by permission of Taylor Devices Europe.

Figure 9.18 A viscous damping device being hoisted at site. Reproduced by permission of Taylor Devices Europe.

Viscous Dampers

Figure 9.16 and 9.17 shows viscous fluid dampers of a conventional piston type that work on the same principle as the well-known automotive dampers. These are widely used, particularly in the USA and their scale can be judged by their installation in buildings. Figure 9.18 shows a damper being hoisted at site.

9.7.4 Analytical Guidelines Currently Available

FEMA 273 entitled *Guidelines for the Seismic Rehabilitation of Buildings* published in 1997 provided guidelines on the use of supplemental devices for the first time. Since then, FEMA 273 has been updated and republished as FEMA 356 in 2000 entitled *Prestandard and Commentary for Seismic Rehabilitation of Buildings*.

9.8 References

Aiken, I.D. and Kelly, J.M. (1990) *Earthquake Simulator Testing and Analytical Studies of Two Energy-Absorbing Systems for Multi-Storey Structures*. EERC, Rept No. UCB/EERC-90/03, Earthquake Engineering Research Center, California.

Anagnostides, G., Hargreaves, A.C. and Wyatt, T.A. (1989) 'Development and applications of energy absorption devices based on friction'. *J. Constr. Steel Res.* **13**: 317–36.

Bathe, K.J., and Wilson, E.L. (1976) *Numerical Methods in Finite Element Analysis*. Prentice Hall, New Jersey.

Bertero, R.D. (1995) 'Inelastic torsion for preliminary seismic design', *J. Struct. Eng.*, ASCE, **121**(8): 1183–89.

Bozorgnia, Y. and Bertero V.V. (2004) *Earthquake Engineering: From Engineering Seismology to Performance-Based Engineering*. CRC Press, Taylor & Francis Group, Boca Raton, USA.

British Standards Institution. (2001) BS 5950: 2001: Structural use of steelwork in building – Part 1: Code of practice for design – Rolled and welded sections. Milton Keynes: BSI.

Chen, W-F. and Atsuta, T. (1977) *Theory of Beam Columns*. McGraw Hill Inc., New York.

Chen, W-F. and Lui, E.M. (2006) *Earthquake Engineering for Structural Design*. CRC Press, Taylor & Francis Group, Boca Raton, USA.

Chopra, A.K. (2001) *Dynamics of Structures: Theory and Applications to Earthquake Engineering*. Prentice Hall, New Jersey.

De La Llera, J.C. and Chopra, A.K., (1995) 'Understanding the inelastic seismic behaviour of symmetric-plan buildings', *Earthq. Eng. Struct. Dyn.* **24**(4): 549–572.

De La Llera, J.C. and Chopra, A.K. (1996) 'Inelastic behaviour of asymmetric multistory buildings'. *J. Struct. Eng.*, ASCE, **122**(6): 597–606.

Dowrick, D.J. (1987) *Earthquake Resistant Design: for Engineers and Architects*. John Wiley & Sons, Chichester, UK.

Federal Emergency Management Agency (FEMA) (2000) *State of the Art Report on Systems Performance of Steel Moment Frames Subject to Earthquake Ground Shaking*. FEMA 355C, prepared the by SAC Joint Venture for the Federal Emergency Management Agency, Washington, DC.

Galambos, T.V. (ed.) (1998) *Guide to Stability Design Criteria for Metal Structures*. John Wiley & Sons, Inc. New York.

Galambos, T.V. (1968). *Structural Members and Frames*. Prentice Hall, New Jersey.

Gupta, A. and Krawinkler, H. (2000) 'Dynamic p-delta effects for flexible inelastic steel structures'. *J. Struct. Eng.*, ASCE, **126**(1): 145–154.

Hahn, G.D. and Liu, X. (1994) 'Torsional response of unsymmetric buildings to incoherent ground motions', *J. Struct. Eng.*, ASCE, **120**(4): 1158–81.

Horne, M.R. (1964a) 'Safe loads on i-section columns in structures designed by plastic theory', *Proc. Inst. Civ. Eng.* **29**, 137–50.

Horne, M.R. (1964b) *The Plastic Design of Columns*. Publication No 23. British Constructional Steelwork Association, London,.

Iemura, H. and Pradono, M.H. (2003) 'Structural control', in *Earthquake Engineering Handbook*, Chen and Scawthorne (eds), CRC Press, New York, Chapter 19.

International Code Council. (1997) *Uniform Building Code (UBC) 1997*. Washington, DC. ICC.

Islam, S. (2001) 'Seismic retrofit of an existing building using energy-dissipation devices'. *Proceedings of the SEAOC (Structural Engineers Association of California) 7th Annual Convention*, San Diego, 287–301.

Lawrence Livermore National Laboratory website (http://www.llnl.gov 18/07/08).

Lou, J.Y.K., Lutes, L.D., and Li, J.J. (1994) 'Active tuned liquid damper for structural control', *Proceedings of the 1st World Conference on Structural Control: Intl Assoc. for Structural Control*, Los Angeles, CA.

Sen, T.K. (1976) *'Inelastic H-column performance at high axial loads'*. Thesis presented to the University of London, England, in partial fulfilment of the requirements of the degree of Doctor of Philosophy.

Taylor Devices Europe. http://www.taylordevices.eu 01/08/08.

Trahair, N.S. (1993) *Flexural-torsional Buckling of Structures*. E&FN Spon, London.

Tsai, K.C. (a) TADAS device; (b) device underload; (c) force-displacement diagram of a device; (d) TADAS devices in the Core pacific city building under construction, Tapei, Taiwan in Chopra (2001).

Uang, C.C., Yu, Q.S., Noel, S. and Gross, J.L. (2000) 'Cyclic testing of steel moment connections rehabilitated with RBS or welded haunch'. *J. Struct. Eng.*, ASCE, **126**(1): 57–68.

Vann, W.P., Thompson, L.E., Whalley, L.E. and Ozier, L.D. (1973) 'Cyclic behaviour of rolled steel members'. *Proceedings of the 5th World Conference on Earthquake Engineering*. Rome, Vol. 1, 1187–93.

Virdi, K.S. and Sen, T.K. (1981) 'Torsion and computed ultimate loads of H-columns'. *J. Struct. Eng.*, ASCE, **127**, No. ST2, 413–26.

Wood, R.H., (1958) 'The stability of tall buildings'. *Proc. Inst. Civ. Eng.* **29**, 69–192.

Wood, R.H. (1974) A *New Approach to Column Design*. Her Majesty's Stationery Office, London.

10

Soil-Structure Interaction Issues

10.1 Introduction

The characteristics of soil surrounding a structure can modify the dynamic response of that structure when it is subjected to ground motion resulting from an earthquake. This is known as soil-structure interaction (SSI). A vast amount of literature compiled over the latter half of the twentieth century has substantiated this fact and helped to shed some light on the complexities of this interaction. This chapter is an overview of the topic wherein some of its most important aspects will be discussed.

10.2 Definition of the Problem

The response of a structure subjected to earthquake forces depends on several factors. The most important of these are (a) the characteristics of the ground motion, (b) the nature of the surrounding soil, and (c) the physical properties of the structure itself. Initial seismological investigations during the 1940s and '50s, which led to the development of response spectra, were based on the assumption that the structure was grounded on firm soil (Housner, 1959). Little was known about soil-structure interaction at that time.

Earthquake body waves are triggered off at a depth below the surface known as the 'hypocentre'. We can visualize that the presence of a large structure can modify the behaviour of ground motion. In this context, the term 'free field motion' is usually used to describe the unimpeded motions of a soil bed subjected to dynamic excitation with no structure present. If the soil is stiff the foundation present will follow the incident wave with little or no SSI. However, this is not true for structures on soft soils.

The foundation motion differs from that in the free-field case due to coupling of the soil and the structure during an earthquake. In other words soil structure interaction takes place. The interaction is the result of (a) scattering of waves due to the presence of the structure, and (b) transfer/radiation of energy from the structure back into the soil produced by the vibration of the structure itself. The outcome of this interaction is that the surface motion characteristics (acceleration, velocity etc.) are significantly different from those in the case of the free-field

Fundamentals of Seismic Loading on Structures Tapan Sen

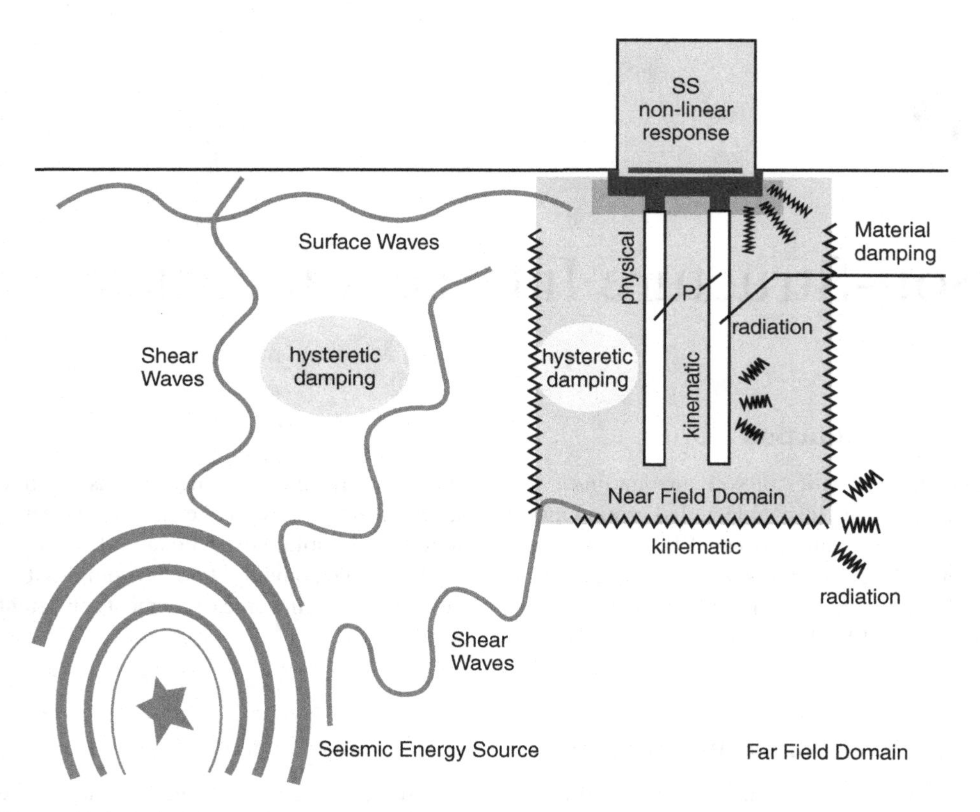

Figure 10.1 Principal mechanism of SSI, SS: Superstructure; P: Piles. Reproduced from Buehler *et al.* (2006) by permission of Dr Michael Buehler.

motion. The schematic diagram shown in Figure 10.1 depicts a simplified version of this very complex mechanism.

10.2.1 Important Features of Soil-Structure Interaction

1. In the case of stiff soils, calculations assuming a fixed base structure are expected to give realistic results. The assumption here is that the foundation present follows the incident wave with little or no SSI.
2. The presence of soft or loose soil can make the overall soil structure system more flexible, thereby increasing the period of the structure. If we refer to the UBC (1994) response spectrum shown in Figure 10.2, this shift in period could lead to a reduced response. This beneficial effect was recognized in the NEHRP 2000 provisions.

   There are a few exceptions (Mexico City, 1985), which indicate substantial increased response at higher time periods. The situation leading to such behaviour is discussed later. Consequently no general conclusions are possible and each case has to be assessed individually.

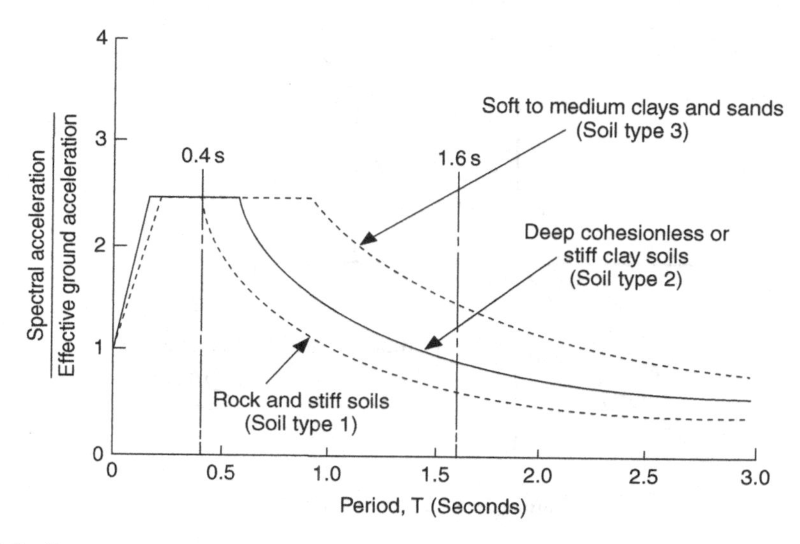

Figure 10.2 Response spectrum UBC 1994. 1994 Uniform Building Code © 1994 Washington DC; International Code Council. Reproduced with permission, all rights reserved.

3. Another aspect which affects SSI is radiation damping. Radiation damping occurs when waves emanating from a seismic source encounter a structural foundation embedded in the soil, and are reflected back into the soil. The scattering of the waves that takes place increases the damping in the soil. It may be viewed as a measure of energy loss from the structure through radiation of waves away from the footing. Radiation damping is very hard to quantify in the field. Radiation damping results in an increase in the effective damping of the entire system. Consequently, this leads to a reduced response of the whole system. In certain situations, where the structure is on a very shallow layer of soil, it is possible to have no waves reflected back in to the soil. In such cases there is only material damping and no radiation damping.

Note on Material Damping

In a travelling wave, part of the energy is always converted into heat. This conversion is accompanied by decrease in the amplitude of the wave. Elastic energy is also dissipated hysteretically; mostly as a result of slippage of grains with respect to each other during cycles of loading. The energy dissipated in a single cycle (assuming a viscoelastic damping model) is shown in the accompanying figure.

As mentioned in Section 10.2, dynamic soil-structure interaction consists of very complex mechanisms. Many of its aspects are in the domain of active research. It should be emphasized that soil-structure interaction must be adequately addressed in the design phase or it may lead to disastrous consequences.

Accounting for the flexibility of the underlying soil complicates the dynamic analysis of the structure considerably. Usually in practice, important features of soil-structure interaction are approximately incorporated in design using appropriate models.

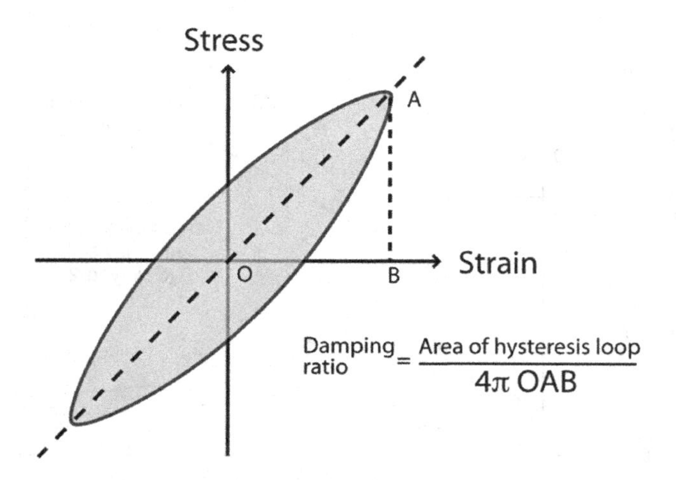

Figure 10.3 Hysteretic damping.

10.2.2 Ground Responses Observed During Earthquakes

In general, there are two types of ground response that are damaging to structures.

1. In one, the *soil amplifies the ground motion* – Loma Prieta, 1989, Mexico City, 1985. Both earthquakes are well documented and their sites were underlain by a variety of different subsurface conditions.
2. In the other, the *soil fails by liquefaction*, as witnessed during the Kobe (1995), and Bhuj (2001) earthquakes.

A brief review of the results of these earthquakes reveals the importance of local site effects on ground motion. The Loma Prieta (1989) and the Mexico City (1985) earthquakes are discussed next. Liquefaction and its effects are discussed later

10.3 Damaging Effects due to Amplification

10.3.1 Mexico City (1985) Earthquake

Effects of Local Site Conditions on Ground Motion

The Mexico City earthquake ($M_S = 8.1$) occurred in 1985. The interesting phenomenon about this earthquake, which generated worldwide interest, is that it caused only moderate damage in the vicinity of its epicentre (near the Pacific coast) but resulted in extensive damage further afield, some 350–360 km from the epicentre, in Mexico City. Fortunately ground motions were recorded at different sites in Mexico City and its vicinity.

Ground motions were recorded at two sites, UNAM (*Universidad Nacional Autonoma de Mexico*) and SCT (*Secretary of Communications and Transportation*) located on Figure 10.4. For the seismic studies that ensued, the city has often been subdivided into three zones (Figure 10.4a). The *Foothill Zone* is characterized by deposits of granular soil and volcanic

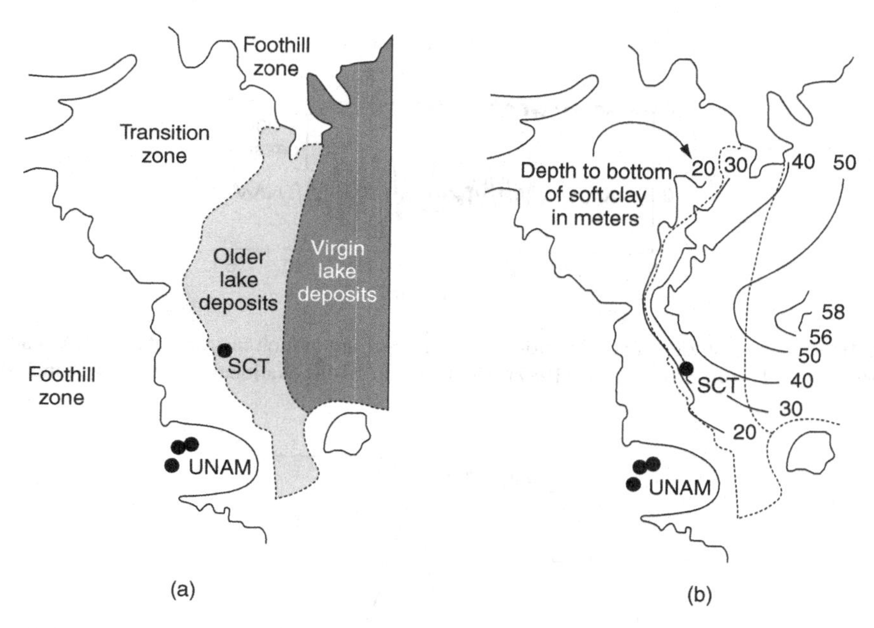

Figure 10.4 Strong motion recording sites and geo-technical make up in Mexico City; (a) location of strong motion instruments in relation to Foothill, Transition and Lake Zones; (b) contours of soft soil thickness. Reproduced from Stone *et al.* (1987). Contribution of the National Institute of Standards and Technology.

fall-off. In the *Lake Zone* there are thick deposits of very soft soil formed over the years. These are deposits due to accompanying rainfall of airborne silt, clay and ash from nearby volcanoes. The soft clay deposits extend to considerable depths (Figure 10.4b). Between the *Foothill Zone* and *Lake Zone* is the *Transition Zone* where the soft soil deposits do not extend to great depths as may be inferred from Figure 10.4b.

The UNAM site was on *basaltic (Oceanic) rock*. Oceanic crust is younger, thinner and heavier than Continental crust (granite). The SCT site was on *soft soil*. Further details of the soil makeup may be found elsewhere (Stone *et al.*, 1987).

The time histories recorded at the two sites are shown in Figure 10.5.

There are notable differences in the peak acceleration recorded at the two sites. From the site measurements of the soil depth and the average shear wave velocity, the natural period of the site was estimated at 2 sec. The computations of response spectra at the two sites from the time histories are due to Romo and Seed (1986) and are shown in Figure 10.6.

The response spectrum is a reflection of the frequency content and the predominant period is again around 2 seconds. The following items coincided at the SCT (soft soil) site:

1. The underlying soft soils had a natural period of about 2 sec;
2. The predominant period of site acceleration was about 2 sec.

As a result of this, structural damage in Mexico City was mixed. Most parts of the *Foot Hill Zone* (rock) suffered hardly any damage. In the *Lake Zone* damage to buildings with a

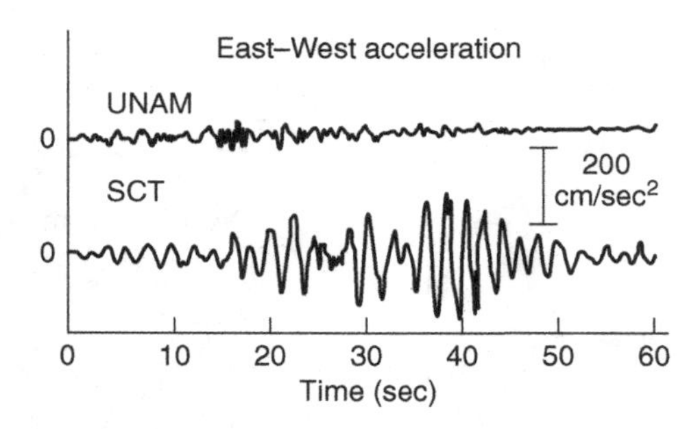

Figure 10.5 Time histories of acceleration recorded by strong motion instruments at UNAM and SCT sites. Reproduced from Stone *et al.* (1987). Contribution of the National Institute of Standards and Technology.

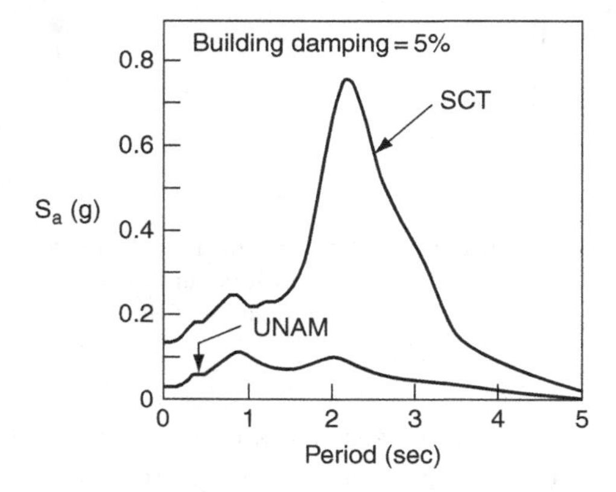

Figure 10.6 Response Spectra computed from recorded motions at UNAM and SCT sites. Reproduced from Romo and Seed (1986) by permission of the American Society of Civil Engineers (ASCE).

natural period of around 2 seconds (not unusual for medium-sized buildings of 10–20 storeys) was severe, whereas damage to taller buildings (more than 30 storeys) and buildings of lesser height (less than 5 storeys) was not major. This was a tragic case of resonance, which produced the widespread damage.

A graphic illustration is shown in Figure 10.7 (a) and (b).

10.3.2 Loma Prieta (1989) Earthquake

Effects of Local Site Conditions on Ground Motion

The Loma Prieta earthquake (1989) struck San Francisco on 17 October, 1989 and had a magnitude of 6.9. Mt. Loma Prieta is located approximately 100 km south of San Francisco and Oakland in the region known as the San Francisco Bay area. The earthquake shares remarkable similarities with the Mexico City (1985) earthquake in the sense that certain areas saw extensive damage while adjacent areas exhibited little damage.

Figure 10.7 (a) Mexico City Earthquake, 19 September 1985. The top floors of the Hotel Continental collapsed. Courtesy of the US Geological Survey. The USGS home page is http://www.usgs.gov 18/07/2007. (b) Mexico City Earthquake, 19 September 1985. Collapsed and damaged upper floors of the Ministry of Telecommunications building. Courtesy of the US Geological Survey. The USGS home page is http://www.usgs.gov 18/07/2007.

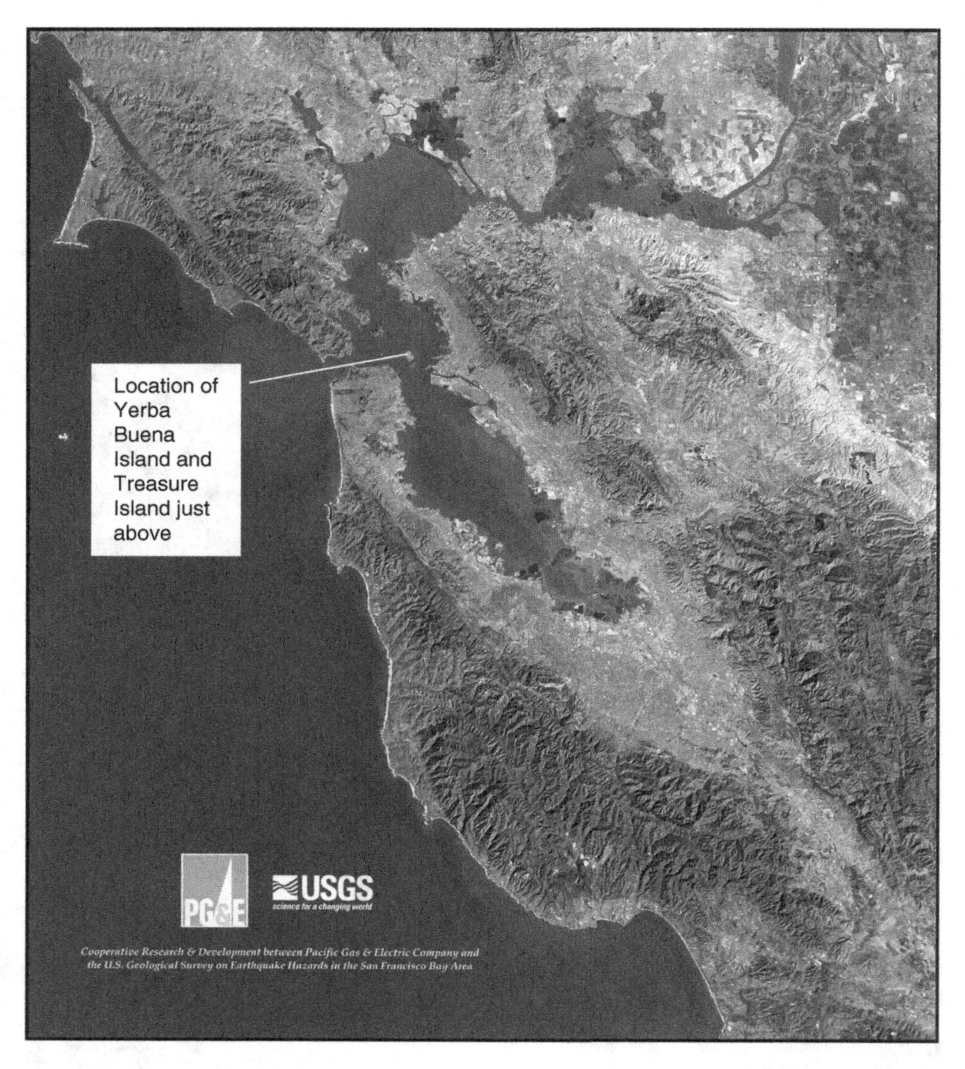

Figure 10.8 Aerial view of the San Francisco Bay area. Courtesy of the U.S. Geological Survey. The USGS home page is http://www.usgs.gov 18/07/2007.

An aerial view of the San Francisco Bay area is shown in Figure 10.8.

For seismic zonation purposes the San Francisco Bay has been divided in three zones (Seed *et al.* 1990):

1. Bay mud.
2. Alluvium.
3. Rock and shallow residual soil.

The San Francisco Bay mud is normally consolidated silty clay and varies from soft near the ground surface to approximately medium stiff at some depth. The thickness of the soil

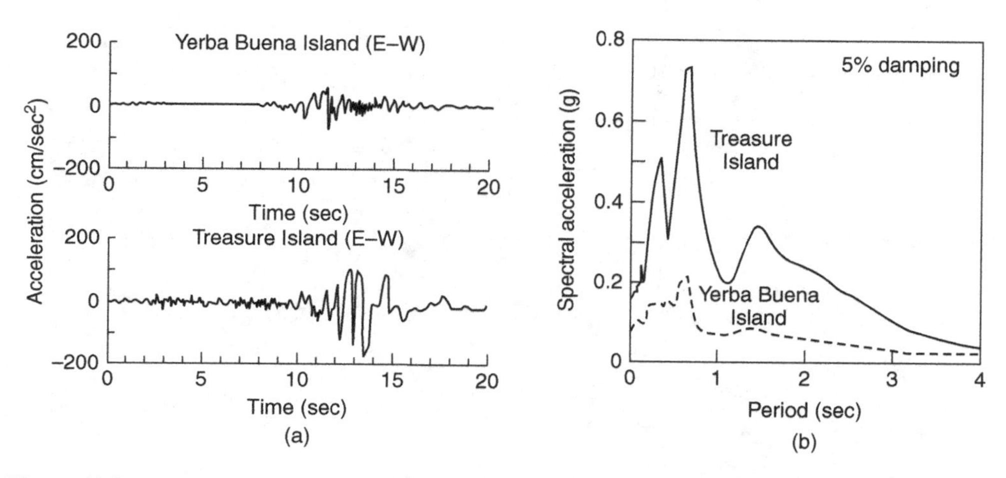

Figure 10.9 Measured peak horizontal accelerations (in g's) in the San Francisco Bay Area during the 1989 Loma Prieta Earthquake. Reproduced from R.B. Seed *et al.* (1990) by permission of the Earthquake Engineering Research Center (EERC).

deposit varies in the Bay mud region. It is fortuitous that the epicentral region and the San Francisco Bay area were well instrumented. As a result peak horizontal accelerations were recorded in several places. A map of peak horizontal accelerations recorded is shown in Figure 10.9.

Treasure Island and Yerba Buena Island are located in the middle of the Bay area in close proximity to each other as shown in Figure 10.9. In the aerial map Treasure Island is just above Yerba Buena Island. The soil composition of the two places is different. Treasure Island is a 400 acre man made hydraulic fill immediately North Northwest of Yerba Buena Island. Yerba Buena Island on the other hand is a rock outcrop. The seismograph on Treasure Island area was underlain by loose sandy soil at a depth of 13 m approximately. The Yerba Buena Island seismograph was directly on the rock.

The peak horizontal accelerations recorded at the two sites are particularly illuminating as shown in Figure 10.9a. This again reinforces the importance of soil composition. Subsequent work by Seed *et al.* (1990) produced the spectral acceleration at the two sites (Figure 10.9b). Amplification of ground motion occurred due to soft soil deposits (Bay mud). This contributed significantly to the damage, as happened during the Mexico City (1985) earthquake. The Northern portion of the I-880 Cypress Viaduct, underlain by San Francisco Bay mud, did not survive and its collapse is shown in Figure 10.10. The Southern portion not underlain by Bay mud, did survive.

A model of the shaking intensity is shown in Figure 10.11.

Mitigating Steps

There have been various studies undertaken (in particular after the Mexico City, 1985, earthquake) to investigate soil-structure interaction. Why was the Mexico City earthquake was so devastating? The epicentre was, after all, 370 km away.

Figure 10.10 Portion of the collapsed Cypress Street Viaduct, Interstate 880, in Oakland, California. Courtesy of the US Geological Survey. The USGS home page is http://www.usgs.gov 18/07/2007.

Estimating the Fundamental Period at the Site

This is one of the quantities affecting the soil amplification. The fundamental period of the site may be estimated from a simple relationship (Kramer, 1996).

$$T_s = \frac{4h}{V_s}$$

where h is the depth of the underlying soil;

V_s is the shear wave velocity;
T_s is the fundamental period of the soil layer.

(at the SCT site in Mexico the shear wave velocity was estimated to be 75 m/sec and the depth of the underlain soil was 35–40 m).

In the case of layered soil with different soil properties, sometimes as an approximation, a weighted average of the shear wave velocities of different layers is used (Kramer, 1996).

Buildings Ought to be Designed away from the Fundamental Period of the Site

Perhaps the fundamental question in designer's mind is *'in which situations should soil-structure interaction be considered?'* Will the rigid base assumption lead to significant errors?

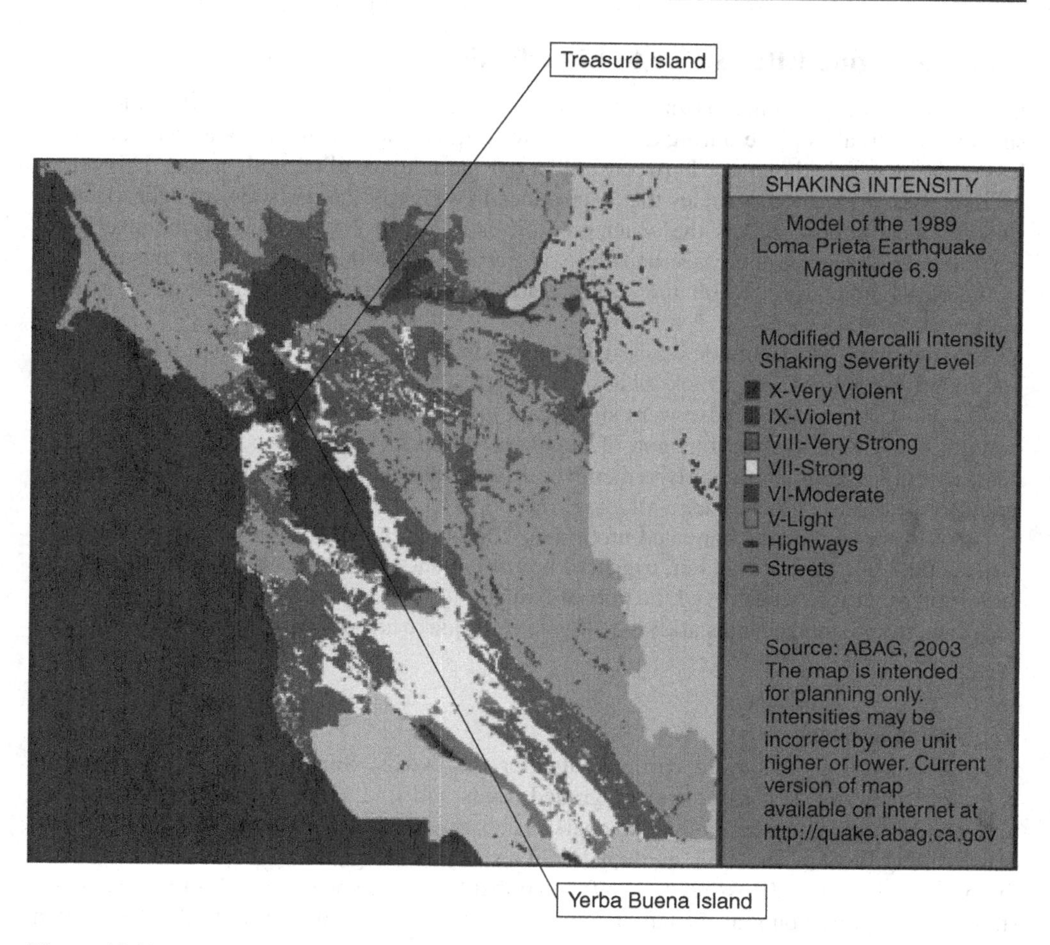

Figure 10.11 Shaking Intensity – model of the Loma Prieta (1989) Earthquake. Reproduced from http://www.abag.ca.gov/cgi-bin/pickmapx.pl 02/08/08 by permission of ABAG Earthquake and Hazards Program.

It is to be noted that the periods of vibration of a given structure increase with the decreasing stiffness of the subsoil. Under such circumstances, evaluating the consequences of soil-structure interaction will require careful assessment of both the seismic input at the bedrock level and soil conditions.

It is known that as the waves travel, with distance, the higher frequencies are filtered out. Only the long periods will remain. Estimating the frequency content of the earthquake arriving at some distance away from the focus is still an active area of research. It often depends upon which part of the fault triggers the earthquake. Extensive reviews of rupture dynamics have been reported by Kostrov and Das (1989) and Scholz (1989). A recent current state-of-the-art review with respect to crustal models has been presented by Madariaga (2006).

10.4 Damaging Effects Due to Liquefaction

Liquefaction is a phenomenon in which the soil loses its strength and stiffness due to the shaking which takes place during an earthquake. Liquefaction occurs in saturated cohesionless soils in which the space between individual particles is filled with water. This water exerts a pressure on soil particles that determines how tightly the particles are held together. Before an earthquake strikes the water pressure is relatively low. During earthquake shaking the water pressure increases and the soil particles lose the bonding that existed prior to the earthquake. As a result the soil strata lose their strength and the result is ground failure.

An earthquake of magnitude 7.5 struck Niigata, in Japan on June 16, 1964. A few months before, an earthquake had struck Alaska on March 27, 1964. The magnitude recorded was 8.4 on the Richter scale. These were significant geotectonic events in modern times, which brought to the fore the phenomenon of liquefaction and its potential to cause damage. The Alaska earthquake is particularly remembered for its savage destructiveness, long duration and also the extent of the damage-affected zone.

The Niigata earthquake damaged more than 2000 houses, but fortuitously few lives were lost. At the same time, a tsunami, triggered by movement of the sea floor associated with the fault rupture, virtually destroyed the port of Niigata.

Scenes from damaged sites after the Niigata earthquake are shown in Figures 10.12(a)–(b)

Notes and Comments

Extensive investigations were carried out after the events. Some of the well-established behaviour patterns such as uniformly graded sands and fine sands, with rounded particles tending to liquefy more easily than do coarse sands or gravelly sands, were very much in evidence during the Alaska, 1964, earthquake. In a study of bridge foundation displacements in the Alaska earthquake (Ross *et al.*, 1969) noted that there were no cases of bridge damage for structures supported on gravels, but there were several cases of damage for bridges supported on sands.

Relative density has an important influence on soil liquefaction. During the Niigata (1964) earthquake, liquefaction was extensive where the relative density of the sand was about 50 %, but did not develop in areas where relative density exceeded about 70 % (Seed and Idriss, 1971). Laboratory tests also support this behaviour pattern.

It is not difficult to comprehend that for soil in a given condition and under a given confining pressure (i.e. pressure applied in a confined state), the vulnerability to liquefaction during an earthquake depends on the magnitude of stresses or strains induced in it by the earthquake; these in turn are related to the intensity of ground shaking. This is borne out by an interesting set of facts on Niigata in Japan (Seed and Idriss, 1971). Records of earthquakes for this city extend well back in time to over 1000 years. The estimated values of maximum ground accelerations for earthquakes affecting the city in the past 370 years are shown in Figure 10.13. It may be seen from this figure that of the previous 25 occasions when the relative loose soil strata of Niigata have been shaken by earthquakes, liquefaction has been observed on only three occasions, and accelerations were in excess of 0.13 g. The damage during the 1964 earthquake was severe when the ground accelerations reported were 0.16 g. During the 22 other occasions ground accelerations ranged from 0.005 g to

Figure 10.12 Scenes from damaged sites (Niigata, 1964): (a) Kawagishi-cho Apartment Niigata, 1964. As a result of ground failure near a river bank (Shinano River), apartment buildings suffered bearing capacity failure and tilted severely. Despite the extreme tilting, the buildings themselves suffered little or no damage. Reproduced from http://www.ngdc.noaa.gov/seg/hazard/slideset/1/1_25_slide.shtml 01/06/08 by permission of the National Geophysical Data Center. Sand boils and ground fissures were visible at various sites in Niigata. Examples of sand boils can be viewed in the Karl V. Steinbrugge Collection of earthquake images at http://nisee.berkeley.edu/elibrary/Image/S3170 03/08/08. (b)Lateral spreading caused the foundations of the Showa Bridge to move laterally (Hamada, 1992) to such an extent, that the simply supported spans became unseated and collapsed. However, recent research and evidence (Bhattcharya *et al.*, 2005; Yoshida *et al.*, 2007) suggest that failure was possibly due to $P-\Delta$ effects. Further investigation is still under way. Showa Bridge Niigata, 1964. Reproduced from http://www.ngdc.noaa.gov/seg/hazard/slideset/4/4_89_slide.shtml 01/06/08 by permission of the National Geophysical Data Center.

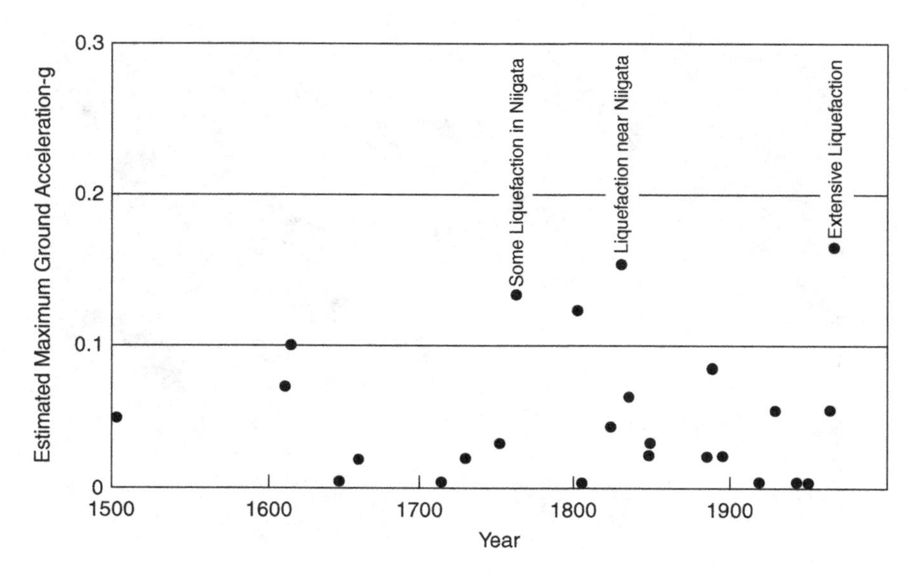

Figure 10.13 Estimated maximum ground acceleration in Niigata. Reproduced from Seed and Idriss (1971) by permission of the American Society of Civil Engineers (ASCE).

0.12 g and there was no evidence of soil liquefaction in the city (Kawasumi, 1968; Seed and Idriss, 1971).

Factors of safety calculated from actual field data for the Niigata earthquake and subsequent analysis point towards liquefaction taking place. The factors of safety plots have been reported by Ishihara (1996).

Another important factor to emerge was the duration of ground shaking which will determine the number of stress or strain cycles to which the soil will be subjected. This was in evidence during the Alaska (1964) earthquake, where landslides were triggered by liquefaction in the city of Anchorage. These slides did not occur until about 90 seconds after the earthquake motions (Seed and Idriss, 1971). This indicated the need for development of a sufficient number of strain cycles for the initiation of liquefaction.

It has been observed that buildings and bridge piers have collapsed with their superstructures being relatively undamaged. Pile failures as a result of liquefaction have been reported as the cause of failure. The pile failures observed during the Niigata earthquake (1964) after excavation are shown in Figure 10.14.

Subsequent analysis in case studies reported (including Hamada, 1992) shows a liquefiable layer of 1.8–8.5 m in depth (Figure 10.15).

A deformed configuration of the pile during the liquefaction process is shown in Figure 10.16.

What are the likely causes of pile failure? In the liquefied layer, the pile has lost the support of the soil and has been reduced to an 'unsupported column'. The initial horizontal movement caused the structure to move sideways. In this kind of situation, the $P-\Delta$ effect will dominate. A failure due to progressive and excessive bending will follow.

Figure 10.14 The pile failure observed during the excavation of the NHK building after the Niigata earthquake (1964). Reproduced from Hamada (1992) by permission of the Multidisciplinary Center for Earthquake Engineering Research (MCEER) formerly known as the National Center for Earthquake Engineering Research (NCEER).

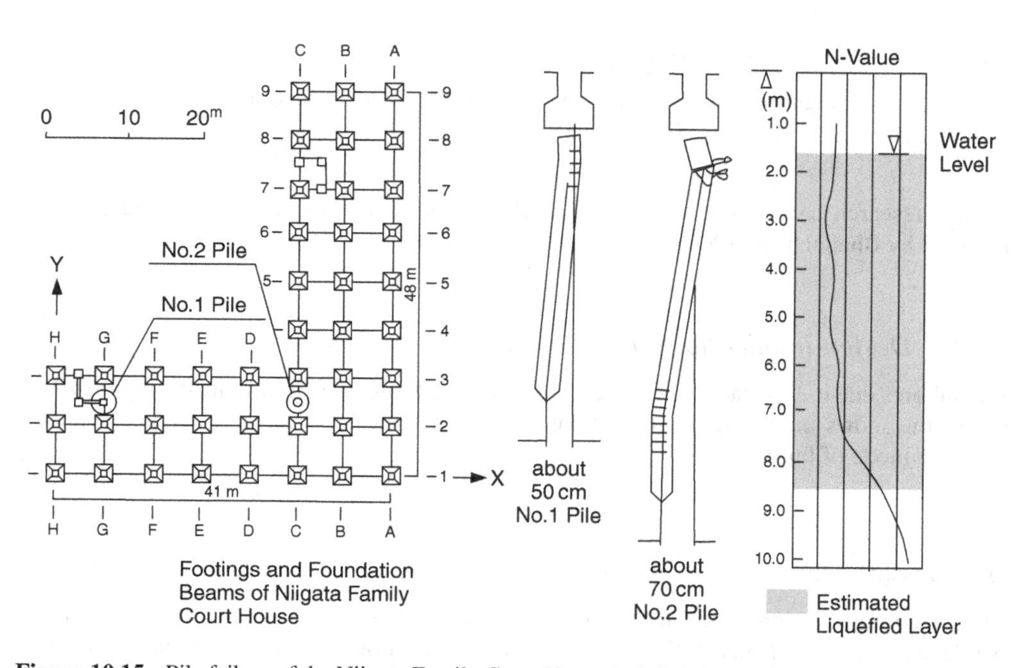

Figure 10.15 Pile failure of the Niigata Family Court House building during 1964 Niigata earthquake, showing the liquefiable layer. Reproduced from Hamada (1992) by permission of the Multidisciplinary Center for Earthquake Engineering Research (MCEER) formerly known as the National Center for Earthquake Engineering Research (NCEER).

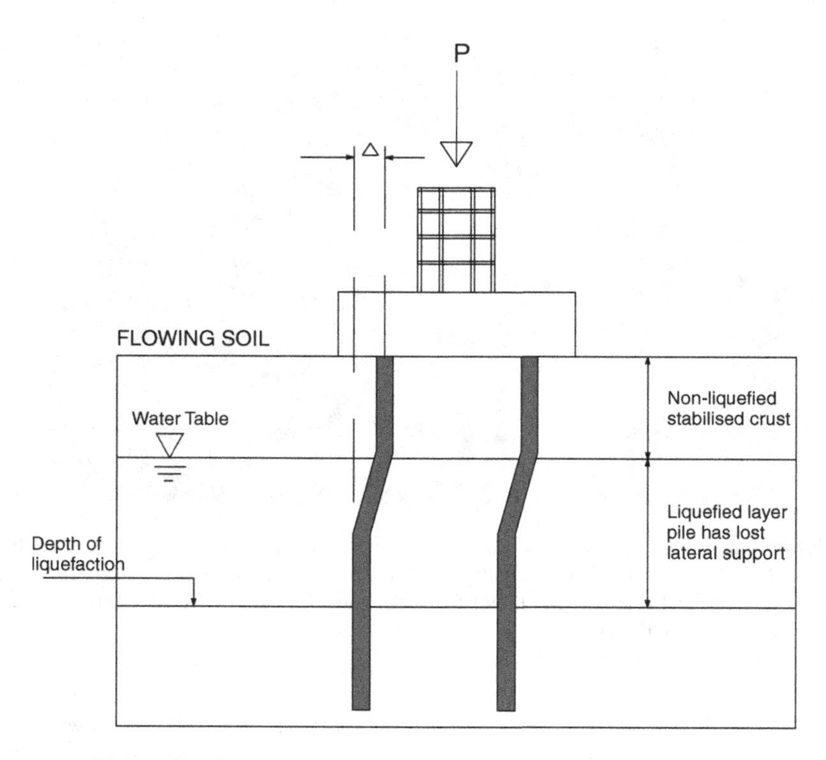

Figure 10.16 Deformed configuration of pile during liquefaction.

New research carried out at Cambridge University supports this type of failure and is reported by Bhattacharya (2003).

10.4.1 Design Implications for Piles due to Liquefaction

Pile failures due to liquefaction were reported after the Kobe (1995) earthquake. Figure 10.17 shows the shows the tilting of large storage tanks after the 1995 Kobe earthquake (Japan). Note the mode of failure.

Design Implications

The governing criterion for the design of piles supporting LNG tanks and other superstructures is to avoid out-of-plane instability due to the combined action of lateral loads and axial loads. This is an important aspect which must be taken into account for the design of piles.

Site Specific Seismic Hazard Analysis (SHA) is important before project implementation and is discussed in Chapter 7.

Figure 10.17 Storage tanks following the 1995 Kobe (Japan) earthquake. Reproduced from http://cee.engr.ucdavis.edu/faculty/boulanger/geo_photo_album/Earthquake%20hazards/Tanks/Tanks%20P1.html 02/03/08 by permission of Prof. Ross W. Boulanger.

10.5 References

Bhattacharya, S. (2003) *Pile Instability during Earthquake Liquefaction*. Ph.D. Thesis, University of Cambridge.

Buehler, M.M., Weinbroer, H. and Rebstock, D. (2006) 'A full soil-foundation-structure interaction approach'. *Proceedings of the 1st European Conference on Earthquake Engineering and Seismology*. Paper no. 421. Geneva, Switzerland.

Hamada, M. (1992) *Large Ground Motions and their Effects on Lifelines: 1964 Niigata Earthquake. Case Studies of Liquefaction and Lifeline Performance during Past Earthquakes*. Technical Report NCEER-92-0001, Vol. 1, Japanese Case Studies, National Center for Earthquake Engineering Research, Buffalo, New York.

Housner, G. (1959) 'Behaviour of structures during earthquakes'. *J. Engg. Mech. Division*, ASCE, **85**, 109–129.

International Code Council (1997) *Uniform Building Code (UBC) 1997*. Washington, DC. ICC.

Ishihara, K. (1996) *Soil Behaviour in Earthquake Geotechnics*. Clarendon Press, Oxford.

Kawasumi, H. (1968) *Historical Earthquakes in the Disturbed Area and Vicinity*. General Report on the Niigata Earthquake of 1964, Electrical Engineering College Press, University of Tokyo.

Kostrov, B. and Das, S. (1989) *Principles of Earthquake Source Mechanics*. Cambridge University Press.

Kramer, S.L. (1996) *Geotechnical Earthquake Engineering*. Prentice Hall, New Jersey.

Madariaga, R. (2006) 'Earthquake dynamics and the prediction of strong ground motion'. *Proceedings of the 1st European Conference on Earthquake Engineering and Seismology*. Paper no. K1b. Geneva, Switzerland.

Penzien, J. Photograph of sand boils at Niigata. Karl V. Steinbrugge Collection. http://nisee.berkeley.edu/elibrary/Image/S3170 03/08/08

Romo, M.P. and Seed, H.B. (1986) 'Analytical modelling of dynamic soil response in the Mexico earthquake of September 19,1985'. *Proceedings of the ASCE International Conference on the Mexico Earthquakes*. Mexico City, 148–162.

Ross, G.A., Seed, H.B. and Migliaccio, R.R. (1969) 'Bridge Foundations in Alaska Earthquake'. *J. Soil Mechanics and Foundations Division*, ASCE, **95** SM4, 1007–1036.

Scholz, C. (1989) *The Mechanics of Earthquake and Faulting*. Cambridge University Press.

Seed, H.B. and Idriss, I.M. (1971) 'Simplified procedure for evaluating soil liquefaction potential'. *J. Soil Mechanics and Foundations Division*, ASCE, **97**(9): 1171–1182. Seed, R.B., Dickenson, S.E., Reimer, M.F., *et al.* (1990) 'Preliminary report on the principal geotechnical aspects of the October 17, 1989 Loma Prieta earthquake,' Report UCB/EERC-90/05, Earthquake Engineering Research Centre, University of California, Berkeley, 137 pp.

Stone, W.C., Yokel, F.Y., Celebi, M., Hanks, T. and Leyendecker, E.V. (1987) *Engineering aspects of the September 19, 1985 Mexico earthquake*. NBS Building Science Series 165, National Bureau of Standards, Washington D.C., 207 pp.

US Geological Survey, Selected images. http://www.usgs.gov 01/08/2008.

Yoshida, N., Tazoh, T., Wakamatsu, K., Yasuda, S., Towhata, I., Nakazawa, H. and Kiku, H. (2007) 'Causes of Showa Bridge collapse in 1964 Niigata Earthquake Based on Eyewitness Testimony'. *Soils and Foundations* **47**(6): 1075–87.

11

Liquefaction

11.1 Definition and Description

In geotechnics the term 'liquefaction' is defined as the transformation of a granular material from a solid to a liquid state as a result of increased pore water pressure and consequent reduction in the effective stress. Increase in pore water pressure takes place due to the tendency of granular materials to compact when subjected to cyclic shear stresses.

Loose to medium dense, dry sand subjected to shaking (i.e. cyclic shearing) would be expected to reduce in volume, as particles are given the energy to drop into inter-particle voids (Figure 11.1a). In this case no liquefaction takes place.

Instead if the soil is saturated and these voids are filled with water, then the water needs to drain out for the reduction in volume. In some cases, depending on the rate of loading, the time scales may be insufficient to allow the water to drain out and sand particles to drop into the gaps during shaking. Such situations may occur with a sudden shock loading or during shaking due to an earthquake. In this case an undrained response with no volume change occurs (Figure 11.1b).

Increase in pore water pressure takes place under such circumstances with consequent decrease in intergranular effective stress (Terzaghi, Peck and Mesri, 1996). The sandy soil reaches a stage where the solid state of the soil transforms into a fluid state with a very high concentration of suspended grains. It then starts to behave like a heavy fluid. Under these conditions complete liquefaction has occurred. The bearing capacity of soil becomes very low and the collapse of the original structure founded on the soil occurs.

Liquefaction as caused by earthquake shaking and its significance from an engineering standpoint are the focus of this chapter.

11.1.1 Geotechnical Aspects of Liquefaction

The prerequisite of occurrence of liquefaction may be conceived as due to triggering of a sudden shock on the soil mass by an earthquake, which may even have a low magnitude. The intensity of the disturbance (shock) necessary for liquefaction of loose sand varies with different gradation of sand and their relative densities. When the relative density of saturated

Fundamentals of Seismic Loading on Structures Tapan Sen

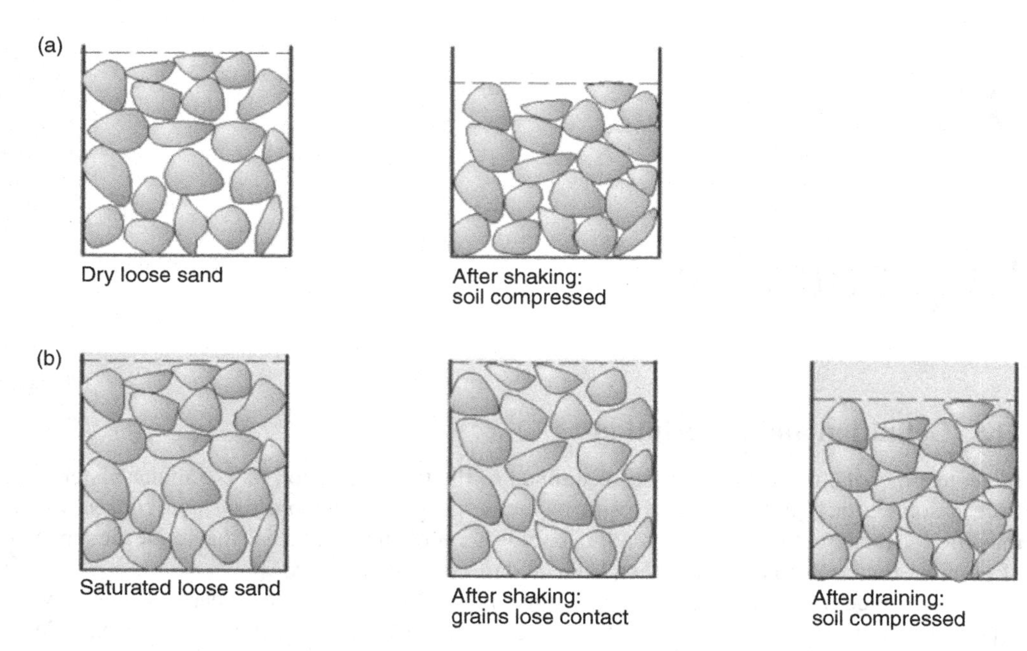

Figure 11.1 Schematic behaviour of loose sand particles under rapid shaking. Reproduced from Brennan *et al.* (2007) by permission of Dr A.J. Brennan.

sand is less than about 40–50 %, liquefaction may occur causing the flow to start. The most unstable sand is made up of rounded grains with effective grain size less than 0.1mm and uniformity coefficient less than 5.

(The *effective size* is the diameter D_{10} that corresponds to $P = 10$ % on the grain size diagram. In other words 10 % of the particles are finer and 90 % coarser than the effective size. The *uniformity coefficient* C_u is equal to D_{60}/D_{10} wherein D_{60} is the grain size corresponding to $P = 60$ % (Terzaghi, Peck and Mesri, 1996).)

Fine sands and coarse silts may be considered most vulnerable to liquefaction. When an isolated footing is placed on a thick layer of saturated fine sand or coarse silt at a very loose density, a sudden inducement of shock or vibration may transform the soil into a liquefied state causing the foundation to fail and thereby resulting in the collapse of the structure.

The liquefaction of saturated sandy soil takes place in the form of loss of foundation support, excessive settlement, lateral spreading or 'landslide' in which heavy liquid soil mass flows down the initial slope initiated by the application of even a sudden small shock or vibration. It finally comes to a state of rest on a flat slope of 4 degrees or less.

Several such occurrences of landslides due to liquefaction have taken place in the past as well as in recent years all over the world, for example as witnessed during the Tokachi-Oki, Miyagi-Oki, Japan, 2003, earthquake (Yasuda et al, 2004). Common geotechnical characteristics of the sands constituting the flow slides in the above cases involved fully saturated grains that were uniform, rounded, and in a loose state of compaction.

It is worthwhile to note certain terminology used to define the states of liquefaction used in other texts.

Liquefaction denotes a condition where a soil will undergo continued deformation at fairly low residual stress or with no remaining residual stress, due to the build-up of high pore water pressure, which reduces the effective confining pressure to a very low value. Pore water pressure build-up, the main cause of liquefaction, may be due to either static or cyclic stress condition imposed on the soil in the initial phase and may be thought of as the start of the liquefaction process.

Initial liquefaction denotes a condition where, during the course of cyclic stress applications, the resultant pore water pressure equals the existing/applied confining pressure on completion of any full stress cycle. This phase, however, does not give any indication of the subsequent deformations in the soil which might follow.

Limited liquefaction with limited strain potential; cyclic mobility or cyclic liquefaction refers to a condition where cyclic stress applications have led to a state of *initial liquefaction* and subsequent stress cycles lead to limited strains within the soil. The soil stabilizes under the applied loads – a phenomenon observed when the pore pressure drops.

Thus, from the above discussions it is appropriate to note:

- Liquefaction is a process in which the strength and stiffness of a soil is reduced by earthquake shaking (or other rapid loading).
- Liquefaction occurs in saturated soils where the space between the individual particles is completely filled with water.
- There is build-up of pore water pressure. Earthquake shaking often generates this increase in pore water pressure.
- Soil starts to behave as a fluid.
- Liquefaction could lead to large settlement, both horizontal and vertical, collapse of buildings, lifting of sewage pipes and broken railways lines (Kobe 1995), Tokachi-Oki, Miyagi-Oki, Japan (2003), Yasuda *et al.* (2004).

11.2 Evaluation of Liquefaction Resistance

11.2.1 Analytical Procedure – Empirical Formulation (Seed and Idriss, 1971)

The Niigata and Alaska earthquakes (1964) triggered off detailed investigations within the seismological profession on liquefaction and ways of predicting its occurrence.

The procedure currently used by many practising engineers is a semi-empirical method (Idriss and Boulanger, 2004; Youd *et al.*, 2001) and will be discussed later.

The first empirical method was due to the pioneering work by Seed and Idriss (1971). It still forms the basic approach and therefore should be studied to gain a better understanding of the subject.

In simple terms, the possibility of liquefaction at a site may be assessed by comparing:

1. The cyclic shear stress induced by earthquake ground motion, (at a 'depth 'h').

against

2. The similarly expressed shear stress required to cause initial liquefaction (at the same depth in question) or a value of stress (or shearing strain) not acceptable in the design.

This will give a measure of the factor of safety and will be discussed further in detail.

Evaluation of Liquefaction Characteristics

The following factors will affect the liquefaction behaviour:

For a given soil condition and under given confining pressure, potential for liquefaction of a layer will depend on the magnitude of the stresses (or strains) and the number of stress cycles induced in it by the earthquake. This will depend on the maximum ground acceleration. It is interesting to note that this fact was established from the records of the Niigata earthquake (1964). Niigata has a long history of maintaining earthquake records. Of the 25 earthquakes reported in the last 370 years in or near Niigata liquefaction occurred only in three earthquakes. The maximum ground acceleration computed in all three earthquakes was over 0.13 g (Seed and Idriss, 1971 – see Chapter 10). In one of the three earthquakes the maximum ground acceleration was 0.16 g and the damage was particularly extensive.

1. Earthquake characteristics:

   (a) Maximum intensity of ground shaking (maximum acceleration) at site (this will be available from the seismic hazard analysis (refer to Chapter 6)).
   (b) Duration of ground shaking.

The duration of ground shaking will determine the number of significant stress cycles to which the soil will be subjected to reach liquefaction. Again, evidence from actual field data obtained from the Anchorage area (Alaska, 1964) shows that liquefaction occurred after 90 seconds of the onset of the shock waves. Prior to this liquefaction was not apparent.

2. Soil conditions (relative density/grain size)

The primary soil condition, which determines the potential for liquefaction is relative density. Relative density is dependent on (a) soil type, characterized by grain size distribution; and (b) void ratio.

For cohesionless soil, such as sand, the soil type is characterized by grain size distribution. As discussed earlier uniformly graded soils are more susceptible to liquefaction than well graded materials. Fine sands tend to liquefy more easily than coarse sand. In some situations loose sand (identified by void ratio) may liquefy, but the same sand in denser formation (smaller void ratio) may not.

3. Initial confining pressure (as measured by triaxial consolidated undrained laboratory test)

There is considerable evidence from laboratory tests to show that the liquefaction potential is reduced with an increase in confining pressure (Seed and Idriss, 1971). Evidence from the Niigata (1964) earthquake reported by Seed and Idriss (1971) also confirms this.

We next consider the method of estimating the number of significant stress cycles.

Step 1

Step 1a

After having established the soil conditions, the design earthquake, the time history of shear stresses induced by earthquake ground motion at different depths can be computed. This may be conveniently obtained from a finite element (FE) analysis program (refer to Chapter 5 for the type of FE analysis required), a finite difference based program (e.g. FLAC) or a one dimensional analysis, eg. SHAKE.

Step 1b

The shear stresses will be irregular. From this irregular time history, we can estimate the average cyclic shear stress ratio (*CSR*). This may be determined from Equation (11.4).

Step 2

From the results obtained from Step 1, plot the variation of *CSR* with depth. This shows the cyclic stress ratio that will be developed during the earthquake. The number of cycles N that the shear stress will be subjected to depends on the magnitude of the earthquake as shown in Table 11.1.

Step 3

From available field or laboratory soil tests on representative samples, conducted under different confining pressures, determine the cyclic resistance ratio (*CRR*) that will have to be developed at the same number of cycles to cause liquefaction.

Note: To distinguish between CSR and CRR

CRR is the cyclic stress required to cause liquefaction (i.e. cyclic resistance ratio) and *CSR* is the cyclic stress ratio induced by earthquake ground motions.

Several tests may be carried out to determine CRR. The options available usually are (1) cyclic triaxial test, (2) cyclic torsional shear test and (3) cyclic simple shear test. All these tests have been discussed by Kramer (1996) and hence will not be discussed here.

Alternatively, field data such as SPT or CPT test data may also be used to determine CRR. This is discussed further in Section 11.4.

Step 4

By comparing the variation of stresses obtained in Step 2 with those obtained in Step 3, we may now determine whether any zone exists within the deposit where liquefaction can be expected to occur i.e. earthquake induced stresses exceed those required to cause liquefaction.

11.2.2 A Simplified Method (Seed and Idriss, 1971)

Seed and Idriss's ground-breaking effort was soon embraced by the engineering community and implemented in practice.

Certain improvements have been proposed over the years (Youd *et al.*, 2001; Idriss and Boulanger, 2004), but the original formulation still forms the backbone of the method currently in use. The Idriss and Boulanger method is discussed in Section 11.4.2.

Estimation of CSR (Cyclic Stress Ratio)

Basic Steps

The schematic for determining the maximum shear stress $\tau_{\max}$ is shown in Figure 11.2.

A simplified method for assessing cyclic shear stress induced at any point within a level ground was first proposed by Seed and Idriss (1971). Referring to Figure 11.2, if the maximum $(\tau_{\max})_r$ horizontal acceleration on the ground surface is $a_{\max}$, then the maximum (peak) shear stress for a rigid column is given by:

$$(\tau_{\max})_t = \frac{a_{\max}}{g} \cdot \gamma_t \cdot h \tag{11.1a}$$

where γ_t = total unit weight of soil
g = acceleration due to gravity
h = depth below ground surface

It is necessary to estimate the equivalent uniform average shear stress. It has been found that the average uniform shear stress is about 65% of the 'peak' cyclic shear stress. Hence, the factor of **0.65** is used to convert the peak cyclic shear stress ratio to a stress ratio that is representative of the most significant cycles over the full duration of the loading. The actual time history of the shear stress would be non-uniform as shown in Figure 11.3.

Thus

$$\tau_{av} = 0.65 \cdot \frac{a_{\max}}{g} \cdot \gamma_t \cdot h \tag{11.2a}$$

Seed and Idriss (1971) further introduced the stress reduction factor r_d as a parameter describing the ratio of cyclic stresses for a flexible column to the cyclic stresses for a rigid column as shown in Figure 11.2. For values of r_d within range of earthquake ground motions and

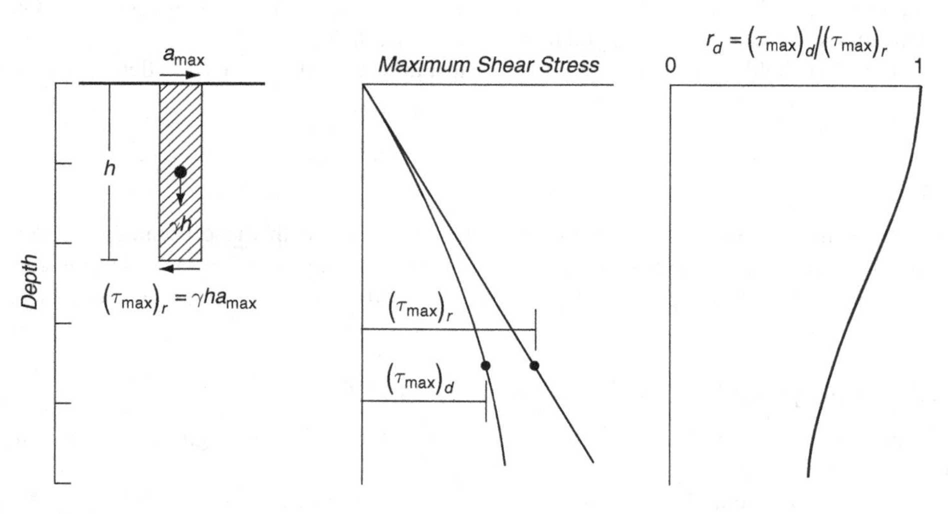

Figure 11.2 Schematic for determining maximum shear stress, $\tau_{\max}$ and the stress coefficient r_d. Reproduced from Seed and Idriss (1971) by permission of the American Society of Civil Engineers (ASCE).

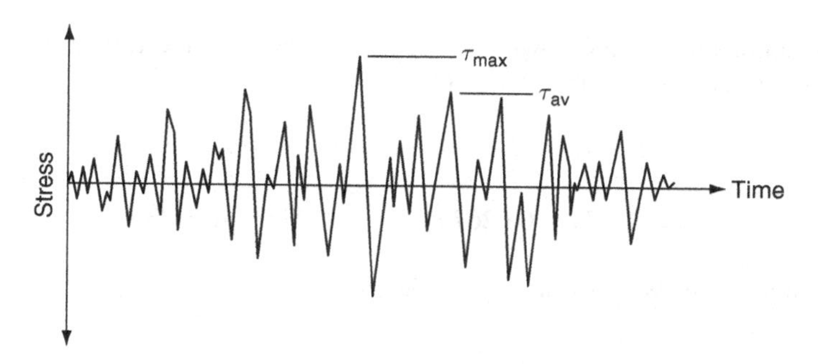

Figure 11.3 Time history of shear stresses during earthquakes for liquefaction analysis. Reproduced from Seed and Idriss (1971) by permission of the American Society of Civil Engineers (ASCE).

soil profiles having sand in the upper 15 *m* approximately, Seed and Idriss suggested an average curve for use as a function of depth. The average curve was intended for all earthquake magnitudes and for all soil profiles.

The shear stresses developed at any point in a soil deposit, during an earthquake, are primarily due to upward propagation of shear waves in the deposit. These stresses can be calculated using analytical procedures and are primarily dependent on earthquake ground motion parameters (e.g. intensity, frequency content etc.).

The variation of r_d with depth is shown in Figure 11.4.

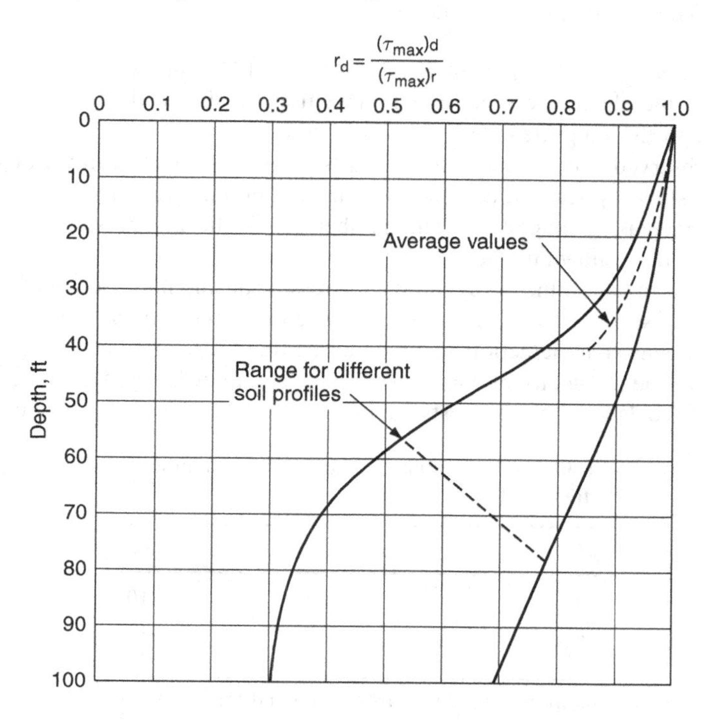

Figure 11.4 Range of values of r_d for different profiles in liquefaction analysis. Reproduced from Seed and Idriss (1971) by permission of the American Society of Civil Engineers (ASCE).

For routine applications, the following equations may be used to estimate average values of r_d (Liao and Whitman, 1986; Youd *et al.*, 2001):

$$r_d = 1.0 - 0.00765h \quad \text{for} \quad h \leq 9.15m \tag{11.2b}$$

$$r_d = 1.174 - 0.0267h \quad \text{for} \quad 9.15m < h < 23m \tag{11.2c}$$

Thus the actual average cyclic shear stress at depth h, $(\tau_{av})_d$

$$(\tau_{av})_d = r_d(\tau_{av}) \tag{11.3}$$

The Seed-Idriss method estimates the cyclic shear stress ratio (CSR) induced by earthquake, at a depth h, below the ground surface from the relation

$$CSR = 0.65 \left\{ \frac{\sigma_{vo} a_{\max}/g}{\sigma'_{vo}} \right\} r_d \tag{11.4}$$

where

r_d is a stress reduction factor that accounts for the flexibility of the soil column
σ_{vo} is the total vertical stress at depth h (r_t.h)
σ'_{vo} is the effective vertical stress at depth h

From the laboratory tests carried out, Seed and Idriss (1971) suggested a simplified method for determining the shear stress causing liquefaction. In the worked out example, which follows, the use of the simplified method is illustrated.

It has been observed that liquefaction is initiated after an adequate number of stress cycles. The number of stress cycles will depend upon the duration of ground shaking. The duration increases with increasing magnitude. The number of significant stress cycles N_c associated with magnitude, M, is shown in Table 11.1.

The next step is to determine the cyclic shear stresses causing liquefaction of a given soil in a number of given stress cycles. This may be done either from estimates of stress conditions known to have caused liquefaction in previous earthquakes, or by means of appropriate laboratory tests. Due to paucity of data from previous earthquakes and also due to the generalized nature of the data, laboratory tests are often preferred. The results of such a laboratory

Table 11.1 Earthquake magnitude significant stress cycles.

M	N_c
7.0	10
7.5	20
8.0	30

Source: Reproduced from Seed and Idriss (1971) by permission of the American Society of Civil Engineers (ASCE).

programme, undertaken at the University of California, Berkeley, are shown in Figures 11.5 and 11.6.

It may be noted, from the above two figures (Figures 11.5 and 11.6 on stress conditions causing liquefaction), that the stress ratio causing liquefaction in 10 cycles is higher than that needed for 30 cycles.

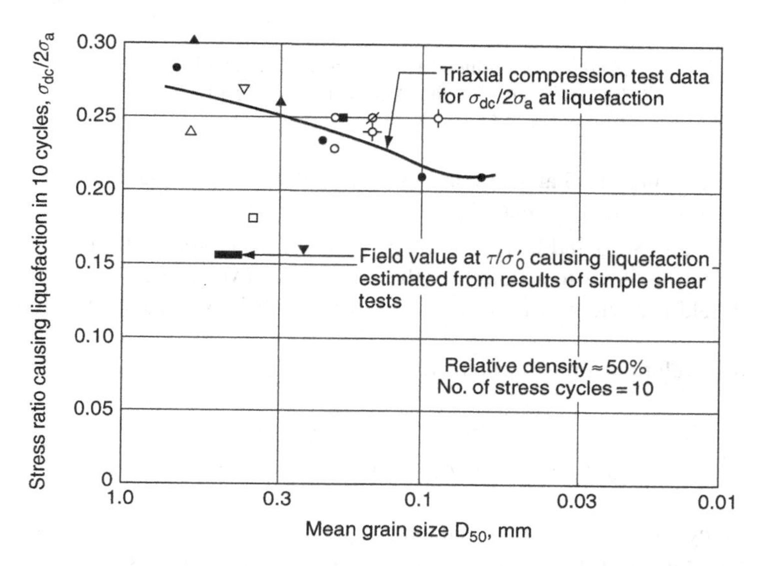

Figure 11.5 Stress conditions causing liquefaction of sands in 10 cycles. Reproduced from Seed and Idriss (1971) by permission of the American Society of Civil Engineers (ASCE).

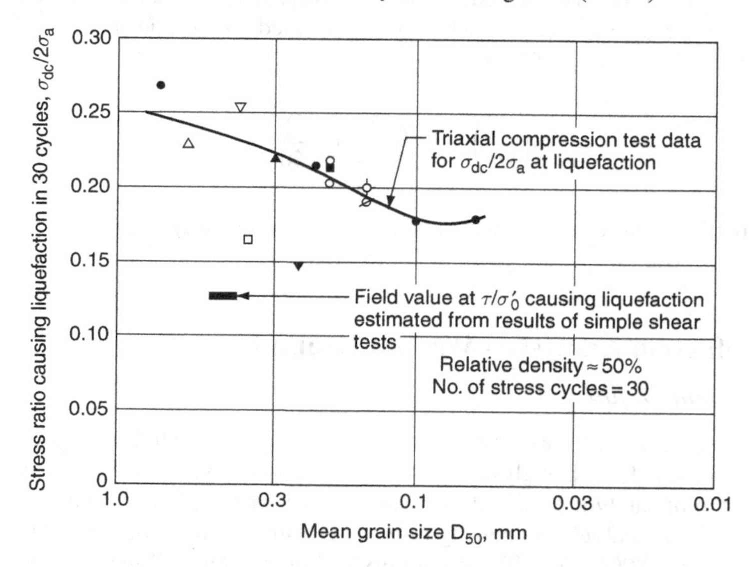

Figure 11.6 Stress conditions causing liquefaction of sands in 30 cycles. Reproduced from Seed and Idriss (1971) by permission of the American Society of Civil Engineers (ASCE).

Table 11.2 Relative density reduction factor.

D_r%	C_r
0–50	0.57
60	0.60
80	0.68

Source: Reproduced from Seed and Idriss (1971) by permission of the American Society of Civil Engineers (ASCE).

Also shown in Figures 11.5 and 11.6 are the appropriate values from the field observations. It is seen that that stress ratio $\frac{\tau}{\sigma_0'}$ is less than $\frac{\sigma_{dc}}{2\sigma_a}$.

To account for this Seed and Idriss (1971) proposed a correction factor C_r.

C_r is the correction factor to account for the difference in values between triaxial compression tests and field behaviour. Values of C_r varying with relative density D_r are presented in Table 11.2.

The suggested relationship is of the form:

$$\left(\frac{\tau}{\sigma_0'}\right)_{lD_r} = \left(\frac{\sigma_{dc}}{2\sigma_a}\right) \cdot C_r \tag{11.5}$$

Note: σ_{dc} is the cyclic deviator stress in the laboratory.

The all round uniform pressure that the soil is subjected to in a triaxial test is denoted here as σ_a and the cyclic stresses applied subsequently is the deviator stress denoted here as σ_{dc}.

The laboratory tests were performed for different grain sizes with a constant relative density of 50 %. Using the appropriate correction factor needed to include the effect of different relative densities, Equation (11.5) becomes

$$\left(\frac{\tau}{\sigma_o'}\right)_{lD_r} = \left(\frac{\sigma_{dc}}{2\sigma_a}\right)_{l50} \frac{D_r}{50} \cdot C_r \tag{11.6}$$

where D_r and 50 denote relative densities D_r and 50% respectively.

The values proposed for C_r are given in Table 11.2.

11.3 Liquefaction Analysis – Worked Example

11.3.1 Problem Definition

In a highly seismic zone, the maximum intensity of ground shaking to be designed for is 0.12 g (derived from seismic hazard analysis). The magnitude is 7.5 on the Richter scale. The site is underlain by 2 m of sandy top soil followed by approximately 15 m of sandy soil. The mean grain size is 0.15 mm and relative density is 47 %. The mass density of top soil is 1650 kg/m^3 and below 2 m it is 1800 kg/m^3. The possibility of liquefaction and the zone of liquefaction need to be assessed. The water table is at 2 m below the ground level.

The soil profile is shown in Figure 11.7.

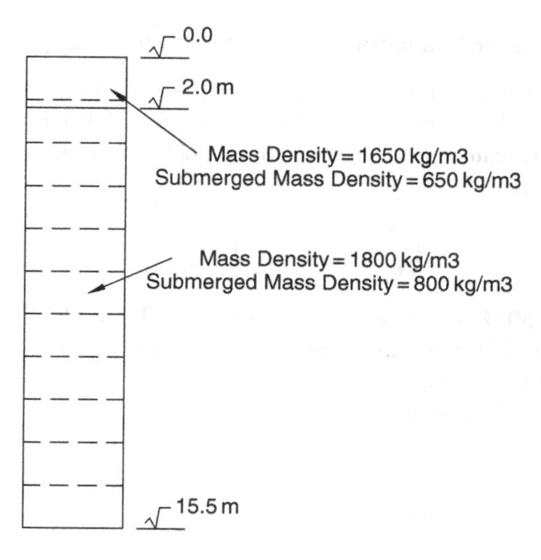

Figure 11.7 Soil profile at site – Ex. 11.3.1.

Use of Simplified Procedure

Evaluate shear stress $(\tau_{av})_d$ at various depths in the deposit

$$(\tau_{av})_d = 0.65 \cdot \frac{a_{\max}}{g} \cdot \gamma_t \cdot h \cdot r_d \qquad [11.2a \text{ and } 11.3]$$

The computed average shear stress is shown in Table 11.3.

Table 11.3 Average shear stress – $(\tau_{av})_d$.

Depth* h m	* γh kN/m^2	$\frac{a_{\max}}{g}$	r_d	$(\tau_{av})_d$ kN/m^2
3.50	58.86	0.12	0.98	4.50
5.00	85.35	0.12	0.97	6.46
6.50	111.89	0.12	0.96	8.37
8.00	138.32	0.12	0.94	10.14
9.50	164.81	0.12	0.93	11.89
11.0	191.28	0.12	0.90	13.43
12.5	217.78	0.12	0.86	14.60
14.0	244.27	0.12	0.82	15.62
15.5	270.76	0.12	0.79	16.79

* h is the depth below the ground surface in m, and γ the mass density, is 1650 kg/m^3 up to 2 m and 1800 kg/m^3 below this depth.

The Simplified Method for Evaluating Shear Stress Causing Liquefaction

Referring to the test data shown in Figures 11.5 and 11.6 together with the values of C_r, we can estimate the stress condition likely to cause liquefaction of different soils in the field. The cyclic stress ratio (CSR) causing liquefaction in the field for a given soil at a relative density D_r may be estimated from

$$\left(\frac{\tau}{\sigma_o'}\right)_{lD_r} = \left(\frac{\sigma_{dc}}{2\sigma_a}\right)_{l50} \frac{D_r}{50} \cdot C_r \qquad [11.6]$$

For relative density 0–50 % the value of C_r is 0.57. (from Table 11.2)

From Figures 11.5 and 11.6 the stress causing liquefaction for soil of D_{50} of 0.15mm is 0.23 for 10 cycles and 0.19 for 30 cycles.

Hence for 20 cycles, we may assume

$$\left(\frac{\sigma_{dc}}{2\sigma_a}\right) = \frac{0.23 + 0.19}{2} = 0.21$$

The value of cyclic shear stress causing liquefaction may now be calculated from

$$\tau = \sigma_o' \times 0.21 \times 0.57 \times \frac{47.0}{50.0} = 0.1125\sigma_w' = k\sigma_o'$$

The calculated shear stress causing liquefaction at various depths is shown in Table 11.4.

The plot of average shear stress induced due to ground intensity of 0.12g and shear stress causing liquefaction is shown in Figure 11.8. The zone of liquefaction is shown.

Table 11.4 Shear stress causing liquefaction.

Depth *h m	**$\sigma_o' = {}^*\gamma_{subm} h$ kN/m^2	k	$\tau = 0.1125 \cdot \sigma_w' = k \cdot \sigma_w'$ kN/m^2
3.50	44.15	0.1125	4.97
5.00	55.92	0.1125	6.29
6.50	67.69	0.1125	7.62
8.00	79.46	0.1125	8.94
9.50	91.23	0.1125	10.27
11.0	103.01	0.1125	11.59
12.5	114.78	0.1125	12.92
14.0	126.55	0.1125	14.24
15.5	138.32	0.1125	15.56

* h is the depth below the top soil in m, and γ, the mass density is 1650 kg/m^3 up to 2 m and γ_{subm}, the submerged mass density, is 800 kg/m^3 below 2 m.

** σ_o' = effective pressure at depth h.

Note: Although the site was underlain with approximately 17 m of sandy soil deposit, corroboration with field tests for the average cyclic shear stress ratio, beyond depth of 15 m, do not exist.

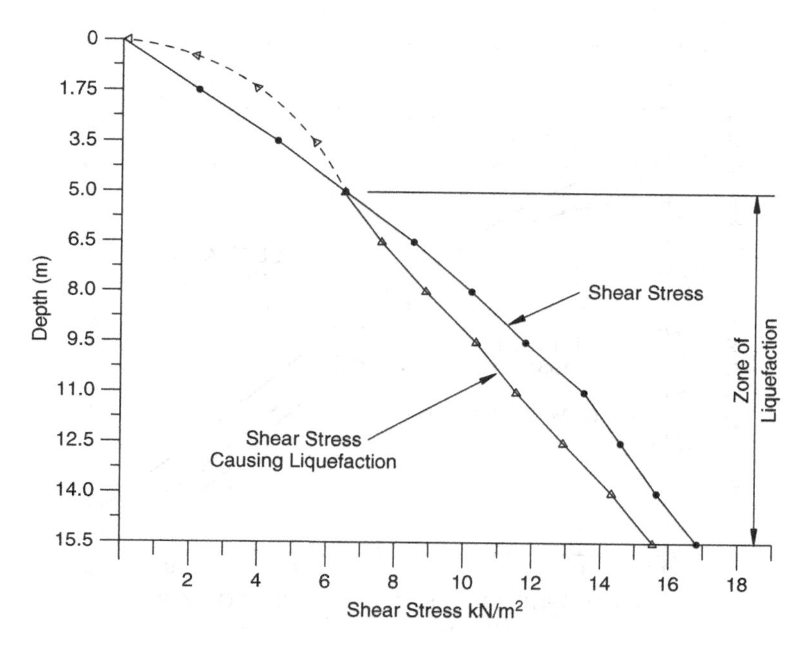

Figure 11.8 Worked example – Ex. 11.3.1 Plot of average shear stress and shear stress causing liquefaction.

11.3.2 Case Study – Using Field Data

Niigata, 1964, Japan

(The following case study is taken from Seed and Idriss, 1982.)

FPS Units
Conversion to SI Units 1m = 39.37 inches.
1kg = 2.204 lbs

This case study (Seed and Idriss, 1982) uses cyclic loading test data to evaluate the liquefaction potential of the sands in the lightly damaged and heavily damaged zones of Niigata, Japan. The elements of such a study were first presented in 1969 using estimated soil properties. Ishihara and Koga presented another study in 1981 based on actual soil test data. The average test results of cyclic triaxial tests performed by these investigators on good quality undisturbed samples of sand taken in depth range 10 to 25 ft from two areas of Niigata, one where no liquefaction occurred in 1964 earthquake and the area of Kawagishi-cho where extensive liquefaction occurred are presented in Figure 11.10. The soil profile in both areas consists of sand to substantial depths below the ground surface with the water table at a depth of 4 ft.

The Niigata earthquake of 1964 had a magnitude of approximately 7.5 and the source of energy release was about 50 km away from the city. From the curves presented in Figure 11.9, the mean value of peak ground acceleration in Niigata would be estimated to be around 0.15 g. It is of interest to note that a value of 0.16 g was actually recorded at Kawagishi-cho and a

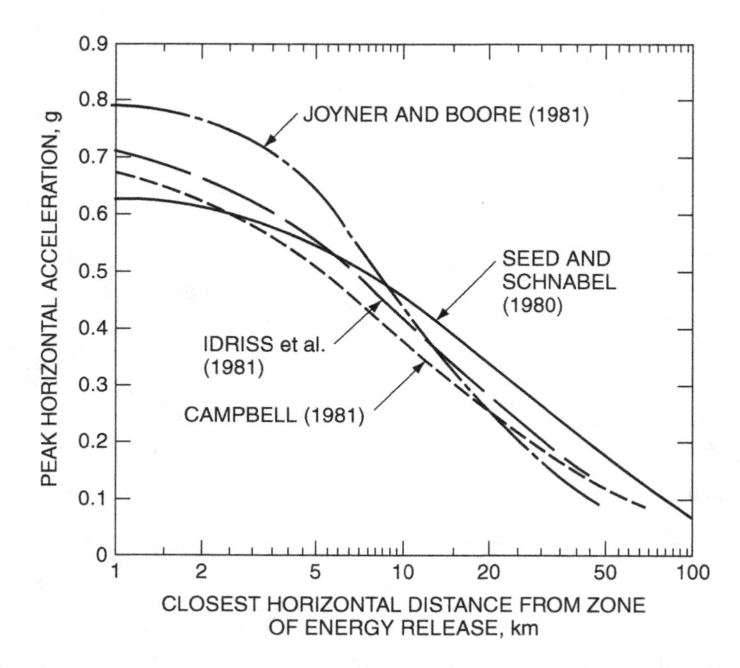

Figure 11.9 Peak ground acceleration curves for stiff soils. Reproduced from Seed and Idriss (1982) by permission of the Earthquake Engineering Research Institute (EERI).

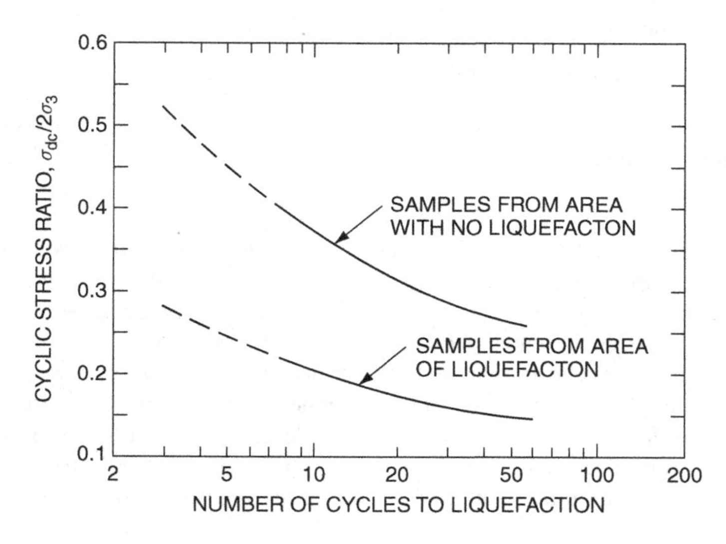

Figure 11.10 Results of cyclic loading triaxial tests on undisturbed samples, Case Study no. 1. Reproduced from Seed and Idriss (1982) by permission of the Earthquake Engineering Research Institute (EERI).

slightly higher value (say as high as 0.18 g) may well have developed in the non-liquefied zones.

For the soil conditions in Niigata, the critical depth for liquefaction was approximately 20 ft. In accord with this fact and utilizing the data presented above, evaluations of the liquefaction potential in the two zones can be made as follows:

Peak ground acceleration $\approx 0.16g$
At a depth 20 ft:

$$\sigma_0 \approx 20 \times 115 = 2300\,\text{psf [density of soil 115 lb/ft}^3]$$

$$\sigma_0' \approx 4 \times 115 + 16 \times 52.5 = 1300\,\text{psf [submerged density of soil} = 52.5\,\text{lb/ft}^3]$$

Cyclic stress ratio developed by earthquake:

$$\left(\frac{\tau_{av}}{\sigma_0'}\right)_d \approx 0.65 \cdot \frac{a_{max}}{g} \cdot \frac{\sigma_0}{\sigma_0'} \cdot r_d$$

$$\approx 0.65 \times 0.16 \times \frac{2300}{1300} \times 0.93 \approx 0.17$$

A magnitude of 7.5 earthquake typically induces about 15 stress cycles, but at Kawagishi-cho liquefaction occurred after about 3 significant cycles. From the data in Figure 11.10, the cyclic stress ratio causing liquefaction in 3 cycles is

$$\tau_{av}/\sigma_0' \approx 0.57 \times 0.28 \approx 0.16.$$

Since the induced stress ratio exceeds that causing liquefaction in 3 stress cycles, it can be concluded that sand in this area is likely to be liquefied early on during the onset of the earthquake, *which is in fact what happened.*

Zone of No Liquefaction

Peak ground acceleration ≈ 0.16 to 0.18 g.
At depth 20 ft

$$\sigma_0 \approx 20 \times 115 = 2300\,\text{psf [density of soil 115 lb/ft}^3]$$

$$\sigma_0' \approx 4 \times 115 + 16 \times 52.5 = 1300\,\text{psf [submerged density of soil} = 52.5\,\text{lb/ft}^3]$$

Cyclic stress ratio developed by earthquake for $a_{max} = 0.18$ g.

$$\left(\frac{\tau_{av}}{\sigma_0'}\right)_d \approx 0.65 \times 0.18 \times \frac{2300}{1300} \times 0.93 \approx 0.19$$

A magnitude of 7.5 earthquake produces 15 equivalent uniform stress cycles. The cyclic stress ratio causing liquefaction in 15 cycles is

$$\tau_{av}/\sigma_0' \approx 0.57 \times 0.34 \approx 0.195$$

based on the data in Figure 11.10.

Thus the soil in this zone would have a small margin of safety against liquefaction at a depth of 20 ft. It may be noted that if the soil does not quite liquefy, it still remains stable. However, the observed stability is not necessarily indicative of a possible imminent threat of catastrophic failure due soil liquefaction.

It can be seen from this case study that the use of a ground motion attenuation curve, together with simple theory, and laboratory test data on undisturbed samples taken in thin-wall tubes could correctly predict the soil behaviour in the Niigata area. Similar conclusions can also be drawn for other medium dense sand deposits elsewhere in the region (Seed and Idriss, 1982).

11.4 SPT Correlation for Assessing Liquefaction

A method for determining the liquefaction induced by cyclic shear stresses from results of laboratory experiments was outlined in the previous section. However, it was recognized that for satisfactory evaluation, recovery of high quality undisturbed samples is required. Unfortunately, in-situ states cannot be easily replicated in the laboratory. Granular soils retrieved by typical drilling techniques are usually too disturbed to yield meaningful results. Specialized sampling techniques like 'ground freezing' are prohibitively expensive for most projects other than the very critical ones. One of the practical options is to use data from field tests. Several popular field tests currently in use are: (1) the standard penetration test (SPT), and (2) the cone penetration test (CPT). The corresponding most popular methods for determining liquefaction potential are the in-situ SPT (Seed *et al.*, 1985) and CPT (Robertson and Campanella, 1985).

In the following sections in this book, the SPT procedure will be discussed followed by a worked example. The references cited have also discussed the CPT procedure but they have not been included here. The reader is encouraged to review the references cited.

A method of determining the liquefaction-inducing cyclic stress ratio as a function of 'N' value (blow counts) in SPT was proposed by Seed *et al.* (1983, 1985). SPT has been the preferred method for field tests amongst engineers and also has the advantage of being supported by a vast database of information on sand deposits where liquefaction has already taken place.

The cyclic stress ratio, believed to have developed at a site, was evaluated on the basis of maximum ground acceleration, after appropriate site adjustments had been made. The cyclic stress ratio, thus obtained, was plotted against SPT 'N' value obtained at the site. If there was any liquefaction observed on the ground surface, the data points were marked on the plot with solid circles. Sites where no evidence of liquefaction on the surface was visible were marked with open circles. Then a line was drawn (conservatively) encompassing the lowest values of the open circles. This line is regarded as the boundary denoting those conditions that lead to the early stages of liquefaction.

11.4.1 Current State of the Art

The method proposed by Seed and Idriss (1971), and Seed *et al.* (1983, 1985), became standard practice throughout North America and other parts of the world. The methodology, empirical in nature, has been modified and improved over the years. To review the developments since the publication of the major papers by Professor Seed and others, a workshop was convened by Professors Youd and Idriss in 1997. The primary purpose was to arrive at a consensus on the updates and augmentations in use and produce a revised 'State of the Art' summary. The revised summary was published in 2001 (Youd *et al.*, 2001).

A major focus of the workshop was on procedures for evaluating cyclic resistance ratio (*CRR*) required to cause liquefaction.

The criteria for evaluation of liquefaction resistance based on SPT have evolved over the years. Those criteria are largely reflected in the *CSR* versus $((N_1)_{60})$ plot reproduced in Figure 11.11.

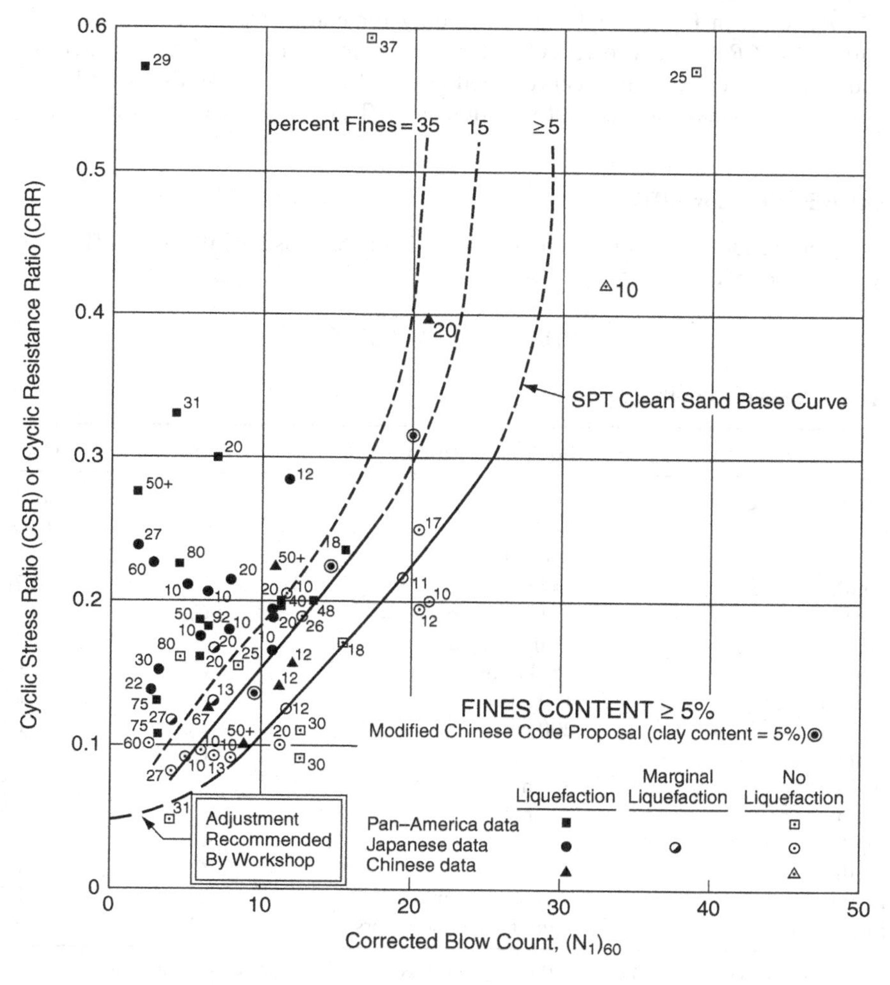

Figure 11.11 SPT Clean Sand (NCEER/NSF Workshop, 2001).

$(N_1)_{60}$ is the SPT blow count normalized to an overburden pressure of approximately 100 kPa (1 ton/sq ft) and a hammer energy ratio or hammer efficiency of 60 %. The normalization factors for these corrections are discussed in detail later (see Table 11.5). Figure 11.11 shows cyclic stress ratios (*CSR*) or cyclic resistance ratios (*CRR*) on the Y coordinate. Figure 11.11 is a graph of calculated *CSR* and corresponding $(N_1)_{60}$ data from sites where liquefaction effects were or were not observed following past earthquakes with magnitudes of approximately 7.5. The *CRR* curves on this graph were conservatively positioned to separate regions with data indicative of liquefaction from regions indicative of non-liquefaction.

The *CRR* curve for fines content < 5 % is the basic penetration criterion for the simplified procedure and is referred to hereafter as the 'SPT clean sand base curve'.

Curves were developed for granular soils with the fines content of 5 % or less, 15 % and 35 % as shown on the plot. The influence of fines content is discussed in Section 11.5.

The *CRR* curves in Figure 11.11 are valid only for magnitude 7.5 earthquakes. Scaling factors to adjust *CRR* curves to reflect differences in number of cycles compared to other magnitudes are addressed in the report (Youd *et al.*, 2001) which are discussed later. Other correction factors to take into account variations in SPT practices are discussed next.

Correction Factors for SPT

Several factors in addition to fines content and grain characteristics influence SPT results and these are detailed in Table 11.5.

$$(N_1)_{60} = N_M C_N C_E C_B C_R C_S \tag{11.7}$$

Table 11.5 Corrections to SPT.

Factor	Equipment variable	Term	Correction
Overburden pressure	—	C_N	$(P_a/\sigma'_{vo})^{0.5}$
Overburden pressure	—	C_N	$C_N \leq 1.7$
Energy ratio	Donut hammer	C_E	0.5–1.0
Energy ratio	Safety hammer	C_E	0.7–1.2
Energy ratio	Automatic-trip Donut-type hammer	C_E	0.8–1.3
Borehole diameter	65–115 mm	C_B	1.0
Borehole diameter	150 mm	C_B	1.05
Borehole diameter	200 mm	C_B	1.15
Rod length	< 3 m	C_R	0.75
Rod length	3–4 m	C_R	0.8
Rod length	4–6 m	C_R	0.85
Rod length	6–10 m	C_R	0.95
Rod length	10–30 m	C_R	1.0
Sampling method	Standard sampler	C_S	1.0
Sampling method	Sampler without liners	C_S	1.1–1.3

Source: (Reproduced from Youd *et al.* (2001) by permission of the American Society of Civil Engineers (ASCE))

where

N_M – measured standard penetration
C_N – is a correction factor for overburden pressure
C_E – correction for hammer energy ratio (ER)
C_B – correction factor for bore hole diameter
C_R – correction factor for rod length
C_S – correction for samplers with or without a liner

A commentary on the use of various factors has been provided by Youd *et al.* (2001) and an extract is reproduced below for the benefit of the readers.

Commentary **Youd *et al.* (2001)**

Because SPT x-values increase with increasing effective overburden stress, an overburden stress correction factor is applied (Seed and Idriss 1982). This factor is commonly calculated from the following equation (Liao and Whitman 1986a):

$$C_N = (P_a/\sigma'_{vo})^{0.5} \qquad (9)$$

where C_N normalizes N_m to an effective overburden pressure (σ'_{vo} of approximately 100 kPa (1 atm) P_a. C_N should not exceed a value of 1.7 [A maximum value of 2.0 was published in the National Center for Earthquake Engineering Research (NCEER) workshop proceedings (Youd and Idriss 1997), but later was reduced to 1.7 by consensus of the workshop participants]. Kayen *et al.* (1992) suggested the following equation, which limits the maximum C_N value to 1.7, and in these writers' opinion, provides a better fit to the original curve specified by Seed and Idriss (1982):

$$C_N = 2.2/(1.2 + \sigma'_{vo}/P_a) \qquad (10)$$

Either equation may be used for routine engineering applications.

The effective overburden pressure σ'_{vo} applied in (9) and (10) should be the overburden pressure at the time of drilling and testing. Although a higher ground-water level might be used for conservatism in the liquefaction resistance calculations, the C_N factor must be based on the stresses present at the time of the testing.

The C_N correction factor was derived from SPT performed in test bins with large sand specimens subjected to various confining pressures (Gibbs and Holtz 1957; Marcuson and Bieganousky 1997a,b). The results of several of these tests are reproduced in Fig. 3 in the form of C_N curves versus effective overburden stress (Castro 1995). These curves indicate considerable scatter of results with no apparent correlation of C_N with soil type or gradation. The curves from looser sands, however, lie in the lower part of the C_N range and are reasonably approximated by (9) and (10) for low effective overburden pressures [200 kPa (<2 tsf)]. The workshop participants endorsed the use of (9) for calculation of C_N, but acknowledged that for overburden pressures >200 kPa (2 tsf) the results are uncertain. Eq. (10) provides a better fit for overburden pressures up to 300 kPa (3 tsf). For pressures >300 kPa (3 tsf), the uncertainty is so great that (9) should not be applied. At these high pressures, which are generally below the depth for which the simplified procedure has been verified, C_N should be estimated by other means.

Another important factor is the energy transferred from the falling hammer to the SPT sampler. An ER of 60 % is generally accepted as the approximate average for U.S. testing practice and as a reference value for energy corrections. The ER delivered to the sampler depends on the type of hammer, anvil, lifting mechanism, and the method of hammer release. Approximate correction factors (C_E = ER/60) to modify the SPT results to a 60 % energy ratio for various types of hammers and anvils are listed in the text as Table 11.5. Because of variations in drilling and testing equipment and differences in testing procedures, a rather wide range in the energy correction factor C_E has been observed as noted in the table. Even when procedures are carefully monitored to conform to established standards, such as ASTM D 1586–99, some variation in C_E may occur because of minor variations in testing procedures. Measured energies at a single site indicate that variations in energy ratio between blows or between tests in a single borehole typically vary by as much as 10 %. The workshop participants recommend measurement of the hammer

Commentary (Continued)

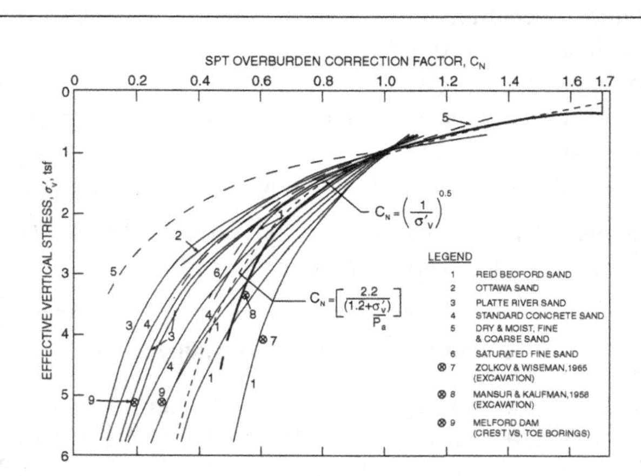

Fig. 3. C_N Curves for Various Sands Based on Field and Laboratory Test Data along with Suggested C_N Curve Determined from Eqs. (9) and (10) (Modified from Castro 1995)

energy frequently at each site where the SPT is used. Where measurements cannot be made, careful observation and notation of the equipment and procedures are required to estimate a C_E value for use in liquefaction resistance calculations. Use of good-quality testing equipment and carefully controlled testing procedures conforming to ASTM D 1586–99 will generally yield more consistent energy ratios and C_E with values from the upper parts of the ranges listed in the text as Table 11.5.

Skempton (1986) suggested and Robertson and Wride (1998) updated correction factors for rod lengths <10 m, borehole diameters outside the recommended interval (65–125 mm), and sampling tubes without liners. Ranges for these correction factors are listed in the text as Table 11.5. For liquefaction resistance calculations and rod lengths <3 m, a C_R of 0.75 should be applied as was done by Seed *et al.* (1985) in formulating the simplified procedure. Although application of rod-length correction factors listed in the text as Table 11.5 will give more precise $(N_1)_{60}$, values, these corrections may be neglected for liquefaction resistance calculations for rod lengths between 3 and 10 m because rod-length corrections were not applied to SPT test data from these depths in compiling the original liquefaction case history databases. Thus rod-length corrections are implicitly incorporated into the empirical SPT procedure.

A final change recommended by workshop participants is the use of revised magnitude scaling factors rather than the original Seed and Idriss (1982) factors to adjust CRR for earthquake magnitudes other than 7.5. Magnitude scaling factors are addressed later in this report.

Reproduced from Youd *et al.* (2001).

Magnitude Scaling Factor

The values of CRR presented in Figure 11.11 are due to an earthquake of moment magnitude 7.5. To adjust the values of *CRR* to some other value of magnitude, magnitude scaling factor (*MSF*) has been introduced. The *MSF* adjusts the values of CRR obtained for a different magnitude, to CRR values at magnitude 7.5. This is because the relevant curves have been developed for magnitude, $M = 7.5$. Hence CRR values need to be adjusted for $M = 7.5$. Thus

$$(\mathrm{CRR})_{\mathrm{M}} = (\mathrm{CRR})_{\mathrm{M}=7.5} \times \mathrm{MSF} \tag{11.8}$$

or

$$MSF = CRR_M / CRR_{M=7.5}.$$

Therefore the lower the earthquake magnitude, the higher the *MSF* and thus the higher the equivalent CRR. Magnitude Scaling Factors proposed by various authors are shown in Figure 11.12.

The relationship proposed by Idriss (1999) is of the form:

$$MSF = 6.9 \exp\left(\frac{-M}{4}\right) - 0.058 \tag{11.9}$$

$$MSF \leq 1.8 \tag{11.10}$$

These equations (11.9 and 11.10) lead to conservative values for magnitudes less than 7.5. Concluding on a conservative note, for magnitudes greater than 7.5 the form

$$MSF = 10^{2.24}/M_w{}^{2.56} \tag{11.11}$$

presented during the workshop may be used.

Magnitude scaling factors proposed by various investigators are shown in Figure 11.12.

Factor of Safety (FS)

The clean sand base or CRR curves in Figure 11.11 apply only to magnitude 7.5 earthquakes. (Note in this section whenever we refer to magnitude we mean moment magnitude.) To adjust the clean sand curves to a magnitude smaller or larger than 7.5, correction factors have been introduced (Seed and Idriss, 1982). These correction factors were termed 'magnitude scaling

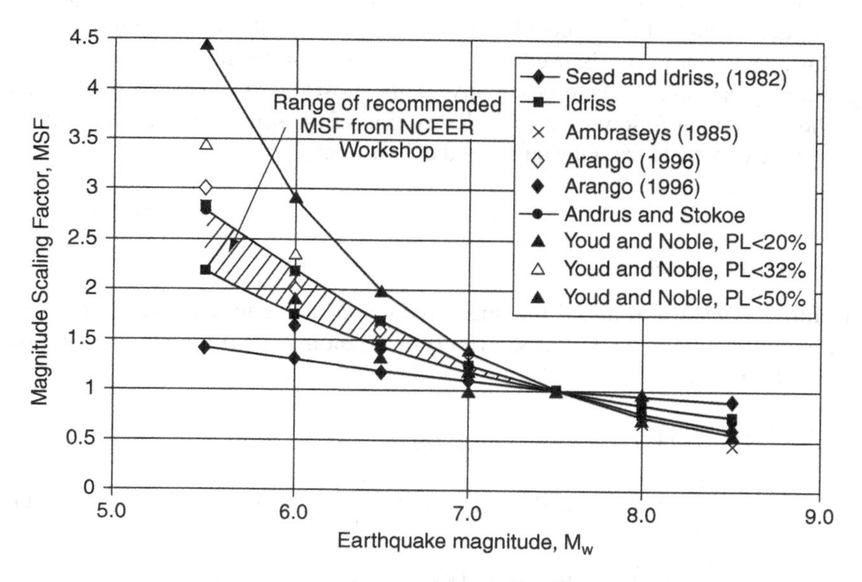

Figure 11.12 Magnitude scaling factor (MSF) values proposed by various investigators. Reproduced from Youd *et al.* (2001) by permission of the American Society of Civil Engineers (ASCE).

factors'. These factors are used to scale up or down on the *CRR* versus $(N_1)_{60}$ plots. Thus the factor of safety may be expressed as

$$FS = \left(\frac{CRR_{7.5}}{CSR}\right) \cdot MSF \tag{11.12}$$

where *CSR* is the calculated cyclic stress ratio generated by the earthquake shaking and $CRR_{7.5}$ is the cyclic resistance ratio for magnitude 7.5 earthquake. $CRR_{7.5}$ is determined from Figure 11.11.

The workshop also recommended procedures for CPT and other field testing methods and these may be found in Youd *et al.* (2001).

11.4.2 Most Recent Work

The most recent work on liquefaction assessment is due to Idriss and Boulanger (2004, 2005, 2006). Idriss and Boulanger have outlined a semi-empirical approach for the determination of various factors covered during the workshop (Youd *et al.*, 2001).

The Seed and Idriss (1971) simplified method, as outlined earlier, is used to determine the cyclic stress ratio (CSR) and has been retained by Idriss and Boulanger.

$$CSR = 0.65\left(\frac{\sigma_{vo}a_{\max}\cdot g}{\sigma'_{vo}}\right)\cdot r_d \qquad [11.4]$$

As explained earlier, $a_{\max}$ is the maximum horizontal acceleration at the ground surface in g's, σ_{vo} is the total vertical stress and σ'_{vo} is the effective vertical stress at depth '*h*'. The parameter r_d is a stress reduction coefficient that accounts for the soil column being flexible (Figure 11.4). The factor 0.65 is used to convert the non-uniform nature of the earthquake generated stress to a stress that is representative of the most significant cycles over the duration of the earthquake. What follows is a brief description of the semi-empirical form introduced in evaluating the constants from the suggested expressions.

Stress Reduction Coefficient

Seed and Idriss (1971) introduced the stress reduction factor to account for the flexibility of the soil column (Figure 11.2). Idriss (1999) in extending the original work proposed the following relationship:

$$\ln(r_d) = \alpha(h) + \beta(h)\cdot M \tag{11.13a}$$

$$\alpha(h) = -1.012 - 1.126\sin\left(\frac{h}{11.73} + 5.133\right) \tag{11.13b}$$

$$\beta(h) = 0.106 + 0.118\sin\left(\frac{h}{11.28} + 5.142\right) \tag{11.13c}$$

where 'h' is the depth in metres and 'M' is the moment magnitude. These equations are considered appropriate up to a depth of 34 m. For 'h' exceeding 34 m the following equation is more appropriate:

$$r_d = 0.12 \exp(0.22M) \tag{11.13d}$$

The above equations should provide a sufficient degree of accuracy for engineering applications.

Normalization of Penetration Resistance

Idriss and Boulanger (2004) recommend that the effective use of SPT blow-count as indices for soil liquefaction characteristics requires that effects of soil density and effective confining stress on penetration resistance be separated. Seed *et al* (1975) included the normalization in sand to an equivalent of $\sigma'_{vo} = 1$ atmosphere ($1P_a \approx 1tsf \approx 100kPa$). This normalization is currently of the form:

$$(N_1)_{60} = C_N(N_{60})$$

where $(N_1)_{60}$ is the SPT N value after correction to an equivalent 60 % hammer efficiency. The recommended relationship for C_N is of the form:

$$C_N = \left(\frac{P_a}{\sigma'_{vo}}\right)^m \tag{11.14}$$

where the exponent m is linearly dependent on D_R.

$$m = 0.784 - 0.521 \cdot D_R \tag{11.15}$$

Equation (11.15) should provide an adequate description of the SPT data over the range of D_R values most relevant to practice.

Modified Standard Penetration – $(N_1)_{60}$ Curve

The modified standard penetration curve (Idriss and Boulanger, 2004) is shown in Figure 11.13.

Provisions in Eurocode 8 (BS-EN 1998-5:2004) Part 5

Recommendations for potentially liquefiable soils are contained in Section 4.1.4 of the code. Empirical charts for simplified liquefaction analyses are provided in Annex B of the code. The charts have been drawn on the basis of field correlations between in-situ measurements and cyclic shear stresses known to have caused liquefaction during past earthquakes. Amongst the most widely used charts are those based on SPT blow-count. The recommended charts in Eurocode 8 for clean and silty sands are similar to those produced by Idriss and Boulanger (2004).

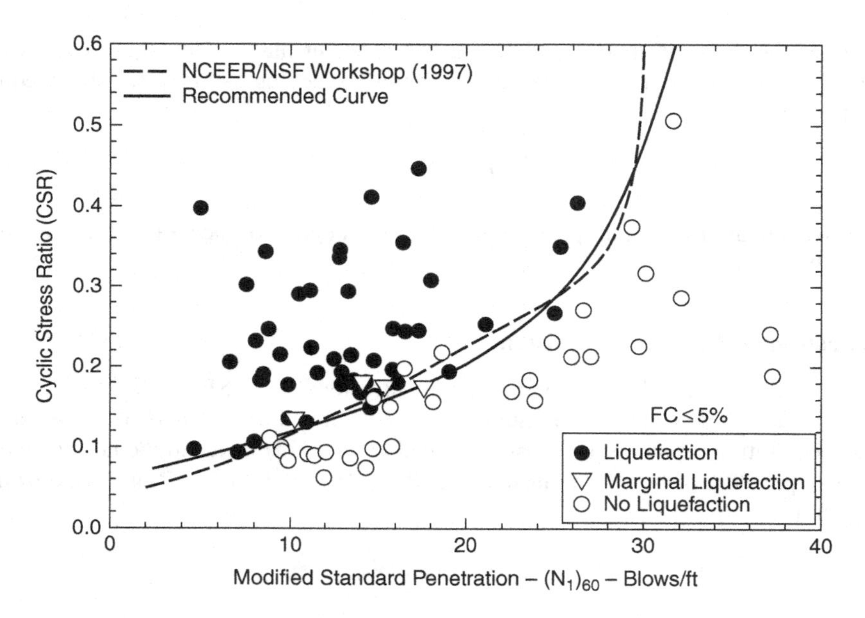

Figure 11.13 Modified Standard Penetration – $(N_1)_{60}$ – Curve. Reproduced from Idriss and Boulanger (2004), courtesy of the International Society for Soil Mechanics and Geotechnical Engineering.

11.4.3 Worked Example with SPT Procedure (Idriss and Boulanger, 2004)

Problem Definition

An area selected as a construction site for buildings is susceptible to liquefaction due to earthquake of magnitude $M_w = 7$.

Evaluate the liquefaction potential of the soil at a depth of 6 m below ground surface, given the following design data:

1. *Design data*

   (a) Ground motion at ground surface by an earthquake magnitude 7 with measured maximum acceleration approximately

$$a_{\max} = 0.16\,\text{g}$$

   (b) The soil at 6 m depth consists of very fine loose sand

$$D_{50} = 0.5$$

   (c) Number of stress cycles = 10
   (d) Water table below ground level = 1.5 m and

$$\text{Mass density of soil } (\gamma) = 1800\ \text{kg/m}^3$$

Calculate by SPT method the possibility of liquefaction at 6 m depth below ground surface.

2. *To establish the correlation between number of blows N and the soil properties by standard penetration tests (SPT)*

Standard penetration tests (SPT) were carried out at site in loose saturated sand at 6 m depth (normal atmospheric pressure)

In the tests a 50 mm tube is driven about 450 mm into the ground below the bottom of a borehole using a drop-hammer weighing 64 kg and falling 760 mm. The first 150 mm of penetration is neglected, in case the soil is disturbed. However, the number of blows required to cause the last 300 mm of penetration forms a guide to the resistance of the stratum. The values are then recorded.

These tests enabled us to formulate a reasonably reliable correlation between the N values and certain soil properties.

The results of the tests showed that the value of N varies between 4 and 10.

For the purpose of evaluation of liquefaction potential we assume an average value of N = 8.

3. *Evaluation of liquefaction potential*

The average blow-count, N = 8

The corrected blow-count factor = (assuming no energy or rod length correction factors) = $C_N \cdot (N)_{60}$

$$C_N = \left(\frac{P_a}{\sigma'_{vo}}\right)^m$$

σ_{vo} is the total vertical stress at depth h

$$= \gamma \cdot h = 18 \times 6 = 108$$

σ'_{vo} is the effective vertical stress

$$= \gamma \cdot 1.5 + (\gamma - 10) \times 4.5 = 18 \times 1.5 + 8 \times 4.5 = 63 \text{ kN/m}^2$$

m = a factor dependent on relative density, $D_r = 0.5$. Thus from Equation (11.15)

$$= 0.784 - 0.521 \times D_r = 0.784 - 0.521 \times 0.5 = 0.52 \; C_N = (100/63)^{0.52} = 1.27$$

$$(N_1)_{60} = 8 \times 1.27 = 10.17 = 10 \text{ (say)}$$

See Figure 11.13 to determine *CRR* for $(N_1)_{60}$

with $(N_1)_{60} = 11$

CRR (cyclic shear resistance ratio) = 0.11 for a magnitude of 7.5

We calculate *MSF* from equations 11.9 and 11.10 (conservative assessment)

$$MSF = 6.9 \exp\left(\frac{-M}{4}\right) - 0.058$$

$$= 6.9 \exp\left(\frac{-7}{4}\right) - 0.058$$

$$= 1.14$$

Therefore $(CRR)_{M=7} = 1.14 \times 0.11 = 0.1254$
We calculate r_d from equations 11.13a to 11.13d:

$$r_d \approx 0.90$$

CSR from equation 11.4.

$$\begin{aligned} CSR &= 0.65 \times (108/63) \times (0.16g/g) \times (0.90) \\ &= 0.16 \end{aligned}$$

This is greater than 0.1254.

So there is the possibility of liquefaction of soil at a depth of 6 m.
i.e. factor of safety $= CRR/CSR = 0.1254/0.16 = 0.78 < 1$.
The computations have been carried in compliance with Eurocode 8 (2004) as well.

11.5 Influence of Fines Content

The NCEER Workshop (Youd et al. *2001) discussed the liquefaction potential in cohesive soils and the influence of fines content.*

In the original development Seed *et al.* (1985) noted an apparent increase in CRR values with increased fines content. Whether this increase was caused by an increase of liquefaction resistance or a decrease in penetration resistance is not clear. Based on empirical data, curves for

% fines = 35 and
% fines = 15

were developed by Seed *et al.* and are shown in Figure 11.11.

The Workshop attendees recommended a revised correction for fines content detailed as follows;

$$(N_1)_{60CS} = \alpha + \beta(N_1)_{60} \tag{11.16}$$

where α and β = Coefficients determined from the following relationships:

$$\alpha = 0 \text{ for } FC \leq 5\% \tag{11.17a}$$

$$\alpha - \exp[1.76 - (190/FC^2)] \text{ for } 5\% < FC < 35\% \tag{11.17b}$$

$$\alpha = 5.0 \text{ for } FC \geq 35\% \tag{11.17c}$$

$$\beta = 1.0 \text{ for } FC \leq 5\% \tag{11.18a}$$

$$\beta = [0.99 + (FC^{1.5}/1{,}000)] \text{ for } 5\% < FC < 35\% \tag{11.18b}$$

$$\beta = 1.2 \text{ for } FC \geq 35\% \tag{11.18c}$$

These equations are recommended for routine liquefaction resistance calculations. A back-calculated curve for a fines content of 35 % is essentially congruent with the 35 %

curve plotted in Figure 11.11. The back-calculated curve for a fines content of 15 % plots to the right of the original 15 % curve.

11.6 Evaluation of Liquefaction Potential of Clay (cohesive) Soil

As discussed in the introduction cohesionless soil (sand) inherits the shear strength by the friction developed between the irregular surfaced graded grains and their closeness. The liquefaction phenomenon in cohesionless soil occurs mainly due to the incidence of high excess pore pressure and consequent decrease in effective stress resulting in the development of high strain and consequently with the collapse of the structure of the sand by the decrement of shear strength even reducing to zero during cyclic or seismic loading. In the laboratory tests it may be defined more precisely that the strain exceeds more than 3% shear strain.

In case of cohesive soil (clay), the structure of the clay and its behaviour are very complex, and somewhat different to that of cohesionless soil. In cohesive soil with grain sizes varying from 0.005 mm to 0.0001 mm, the grains are assumed to touch each other.

Questions regarding the potential liquefiability of finer, 'cohesive' soils (mainly silts and silty clays) have been a major focus for geotechnical specialists in recent years.

Traditional conditions for liquefiability of cohesive soils were based on the modified Chinese criteria and were introduced by Seed and Idriss (1982), i.e.

Percent finer than 0.005 mm < 15
Liquid Limit (LL) < 37
Plasticity Index (PI) < 12
Water Content > 0.85 LL

These criteria refer to liquefaction due to pore pressure development and do not address other important issues such as loss of soil strength due to remoulding of soil under cyclic loading due to 'sensitivity' and other such issues. These criteria have been the most widely used for defining and identifying potentially liquefiable soils for the past two decades.

11.6.1 Fresh Evidence of Liquefaction of Cohesive Soils

Two recent major earthquakes (Kocaeli, Turkey and Chi-Chi, Taiwan, 1999) have provided fresh new evidence of liquefaction-induced damage including settlements and/or partial or complete bearing failure at sites with cohesive soils. At all the sites in Turkey and Taiwan failures occurred *with soil types which appeared to be more 'cohesive' than would be expected based on the modified Chinese criteria.*

Liquefaction would be unexpected based on current understanding. Post earthquake studies (Bray *et al.*, 2000; Sancio *et al.*, 2002, and Sancio *et al.*, 2003) suggest that current understanding needs to be reviewed.

It is suggested that *it is not the percentage of 'clay-sized' particles that is important but the percentage of clay minerals present in the soil.* Fine quartz particles may sometimes be smaller than either 0.005–0.002 mm, but if largely non-plastic, these soil particles respond as cohesionless material in terms of liquefaction under cyclic loading (Seed *et al.*, 2003).

Figure 11.14 represents interim recommendations regarding 'liquefiability' of soils with significant fines contents. These may evolve further, based on work in progress, but for the

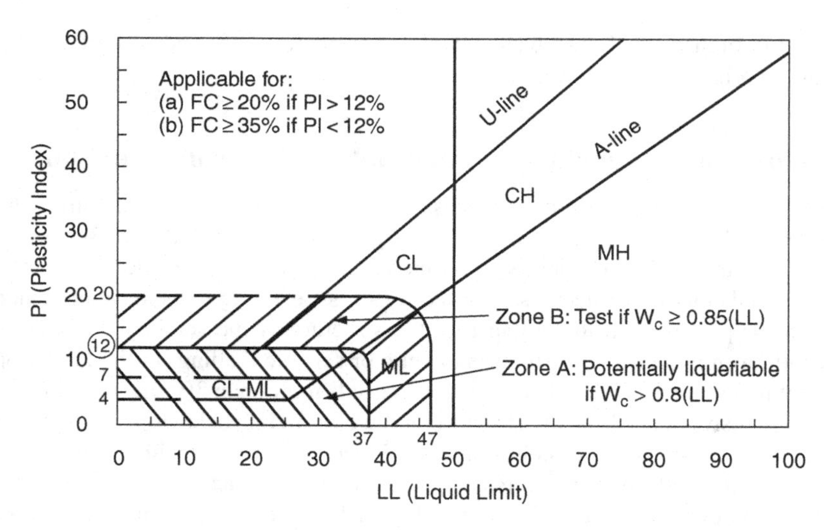

Figure 11.14 Recommendations regarding assessment of liquefiable soil types. Reproduced from Seed *et al.* (2003) by permission of the Earthquake Engineering Research Center (EERC).

time being this depicts a good summary of what is currently understood. For soils with sufficient fines content, where the fines separate the coarser particles and control the overall behaviour, the recommendations are: (1) soils within zone A are considered to be potentially susceptible to 'classic' cyclically induced liquefaction; (2) soils within zone B may be liquefiable; and (3) soils in zone C (not embedded within zones A or B) are generally susceptible to 'classic' cyclically induced liquefaction, but should be checked for potential sensitivity (loss of strength with remoulding or monotonic accumulation of shear deformation), Seed *et al.*, 2003.

A comprehensive picture regarding the potential for liquefiability of finer 'cohesive' soils is slowly beginning to emerge.

11.7 Construction of Foundations of Structures in the Earthquake Zones Susceptible to Liquefaction

The following procedures as suggested by D'Appolonia (1970) may be adopted in the construction of structural foundations:

1. Soil Replacement Method

- Remove the sandy soil stratum susceptible to liquefaction encountered in the foundation excavation.
- Replace it with non-liquefiable granular soil of low compressibility and high shear strength.
- Compact the soil with suitable compaction equipment to reach to a minimum desired relative density of 85 %. Large scale field tests should be carried out during the compaction operation.

- The area of replacement and compaction should extend to at least one-half the average width of the foundation of buildings in all directions.
- In case of machinery foundation, the depth of the replacement and compaction of soil should be carried out for a depth of 1.5 times the average width of foundation.

2. Vertical Drains Principle

Vertical drains principle is one of the options for alleviation of liquefaction problems. Vertical drains can be installed through liquefiable soils of either coarse or permeable type. The principle is that excess pore pressure is allowed to dissipate rapidly before reaching liquefaction conditions. Even if liquefaction occurs, fast dissipation after the shaking can rapidly reduce excess pore pressure and hence reduce the time spent in the unstable liquefied state.

A comprehensive study has been carried out by Onoue (1988). Other studies on the subject have been reported by Brennan (2004), and Brennan and Madabhushi (2002, 2005, 2006).

3. Vibroflotation method

- Several holes of required designed diameter in the foundation area are drilled through the susceptible liquefiable sand strata down to the dense soil stratum of adequate shear strength.
- The holes are then filled with a mixture of coarse grained sand and gravel.
- The fill material is densely compacted by means of vibroflotation method to form stone columns of adequate load bearing capacity to support the foundation of structures, as suggested by D'Appolonia, Miller and Ware (1955).

A design example of the vibroflotation method follows.

Design Example – Vibroflotation Method

In an industrial process plant in a seismically active zone, heavily loaded flat bottomed large diameter tanks were used for the chemical process requirements. The circular raft foundations for the tanks were to be supported directly on soil. The soil was composed of sand of considerable depth throughout the entire process area.

Given design data on foundation

(See Figure 11.15.)

- Total vertical load on foundation — $Gk = 45{,}000\,\text{kN}$
- Horizontal seismic load on tank — $Ek = 1136\,\text{kN}$
- Horizontal wind load on tank — $Wk = 100\text{kN}$

(Both the horizontal forces are assumed acting at half the tank height)

- Diameter of tank — $D = 16.8\text{m}$
- Height of tank from foundation level — $H = 12.8\text{m}$.
- Water Table: at Ground level
- Geotechnical data.

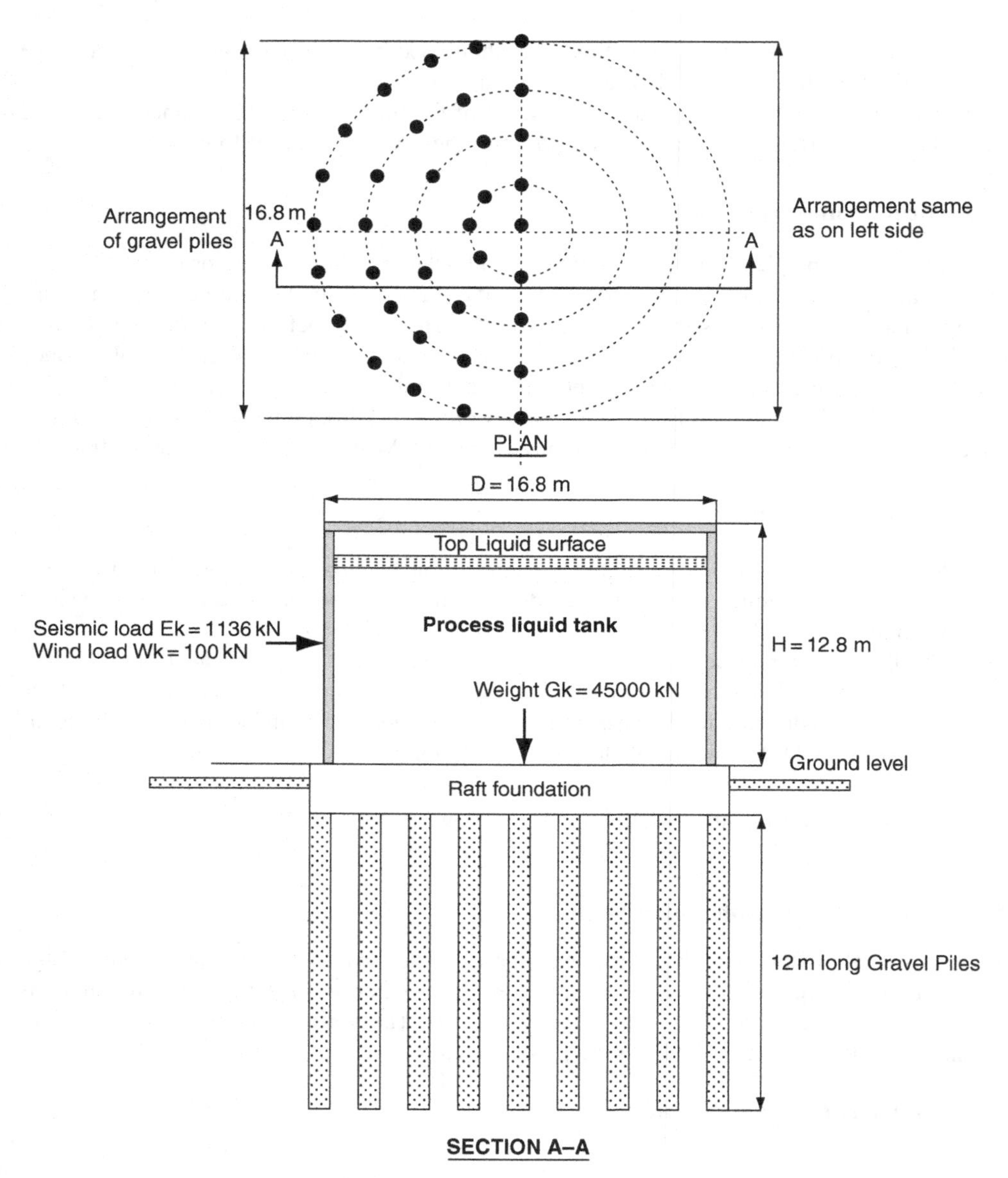

Figure 11.15 Liquid storage tank foundation.

Geotechnical data

The results of geotechnical field investigations and laboratory tests showed that the sand was poorly graded and the angle of shear resistance of soil low. The relative density of sand was between 0.6 and 0.65. The process site was in the seismic zone. So, there might have been the possibility of liquefaction in the sand stratum below the foundation due to a seismic event.

In normal conditions the allowable bearing capacity was limited to 200–240 kN/m^2 and even reduced to 150–175 kN/m^2 taking into account the development of a large bulb of pressure in the soil under the large diameter foundation raft subjected to high contact pressure.

Therefore to increase the allowable bearing capacity of soil and to eliminate the possibility of occurrence of the liquefaction, the compacted stone (gravel) piles were installed by driving hollow steel mandrels with false bottoms to the required depth of between 10 m and 12 m and at 2.0 m spacing all around the foundation area (see plan, Figure 11.5), removing the material from the mandrel and replacing with granular well-graded sand grain sizes (including gravel) suitable for well compaction and densifying the soil by vibroflotation compaction. After compaction the mandrels were withdrawn. Thus, the whole foundation area was covered with clusters of stone (gravel) columns ready to take up uniformly distributed loads over the entire area.

As a result of the application of vibroflotation, the allowable bearing capacity of soil was substantially increased to 300–350 kN/m^2 with increased internal angle of shearing resistance and with relative density reaching more than 75 %.

The important thing to note is that the gravel piles were placed within the zone of influence of each of the piles determined from experiments at the time. Further laboratory tests revealed improved liquefaction resistance in the surrounding areas and engineering judgement dictated further compaction being unnecessary. Also strength and compressibility improved with time and several such cases were reported by Lukas (1997). The measurements were carried over extended periods of time after completion.

Calculations of actual soil pressure due to (vertical + seismic + wind) loadings on foundation

Pressure on soil due to vertical load = Gk/A (where 'A' is the base area)

$$= 45,000/(\pi D^2/4) = 45,000/(\pi 16.8^2/4) = 203 \text{ kN/m}^2$$

Moment due to (seismic + wind) loads acting at half the height of tank, M_{sw}

$$= 1236 \times 12.8/2 = 7910 \text{ KN} \cdot \text{m}$$

Section modulus of base = $Z_x = \pi \cdot D^3/32 = 466 \text{ m}^3$
(where D = diameter of raft assumed the same as diameter of tank)
pressure due to (seismic + wind) load = $\pm 7910/466 = \pm 17 \text{ kN/m}^2$
Total maximum pressure on soil $p_{max} = 203 + 17 = 220 \text{ kN/m}^2$
$< 300 \text{ kN/m}^2$ (allowable)

Since the spacing of the gravel piles was within the zone of influence of the surrounding neighbouring piles, liquefaction was not expected and confirmation followed from further tests.

11.8 References

Bray, J.D. and Stewart, J.P., Co-ordinators (2000), 'Damage Patterns and Foundation performance in Adapazari, Kocaeli, Turkey, Earthquake of August 17, 1999'. Reconaissance Report, Youd, T.L. Bardet, J.P. Bray, J.D. eds. *Earthquake Spectra,* Supplement A, Vol. 16, 163–189.

Brennan, A.J. and Madabhushi, S.P.G. (2002) 'Effectiveness of vertical drains in mitigation of liquefaction', *Soil Dynamics and Earthquake Engineering*, Elsevier Science, **22**(9–12): 1059–1065.

Brennan, A.J. and Madabhushi, S.P.G. (2005) 'Liquefaction and drainage in stratified soil', *J. Geotechnical and Geoenvironmental Engg*, ASCE 131(7): 876–885.

Brennan, A.J. and Madabhushi, S.P.G. (2006) 'Liquefaction remediation by vertical drains with varying penetration depths', *Soil Dynamics and Earthquake Engineering*, Elsevier Science, **26**(9–12): 469–475.

Brennan, A.J., Govindaraju, L., Bhattcharya, S. (2007),'Liquefaction – susceptibility and remediation'. Proceedings of Intl Workshop on Earthquake and Geotechnical Engineering, Ed. S. Bhattacharya National Information Centre of Earthquake Engineering', NICEE, IIT, Kanpur, India.

Brennan, A.J. (2004), 'Vertical drains as a countermeasure to earthquake-induced soil liquefaction'. Ph.D. Thesis, University of Cambridge, UK.

British Standards Institution. (2005). BS EN 1998-1: 2005 Eurocode 8: *Design of structures for earthquake resistance Part 5: Foundations, retaining structures and geotechnical aspects*. Milton Keynes: BSI.

Castro, G. (1995), 'Empirical methods in liquefaction evaluation'. *Primer Ciclo de Conferencias Internationales*, Leonardo Zeevaert, Universidad Nacional Autonoma de Mexico City.

D'Appolonia, E. (1970), 'Dynamic loadings'. *J. Soil Mechanics and Foundations Division*, ASCE, Vol. 95, SM1.

D'Appolonia, E., Miller, C.E. and Ware, T.M. (1955). 'Sand compaction by vibroflotation'. *Transactions of ASCE*, Vol. 120, 154–168.

Gibbs, H.J. and Holtz, W.G. (1997), 'Research on determining the density of sand by spoon penetration testing'. *Proceedings of the 4th International Conference on Soil Mechanics and Foundation Engineering*. Vol. 1, 35–39.

Idriss, I.M. (1999), 'Earthquake ground motions at soft soil sites', Proceedings of the 2nd Intl Conf. on Recent Advances in Geotechnical Earthquake Engg and Soil Dyn., Vol. 3, 2265–2271.

Idriss, I.M., Boulanger, R.W. (2004), 'Semi-empirical procedures for evaluating liquefaction potential during earthquakes', Proceeding of the 3rd Int. Conf. on Earthquake Geotechnical Engg, University of California, Berkeley.

Idriss, I.M. and Boulanger, R.W. (2005), 'Evaluating liquefaction potential, consequences and mitigation'. Intl Conf. Indian Geotechnical Society, India, 3–25.

Idriss, I.M. and Boulanger, R.W. (2006), 'Semi-empirical procedures for evaluating liquefaction potential during earthquakes'. *Soil Dyn. Earthq. Engg,* Vol. 26, 115–130.

Ishihara, K. and Koga, Y. (1981), 'Case studies of liquefaction in the 1964 Niigata Earthquake'. *Soils and Foundations*, Vol. 21 No. 3, 35–52.

Kramer, S.L. (1996), *Geotechnical earthquake engineering*. Prentice Hall, New Jersey.

Kayen, R.E., Mitchell, J.K., Seed, R.V., Lodge, A., Nisho, S. and Coutinho, R. (1992), 'Evaluation of SPT-, CPT-, and shear-wave based methods for liquefaction potential using Loma Prieta data'. *Proceedings of the. 4th Japan-US Workshop on Earthquake Resistant Design of Lifeline Facilities. and Countermeasures for Soil Liquefaction,* Vol. 1, 177–204.

Liao, S. and Whitman, R.V. (1986), 'Overburden correction factors for SPT in sand'. *J. Geotech. Engg*, ASCE, Vol. 112 No. 3, 373–377.

Lukas, R.G. (1997), 'Delayed soil improvement after dynamic compaction', Geotechnical Special Publication , No. 69, Ground Improvement, Ground Reinforcement, Ground Treatment, 409–420.

Marcuson, W.F. III. and Bieganousky, W.A. (1997a), 'Laboratory standard penetration tests on fine sands', *J. Geotech Eng. Div.* (ASCE), 103(6), 565–588.

Marcuson, W.F. III. and Bieganousky, W.A. (1997b), 'SPT and relative density in course sands'. *J. Geotech Eng. Div.* (ASCE), 103(11), 1295–1309.

Onoue, A. (1988), 'Diagrams considering well resistance for designing spacing ratio of gravel drains'. *Soils and Foundation*, Vol. 28 No. 3, 160–168.

Robertson, P.K. and Campanella, R.G. (1985), 'Liquefaction potential of sands using the CPT'. *J. Geotech. Engg*, ASCE, Vol. 111 No. 3, 384–403.

Robertson, P.K. and Wride, C.E. (1998), 'Evaluating cyclic liquefaction potential using cone penetration test'. *Canadian Geotechnical J.,* 27(1), 151–158.

Sancio, R.B., Bray, J.D., Stewart, J.P., Youd, T.L., Durgunoglu, H.T., Onlap, A., Seed, R.B., Christencen, C., Baturay, M.B. and Karadayilar, T. (2002), 'Correlation between ground failure and soil condition in Adapazari, Turkey'. *Soil Dyn. Earthq. Engg*, Vol. 22, 1093–1102.

Sancio, R.B., Bray, J.D., Reimar, M.F. and Durgunoglu, H.T. (2003), 'An assessment of liquefaction susceptibility of Adapazari silt', Paper 172, Proceedings of the Pacific Conference on Earthquake Engineering, New Zealand.

Seed, H.B., Mori, K. and Chan, C.K. (1975), 'Influence of seismic history on the liquefaction characteristics of sands'. EERC Report No. 75-25, Earthquake Engineering Research Center, University of California, Berkeley.

Seed, H.B., Idriss, I.M. and Avango, I. (1983), 'Evaluation of liquefaction potential using field performance data'. *J. Geotechnical Engg,* ASCE, Vol. 109, GT3, 458–482.

Seed, H.B., Tokimatsu, K., Harder, L.F. and Chung, R.M. (1985), 'Influence of SPT procedures in soil liquefaction evaluations, *J. Geotechnical Engg*, ASCE, Vol. III No. 12, 1225–45.

Seed, R.B., Cetin, K.O., Moss, R.E.S., Kammerer, A.M., Wu, J., Pestana, J.M. (2003), 'Recent advances in soil liquefaction engineering: A unified and consistent framework', *EERC* Report no. 2003–06, Earthquake Engineering Research Center, University of California, Berkeley.

Seed, H.B. and Idriss, I.M. (1971), 'Simplified procedure for evaluating soil liquefaction potential'. *J. Soil Mechanics and Foundations Division*, ASCE, Vol. 97 No. 9, 1171–1182.

Seed, H.B. and Idriss, I.M. (1982), *Ground motion and soil liquefaction during earthquakes*. Earthquake Engineering Research Institute, *Engineering* Monographs on Earthquake Criteria, Structural Design and Strong Motion Records.

Seed, H.B., Tokimatsu, K., Harder, L.F. and Chung, R.M. (1985), 'The influence of SPT procedures in soil liquefaction resistance evaluations', *J. Geotechnical Engineering Division*, ASCE, 111(12), 1425–1445.

Skempton, A.K. (1986), 'Standard penetration test procedure and the effects in sands of overburden pressure, relative density, particle size, aging and overconsolidation' *Geotechnique,* 36(3), 425–447.

Terzaghi, K., Peck, R.B. and Mesri, G. (1996), *Soil Mechanics in Engineering Practice*, John Wiley and Sons, Inc., New York.

Yasuda, S., Morimoto, I., Kiku, H. and Tanaka, T. (2004) 'Reconnaissance Report on the Damage caused by Three Japanese Earthquakes, 2003'. *3rd Intern. Conference on Earthquake Geotechnical Engineering (ICEGE),*

Youd, T.L, Idriss, I.M. Eds. (1997), *Proceedings of the NCEER Workshop on evaluation of the liquefaction resistance of soils*. National Center for Earthquake Engineering Research, Buffalo, New York.

Youd, T.L., Idriss, I.M., Andrus, R.D., Arango, I., Castro, G., Christian, J.T., Dobry, R., Finn, W.D.L., Harder, L.F., Hynes, M.E., Ishihara, K., Koester, J.P., Liao, S.S.C., Marcuson III, W.F., Martin, G.R., Mitchell, J.K., Moriwaki, Y., Power, M.S., Robertson, P.K., Seed, R.B., Stokoe II, K.H. (2001), 'Liquefaction resistance of soils: Summary report from the 1996 NCEER and 1998 NCEER/NSF workshops on evaluation of liquefaction resistance of soils'. *J. Geotech. and Geoenv. Engg,* Vol. 127 No.10, Oct. 2001.

12

Performance Based Seismic Engineering – An Introduction

12.1 Preamble

Performance based seismic engineering (PBSE) is an innovative concept which has gained prominence as an approach allowing more transparent choices about how an engineered structure should perform in the event of an earthquake. Greater flexibility, with a non-prescriptive approach and the involvement of the stakeholder is at the core of this methodology.

Development of a fully performance based design approach within the framework of civil engineering is the preferred way forward in earthquake engineering. This is the path that the engineering community, the federal code agencies and other national code drafting authorities have decided to follow in the near future.

> The concept, if not new in theory is certainly most up-to-date in terms of practical implementation, according to which the design, evaluation, construction and maintenance of civil engineering structures, have to be devised and carried out in such a way that these facilities are guaranteed to meet a number of diverse performance objectives during their life cycle, tailored to the needs of the owners-users and the society as a whole. (Krawinkler, 1999)

Perhaps Krawinkler's most powerful observation, providing the basis for further progress is that the system be 'tailored to the needs of the owners-users'. Viewed from this angle, performance based seismic engineering is a new method that is slowly emerging in scope. It is a more sophisticated approach to disaster scenario planning than contemplated before, which allows for specific tailor-made designs of the system to meet requirements laid down by the owners-users.

May's 2002 report discusses the characteristics of introducing innovation. With reference to introducing PBSE to the engineering profession, the five attributes of relevance identified in the report are: (a) relative advantage; (b) compatibility; (c) complexity; (d) trialability; and (e) observability. 'Relative advantage' is the degree to which an innovation is perceived to be better than the idea it supersedes. 'Compatibility' is the degree to which an innovation is perceived as consistent with the existing values, past experiences and needs of potential adopters. 'Complexity' is the degree to which an innovation is perceived as relatively difficult to understand and use. 'Trialability' is the degree to which an innovation may be experimented with on a limited basis. 'Observability' is the degree to which the results of an innovation are visible to others. The attributes above help us to understand PBSE in the present context.

In its most simplistic terms PBSE allows for the design objectives selected by owners or other relevant decision makers to be met. The performance approach is different in content and thinking from anything currently in place. It is essentially thinking in terms of the ends rather than meeting prescribed limits, with a focus on the way the structure is supposed to respond. The design process becomes one of demonstrating through rational means, which may consist of calculations, finite element modelling of structure and joints, prototype testing or a combination of these, that the completed design will be capable of performing in the desired way.

'It is rather the focus on determining the quantitative nature of these objectives and the attempt to provide reliability to the predictions of the performance, both from the side of capacity and from that of demand, which sets the contemporary effort apart from older practice' (Krawinkler, 1999).

Area of Concern

Thus explicitly defining and quantifying the various aspects of the performance of the structure has become a key issue. While these objectives are being defined, an area of concern that emerges is that of developing effective approaches to ensure quality of construction, e.g. that the detailing developed at the joints is capable of meeting the ductility requirements set forth. Experience after the Northridge (1994) earthquake showed this could indeed be the weak link in the process, as a lack of quality control and poor detailing can affect the performance of the final product negatively (Negro and Mola, 2006).

12.2 Background to Current Developments

It may be argued that the first steps toward a 'limited' performance based design were taken in the 1960s when SEAOC published the following performance objectives:

1. No damage for minor levels of ground motion.
2. No structural damage but some non-structural damage for moderate ground motions.
3. Structural and non-structural damage but no collapse for major levels of ground motion (which includes the strongest credible earthquake at the site).

These performance objectives were ultimately incorporated in the Uniform Building Code (UBC) and have served the engineering community and society. Despite the acceptance and recognition of these SEAOC objectives in modern codes, life safety has been the principal concern in seismic design and it continues to be so. Explicit assessment of the expected performance of various components is not carried out, as suitable methods of doing so are not available.

Experiences from two key earthquakes, Loma Prieta (1989) and Northridge (1994) have been instrumental in promoting the development of PBSE. These events illustrated the urgency of a revised approach towards engineered structures with recognition of the tremendous financial and economic stakes of urban earthquakes. Facility owners and businesses were awakened to the realization that, even when buildings are built to modern code specifications, there can be months of interruption and disruption following an earthquake, sometimes leading to financial ruin. Seismic regulations in place focused on preventing loss of life and this objective was met to a large extent in the two earthquakes, but the financial losses resulting from downtime of facilities due to non-structural damage were significant. The losses were far greater than society, engineers and public policy makers were willing to accept for such moderate earthquake events. Interestingly, the loss due to non-structural damage and loss of service was far greater than due to structural damage. A need was identified for new building design, which in addition to preventing loss of life could also better meet society's expectations.

12.2.1 Efforts in the USA

Closely following the Loma Prieta (1989) and Northridge (1994) events, a devastating earthquake struck Kobe in 1995. As a result of observed damages in the two earlier earthquakes in the USA and the aftermath of Kobe in Japan, a national effort was launched in the States to revaluate existing design methods which though focusing on preventing loss of life did not explicitly address non-life threatening damage to facilities. This revaluation is an ongoing process; published documents include ATC 40 (1996), FEMA 350 and 356 (2000) and these address key aspects embodied in the concept of PBSE. Two documents in particular, SEAOC Vision (2000) and NEHRP Guidelines (1997), show the first attempts to develop more quantitative definitions of performance objectives, including specific terminology to describe different levels of damage to structural and non-structural elements. A range of four performance levels was thus introduced starting with (1) Operational (called Fully Functional in Vision 2000); to the next level (2) Immediate Occupancy; to (3) Life Safety; followed by (4) Collapse Prevention. A qualitative description of each performance was provided.

In addition different levels of ground intensity were considered in Vision 2000. The ground motions were identified for Frequent, Rare or Very Rare events corresponding to 50 %, 10 % and 2 % Probability of Exceedance in 50 years. The performance based approach enabled a choice to be made; a given performance level with a given earthquake intensity. SEAOC (1999) consolidated the various recommendations and published 'Tentative Guidelines for Performance Based Seismic Engineering'.

FEMA along with ATC spearheaded the development efforts in the USA and has planned a multi-agency development programme whose general approach is outlined in FEMA 349 (2000) 'Action Plan for Performance Based Seismic Design'. As a part of the initiatives undertaken, a new project has been launched by ATC to develop the next generation of PBSE procedures (ATC 58).

12.2.2 Efforts in Japan

Similar efforts are taking place in Japan. The Proceedings of the Bi-Annual US-Japan Workshop on Performance Based Earthquake Engineering provides recent trends in Japan (PEER Report 2000/10).

12.2.3 Efforts in Europe

Eurocode 8 is also based on the dual-design philosophy (see Chapter 8). The code specifies that the purpose of the design of structures in the event of a major earthquake is that human lives are protected and damage is limited. Thus life safety in the event of major ground shaking and guaranteeing limited damages to structural members and disruption of facilities during more frequent earthquakes is vital. The standard structural design procedure for collapse prevention performance level is force-based – a linear analysis for 5 % damped elastic spectrum reduced by the behaviour factor, q. This is similar to the Uniform Building Code (UBC); the elastic spectrum is reduced by strength factor R.

In the latest version of Eurocode 8 though, it is possible to implement a fully performance based approach. Unless a country objects to that in its National Annexes, it is allowed to design without employing the q factors, but directly on the basis of non-linear analysis with member verification on the basis of deformations. In this way Eurocode 8 paved the way for full displacement and deformation-based design in the next generation (Negro and Mola, 2006).

The displacement-based approach is the preferred way forward amongst researchers and code providers for the development of PBSE.

In this chapter . . .

The remainder of this chapter focuses on seismic performance and hazard levels as laid down in SEAOC 1999 (Section 12.3). Second generation PBSE objectives are introduced in section 12.4.

12.3 Performance-Based Methodology

The International Code Council (ICC), responsible for developing the International Building Code has in place a standing performance based code committee to focus and develop a performance based code. The result is a performance based code published under the title 'ICC Performance Code for Buildings and Facilities' (2001). Its stated aim is 'to provide appropriate health, safety, welfare and social and economic value, while performing innovative, flexible and responsive solutions that optimize the expenditure and consumption of resources'.

Also of relevance here are developments in performance based approaches in the fire safety community. The National Fire Protection Association (NFPA) has just completed a performance based design option (NFPA 5000, Building Code, 2003). Writing in the NFPA Journal, Harrington (2002) comments:

> The PBD (performance based design) option offers designers more flexibility and requires a greater level of sophistication than prescriptive design. Using the PBD option designers will have to prove to the authority having jurisdiction (AHD) that a building's design meets the code's goals and objectives using scientific methods, including computer and physical models that address the various design-hazard scenarios. And some of the scenarios are extremely challenging [. . .] By incorporating the PBD option in to our new Building Code, NFPA shows it's facing the twenty-first century headlong.

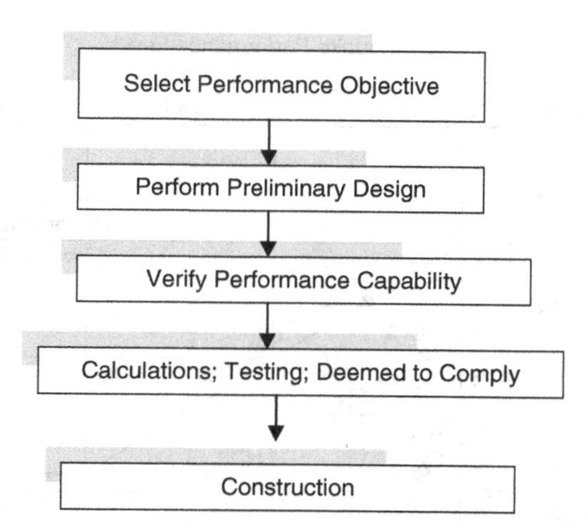

Figure 12.1 PBSE concept. After the International Code Council, 2001.

Developments within the seismic community are equally forward thinking.

Figure 12.1 shows the process of performance based design as envisaged by the ICC. In addition to verification that a design will be capable of meeting the selected performance objectives by the usual means of analysis, calculation, prototype testing etc., it also allows verification through conforming to relevant standards.

While the last two items in Figure 12.1 are self-explanatory, the first steps are elaborated further.

12.3.1 Performance Objectives

A performance objective stated simply is the desired structural performance when subjected to a strong ground motion of specified severity. A seismic performance objective essentially consists of two parts – (1) defining a damage state and (2) a level of hazard. For example, Life Safety and Immediate Occupancy (see below) are descriptors of damage states. The standard performance objectives matrix, as set out in SEAOC (1999) is shown in Figure 12.2.

From the matrix shown above it is easy to deduce that a performance objective may be a single-level performance objective or a multiple-level performance objective. A single-level performance objective considers a building performance for one level of ground motion. A multiple level performance objective, on the other hand, considers damage states for several states of ground motion. Defined in this way, a performance objective addresses a specified risk or a set of risks. A performance objective should be selected based on several factors. For example for a building, it may include:

1. building's occupancy;
2. the importance of functions being carried out in the building (for example, an expensively equipped hospital);
3. considerations relating to building damage repair/equipment replacement;
4. costs;
5. costs related business interruption.

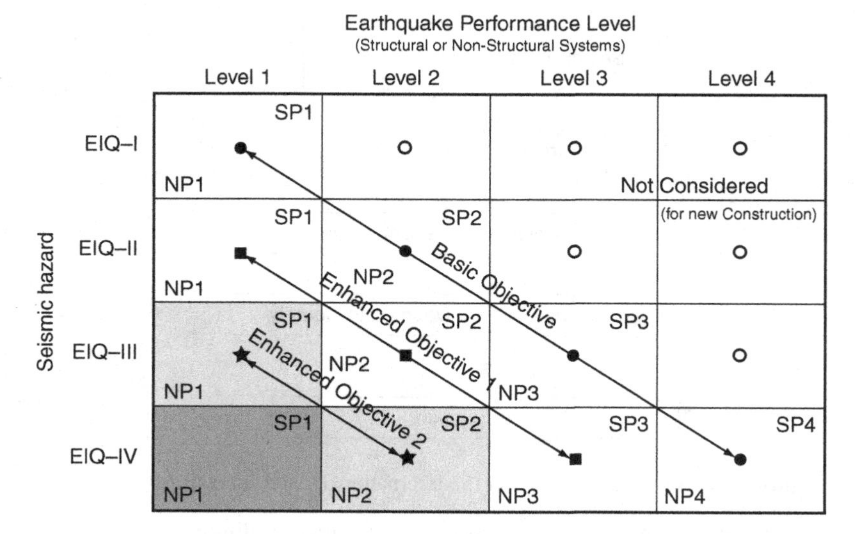

Commentary:

1. The performance objective matrix illustrates the typical performance levels to be achieved at each hazard level to meet the three defined performance objectives.
2. Meeting one performance level at one hazard level does not indicate that the specified performance objective is met. Performance should be checked at each hazard level to verify that the selected objective is met.
3. The performance objectives illustrated above indicate the same performance objectives for both structural and non-structural systems. Alternatively, separate and different objectives can be selected for the structural and non-structural systems.

Figure 12.2 Typical seismic performance objectives for buildings. Reproduced from SEAOC (1999) by permission of the Structural Engineers Association of California.

Remembering that each building and building owner are unique, the performance objectives are selected by the owner as being most appropriate for that project. The designer while trying to meet the specified objectives must also meet or exceed the minimum requirements of the applicable codes.

12.3.2 Performance Levels

System performance levels may be defined in terms of the structural and non-structural performance of a building at a specified hazard. SEAOC (1999) introduces the term inelastic displacement demand ratio (IDDR). This is shown in Figure 12.3 where an illustration of an inelastic displacement capacity curve is given.

Structural performance (SP) levels are defined in terms of structural system displacement response as measured by the IDDR (see Figure 12.3). Corresponding non-structural performance (NP) levels are defined in terms of expected loss-to-value ratios for the non-structural systems. Building performance may be described in terms of combinations of

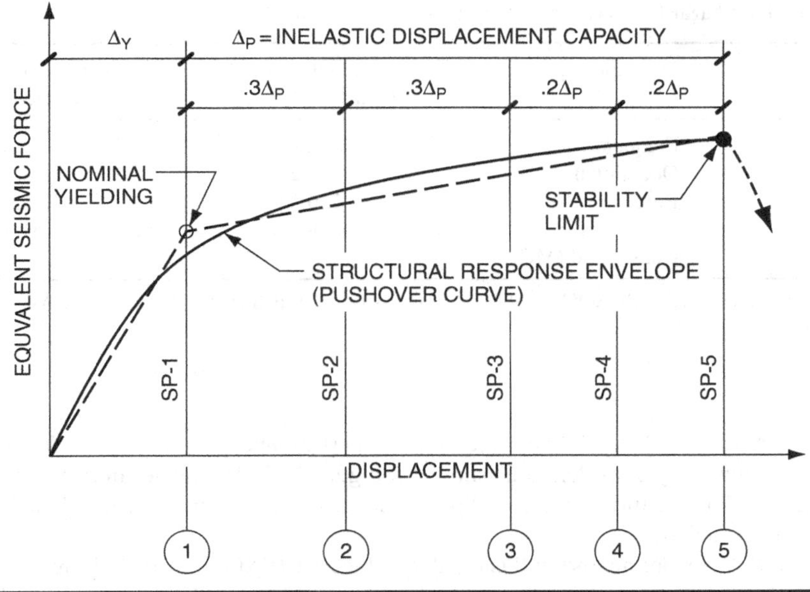

Structural Performance Level	Qualitative Description	System Displacement Limit	Inelastic Displacement Demand Ratio (IDDR)	Non-structural Performance Level	Non-structural Damage Ratio
SP1	Operational	Δ_Y	0%	NP1	0%–10%
SP2	Occuplable	$\Delta_Y + .3\Delta_P$	30%	NP2	5%–30%
SP3	Life safe	$\Delta_Y + .6\Delta_P$	60%	NP3	20%–50%
SP4	Near collapse	$\Delta_Y + .8\Delta_P$	80%	NP4	40%–80%
SP5	Collapsed	$\Delta_Y + \Delta_P$	100%	NP5	>70%

Figure 12.3 Illustration of structural performance levels and seismic response curve for structural systems. Reproduced from SEAOC (1999) by permission of the Structural Engineers Association of California.

structural performance (SP) and non-structural performance (NP) levels. For example, a building performance target may be SP1-NP3, meaning Structural Performance level 1 and Non-Structural Performance level 3 (SEAOC, 1995). Qualitative descriptions of structural and non-structural performance may be found in SEAOC (1995). Generally buildings that meet performance levels SP1-NP1 (Level 1) at the operational level earthquake remain functional. At the other end of the spectrum, buildings at SP5-NP5 (Level 5) may have total failure. The owner/stakeholder involvement is at the heart of PBSE.

12.3.3 Performance Objectives

Performance objectives are basically tied in with performance levels to be achieved at several specified seismic hazard levels (i.e. design earthquake levels – see Table 12.1).

Table 12.1 Four hazard levels specified in SEAOC (1995) guidelines levels.

Hazard class	Description	Mean return period (years)	Annual probability of exceedance (%)
EQ-I	Frequent	25	4.0
EQ-II	Occasional	72	1.4
EQ-III	Rare	250–800	0.12–.4
EQ-IV	Maximum Considered (MCE)	800–2500	0.04–0.12

Source: Reproduced from SEAOC (1999) by permission of the Structural Engineers Association of California.

In the SEAOC (1999) guidelines three typical performance objectives are defined based on common occupancy considerations shown in Figure 12.3. The owner in consultation with the engineer will define further enhanced performance objectives, which exceed the minimum requirements of the code.

SEAOC (1999) recommends the Basic Safety Objective (BSO) for PBSE design of standard occupancy buildings. In terms of structural systems, it is similar to the implied performance objectives in existing codes. In the case of non-structural systems it exceeds the implied objective of current codes by systematically addressing the non-structural systems. The BSO is similar to the implied performance objectives for school buildings in the State of California. This would correspond to SP1-NP1 performance in hazard classification EQ-I (see Figure 12.2).

12.3.4 Hazard Levels (design ground motions)

The selection of a set of performance objectives involves specification of a hazard level, defining the expected earthquake intensity for a given location. Hazard levels are usually specified in terms of a response spectrum. Ground motion time history analyses for the project are introduced later (see Chapter 6 and 7 for a fuller discussion).

In the SEAOC (1995) guidelines, four hazard levels are specified and are shown in Table 12.1.

Design ground motions should be based either on site-specific seismic hazard analysis or on generation of response spectra specified in the relevant codes.

The generation of site-specific hazard spectra has been discussed in Chapter 6 and uniform hazard spectra in Chapter 7. It was seen that generating a spectrum requires the definition of several parameters, most of which pertain to site characteristics.

Site-specific hazard analysis is recommended for very soft soils that are vulnerable to failure due to cyclic degradation during a strong motion earthquake. Certain types of non-cohesive soils are vulnerable to liquefaction and should be specially investigated. Site-specific studies are also recommended for structures located near an active fault.

SEAOC (1999) provides guidelines for establishing the design ground motions for the hazard levels indicated in Table 12.1 where site-specific studies are not available.

12.4 Current Analysis Procedures

Evaluation of engineering demand parameters (EDPs) on which seismic performance may be based is a key step in the whole PBSE approach. It requires an analytical tool to assess what happens to a building in terms of damage when subjected to a ground shaking that that exceeds the elastic capacity of the structural system.

Fundamentally, the analytical exercise should enable a comparison of a demand parameter with a corresponding capacity parameter for a building.

Base shear (total horizontal force at the support level of the building) has been the traditional parameter used in the design of buildings. The base shear force demand that would be generated by the input ground motion is compared against the base shear force capacity of the building. The design earthquake forces are reduced by a design factor '*R*' obtained from the codes to keep the design process in the elastic range. A schematic of the design process (ATC 1996) is shown in Figure 12.4.

When working in the inelastic range, the generalized representation of capacity as a force makes it difficult to investigate the actual damage to the specific parts of the building, since damage is progressive. In the process some parts of the building yield while redistribution of forces take place; inelastic demands vary from component to component. Thus component damage/damage to specific parts of the building would be sensitive to displacements and cannot be investigated rationally through calculated elastic forces. Short of non-linear dynamic time history analysis (which requires availability of ground motion records that are representative of the hazard at the site), presently employed procedures have to rely on simplified approaches.

A few years after the Loma Prieta (1989) earthquake, an association of the Applied Technology Council (ATC), the American Society of Civil Engineers (ASCE) and the Building Seismic Safety Council (BSSC) began work on the development of FEMA (Federal Emergency Management Agency) 273, NEHRP (National Earthquake Hazard Reduction Program) Rehabilitation Guidelines. The publication was specifically focused on: (a) providing performance based design criteria and (b) procedures for seismic upgrade of existing buildings. Almost at the same time the Applied Technology Council began developing a similar document (ATC 40) for rehabilitation of concrete buildings in the California region at the behest of the California Seismic Safety Commission.

The prevailing thought veered towards adopting a simplified procedure to achieve the goals of PBSE. Both FEMA 273 and ATC 40 adopted an approach to performance evaluation based on non-linear static analysis known as 'pushover' analysis.

In 'pushover' analysis the structure is idealized by an assembly of component models capable of representing the structural behaviour under seismic loading. A load pattern (under the presence of gravity loading) is applied; the structure is gradually pushed monotonically (in small increments) to large inelastic deformations until a target value is reached at a reference point of the structure – usually the centre of mass at the roof level. A schematic of the pushover analysis is shown in Figure 12.5

Non-linear static procedures are a departure from the traditional approach in several ways. To start with, the basic demand and capacity parameter for analysis is the lateral displacement of the building. A schematic representation of the capacity curve identifying some of the damage states is shown in Figure 12.6.

Referring to the curve, a point on the curve defines a specific damage state for the structure, since the deformation of all the components can be related to the global displacement of

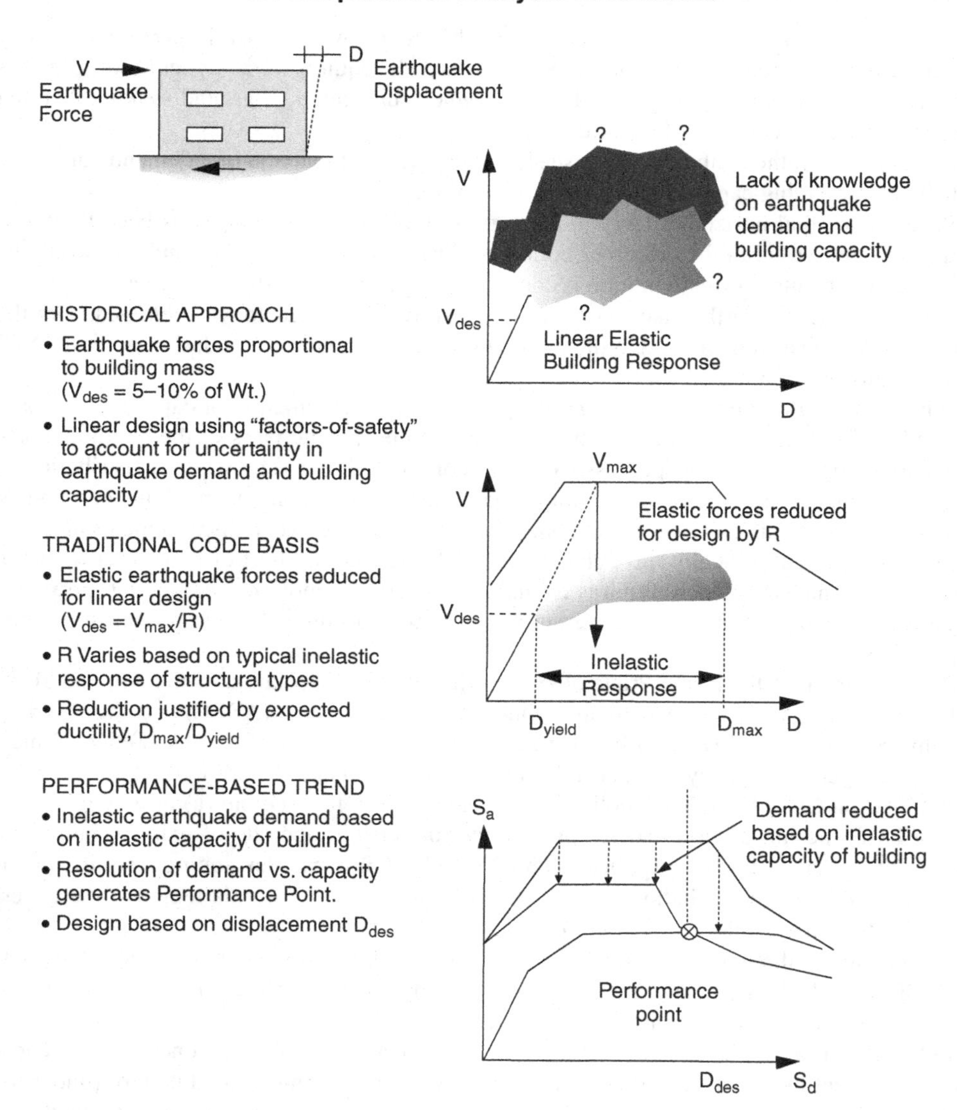

Figure 12.4 Evolution of seismic design (ATC 1996). Reproduced from ATC 40 (1996) by permission of the Applied Technology Council (ATC).

the structure. By correlating this capacity curve to the seismic demand generated by a specific earthquake (identified at the hazard level definition stage) a point can be found on the capacity curve that would identify the damage state. This is called a performance point. The location of the performance point relative to the performance level defined by the capacity curve indicates whether the performance objective has been met.

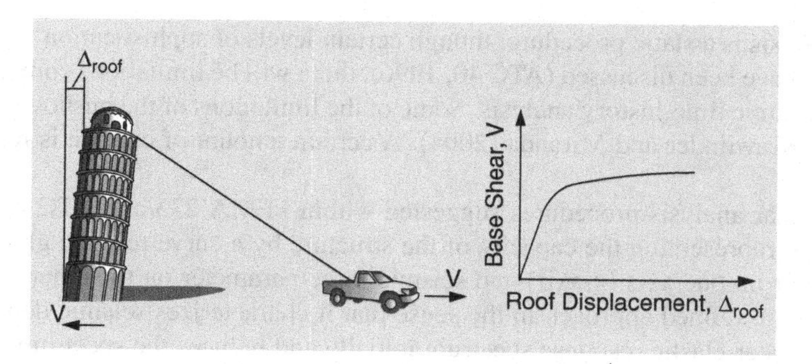

Figure 12.5 Schematic of 'pushover' analysis. Reproduced from ATC 40 (1996) by permission of the Applied Technology Council (ATC).

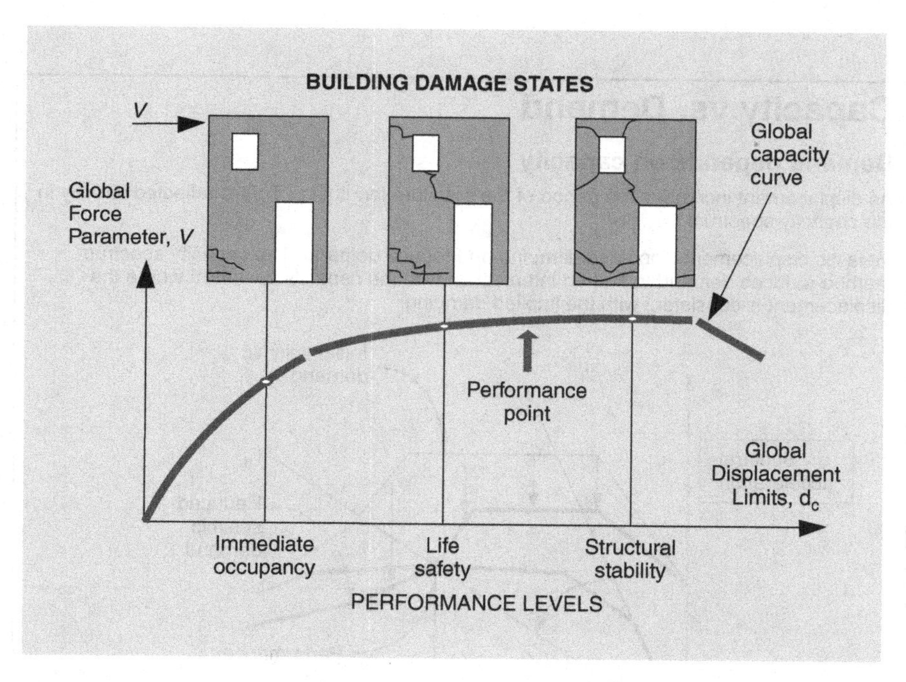

Figure 12.6 Building capacity curve and global displacement capacities d_c for various performance levels. Reproduced from Comartin *et al.* (2000) by permission of the Earthquake Engineering Research Institute (EERI).

12.4.1 ATC Commentary

The capacity curve is generally constructed to represent the first mode response of the structure based on the assumption that the fundamental mode of vibration is the predominant response of the structure. This is generally valid for buildings with fundamental periods of vibration up to one second. For more flexible buildings with a fundamental period greater than one second, the analyst should consider addressing higher mode effects in the analysis. (ATC 40, 1996)

Pushover analysis is a static procedure; though certain levels of sophistication for more flexible structures have been discussed (ATC 40, 1996), there will be limitations compared to a 3D non-linear dynamic time-history analysis. Some of the limitations of the pushover analysis are discussed by Krawinkler and Miranda (2004). A certain amount of caution is recommended in its use.

The non-linear analysis procedures suggested within FEMA 273 and ATC 40 are similar with respect to representing the capacity of the structure by a curve relating global displacement parameter on one axis (x-axis) and seismic force parameter on the other axis. ATC 40 advocates a more refined approach in the sense that it characterizes seismic demand using a 5 % damped linear-elastic response spectrum initially and reduces the spectrum to reflect the effects of energy dissipation to estimate the inelastic displacement demand (Comartin *et al.*, 2000). This is identified as the Capacity Spectrum method and a schematic of the process is shown in Figure 12.7.

Capacity vs. Demand

Demand depends on capacity

As displacement increases the period of the structure lengthens. This is reflected directly in the capacity spectrum.

Inelastic displacements increase damping and reduce demand. The capacity spectrum method reduces demand to find an intersection with the capacity spectrum where the displacement is consistent with the implied damping.

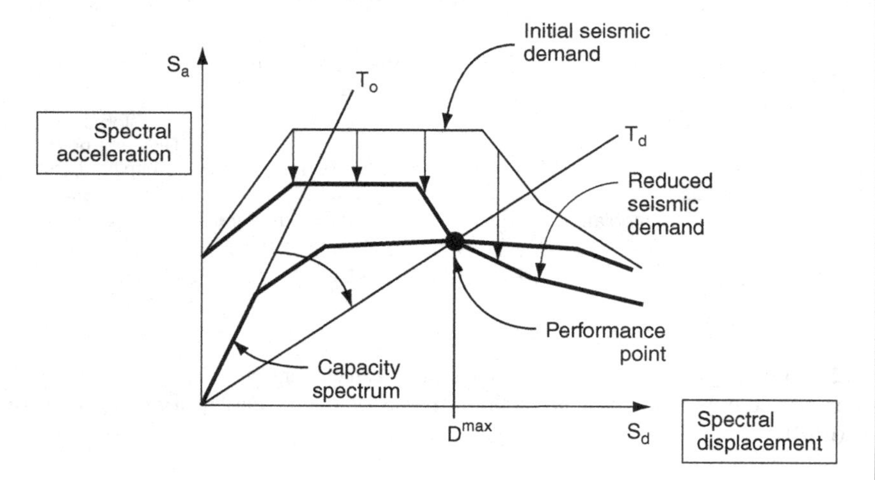

At the performance point; capacity and demand are equal.

The maximum displacement implies a unique damage state of the building related directly to a specific earthquake or intensity of ground shaking. The damage state comprises deformations for all elements in the structure. Comparison with acceptability criteria for the desired performance goal leads to the identification of deficiencies for individual elements.

Figure 12.7 Capacity Spectrum Method. Reproduced from ATC 40 (1996) by permission of the Applied Technology Council (ATC).

12.4.2 Displacement Design Procedures

From the foregoing it emerges that development of suitable displacement-based-design (DBD) methods is central to the further development of PBSE. An easy-to-follow DBD method is included in SEAOC (1999).

Contributions to DBD methods proposed recently have been numerous. Some of the contributors include: Moehle (1992), Kowalski *et al.* (1995), Panagiotakos and Fardis (1999), Chopra and Goel (1999), Kappos and Manafpour (2001), Aschheim and Black (2000), Chopra and Goel (2001), Priestley and Kowalsky (2000), and Humar *et al.* (2006).

Sullivan *et al.* (2003) presented the findings of a study that uses eight different DBD methods for seismic design. Five different case studies with identical sets of design parameters were undertaken for this exercise. Limitations of each of the methods were identified and reported. The study also showed that despite all the DBD methods using the same set of design parameters, a large variation in design strength is obtained. Performance of each method was compared with a non-linear time history analysis method. The performance assessment indicated that each of the eight methods provided designs that ensure limit states are not exceeded.

A matrix of design procedures proposed is shown in Table 12.2.

Table 12.2 Matrix of design procedures.

	Deformation – Calculation Based (DCB)[3]	Iterative Deformation – Specification Based (IDSB)[4]	Direct Deformation – Specification Based (DDSB)[5]
Response Spectra Initial Stiffness Method	Moehle (1992) FEMA (1997) UBC Panagiotakos and Fardis[1,2] (1999) Albanesi *et al.* (2000) Fajfar (2000)	Browning[1] (2001)	SEAOC (1999) Aschheim and Black (2000) Chopra and Goel (2001)
Response Spectra Secant Stiffness Based	Freeman (1978) ATC (1996) Paret *et al.* (1996) Chopra and Goel	Gulkan and Sozen (1974) Humar *et al.* (2006)	Kowalsky *et al.* (1995) Priestley and Kowalsky[1] (2000)
Direct Integration: Time History Analysis Based	Kappos and Manafpour[2] (2001)	N/A	N/A

Source: After Sullivan *et al.*, 2003.

Notes

[1] Method has been developed for particular structural types and not intended for application to other structural types.

[2] Method has been developed with specific limit states in mind that must be checked during design.

[3] DCB – Deformation Calculation Based: calculation of the expected maximum displacement for an already designed structural system.

[4] IDSB – Iterative Deformation-Specification Based: involves analysis of an already designed system for the maximum limit displacement. Structural changes are introduced such that the analytical displacements are less than the limit imposed. Involves iteration.

[5] DDSB – Direct Deformation-Specification Based: sets a pre-defined target displacement

12.5 Second Generation Tools for PBSE

Aspects of 'First Generation Tools' for PBSE have been broadly covered in the previous sections of this chapter, including the beginnings of a more client oriented design process. First Generation Tools' accounted for:

- Performance levels and objectives
- Hazard computations
  (Hazard level is identified for each performance level)
- Deformation-based analysis
  (non-linear static pushover analysis being pursued by engineers; displacement-based design methods emerging)
- Deformation based acceptance criteria
- Successfully used for retrofit of existing buildings

12.5.1 Targets for 'Second Generation Tools'

More pertinent questions from the viewpoint of the owner/stakeholder regarding the performance of the building and the damage state remain to be answered. Examples of possible questions include:

1. How much will it cost to repair?
2. How much of the costly equipment will be written off?
3. How long will the building remain vacant?
   (long downtime can hurt the business significantly)
4. Scale of human casualties?
   (an important consideration for the community)

These types of questions prompted the development of 'Second Generation Tools' ATC-58 (Phase 2), capable of a more holistic performance assessment of structures using a risk-based approach. In particular, probabilistic methods allow for a quantitative evaluation of the performance of a structural system under seismic loading. Furthermore these methods can also be extended to cover life-cycle assessments of the structure, providing a decision-making tool for all stakeholders involved.

Another benefit of using a quantitative approach is the capability of modelling the uncertainties inherent with the randomness of the seismic action (aleatory uncertainty) and those associated with the lack of knowledge of certain parameters (epistemic uncertainty Tekie and Ellingwood, 2003). The likelihood of hazard occurrence, performance goals and cost benefits may be combined to assess the safety over several performance levels or limit states.

Next generation tools will focus on:

- loss computations
- downtime business loss
- damage states
- fragility curves

- clear definitions of hazard levels at each performance level
  (site dependent hazard analysis to be encouraged)
- more refined non-linear analysis procedures.

Loss estimation is a serious business item which will be scrutinized by the stakeholders/owners and play a major part in the decision-making process. A useful parameter for loss estimation is the expected annual loss, i.e. the average loss that occurs every year in a building. Owners, lending institutions and insurers can then quantitatively compare with business profit and overheads, for example. Hence quantitative assessment of economic losses will constitute vital information for the owners/stakeholders, leading to more rational decision-making than is currently possible. Mathematical models of loss estimation have been discussed by Krawinkler and Miranda (2004).

Downtime losses are as important to owners as the economic losses discussed previously and are also difficult to quantify. Downtime losses can be due to a variety of factors, including loss of manufacturing capacity, loss of rent, loss of direct income, and inability to fulfil prior commitments thereby leading to loss. All of these can be significant. Quantification of downtime loss is an area of future research. In an interesting study on this type of loss in a premier educational campus, Comerio (2000) reported the effects of three earthquake scenarios at the University of Berkeley, California.

The development of quantitative methods for measuring performance has led to the use of fragility curves as a graphical tool for presenting the results of an analysis. Fragility functions are created to calculate the probability of damage to facility components when subjected to a given force, deformation or other engineering demand parameter. Damage data is required for creating the fragility functions. Damage data may be empirical, analytical or from expert opinion. The data would come with the knowledge of the engineering demand parameter to which the components are subjected, as well as the associated damage state. One of the ways of representing damage would be in terms of required repairs and other consequences. It may be imagined that creating fragility functions for a high valued hospital equipped with costly equipment to be protected would be a complicated process. Fragility functions and their derivation have been discussed by Porter *et al.* (2007).

12.5.2 Prognosis

At this point in time it is difficult to make a prognosis or gauge the speed with which innovations in performance based seismic engineering will be adopted and implemented. Performance objectives are selected by the owners/stakeholders. Viewed in this context performance based seismic engineering is a bold and innovative step forward; however, many uncertainties lie ahead. One certainty is that development of a fully performance based design approach within the framework of civil engineering, will certainly change the way structures are designed in the future. Concepts for the next generation tools have been formulated but rigorous methods and techniques are still largely on the drawing board. May (2002) observes that

> the promise of performance-based seismic engineering requires more than the development of sound methodologies and analytical tools. Such advances will be left on the conceptual drawing board unless they are adopted by the engineering profession and are effectively used to inform seismic

safety decisions. Recognizing this, it is important to remember that the adoption of new methods and tools is not automatic. The availability of a methodology or tool does not guarantee that it will be effectively employed. In short, it is a long way from the research laboratory to actual practice.

The Federal Emergency Management Agency has released a 10-year action plan (FEMA 349, 2000) for promoting PBSE and along with it hopes of a better future.

12.6 References

Albanesi, T., Nuti, C. and Vanzi, I. (2000) 'A simplified procedure to assess the seismic response of non-linear structures', *Earthquake Spectra* **16**(4): 715–34.

Applied Technology Council (2004) *ATC-58 Task Report Phase 2, Task 2.2. Engineering demand parameters for structural framing systems.* ATC, Washington, DC.

Aschheim, M.A. and Black, E.F. (2000) 'Yield point spectra for seismic design and rehabilitation'. *Earthquake Spectra* **16**(2): 317–36.

ATC-40 (1996) *Seismic Evaluation and Retrofit of Concrete Buildings.* Applied Technology Council, Redwood City, C.A.

British Standards Institution (2005) BS EN 1998-1: 2005 Eurocode 8: *Design of Structures for Earthquake Resistance Part 5: Foundations, Retaining Structures and Geotechnical Aspects.* Milton Keynes: BSI.

Browning, J.P. (2001) 'Proportioning of earthquake-resistant RC building structures', *J. Struct. Div., ASCE,* **127**(2), 145–151.

Chopra, A.K. and Goel, R.K. (1999) 'Capacity-demand-spectrum methods based on inelastic design spectrum'. *Earthquake Spectra* **15**(4): 637–56.

Chopra, A.K. and Goel, R.K. (2001) 'Direct-displacement-based design: Use of inelastic vs. elastic design spectra'. *Earthquake Spectra* **17**(1): 47–65.

Chopra, A.K. and Goel, R.K. (2002) 'A modal pushover analysis procedure for estimating seismic demands for buildings'. *Earthquake Engg Struct. Dyn.* **31**: 561–82.

Comartin, C.D., Niewiarowski, R.W., Freeman, S.A. and Turner, F.M. (2000) 'Seismic evaluation and retrofit of concrete buildings: A practical overview of the ATC 40 document'. *Earthquake Spectra* **16**(1): 241–261.

Comerio, M.C. (2000) *The Economic Benefits of a Disaster Resistant University: Earthquake Loss Estimation for U.C. Berkeley.* Report No. WP-2000-02, Institute of Urban Regional Development, University of California, Berkeley.

Fajfar, P. (2000) 'A nonlinear analysis method for performance-based seismic design', *Earthquake Spectra* **16**(3): 573–92.

FEMA 273/274 (1997) *NEHRP Guidelines for the Seismic Rehabilitation of Buildings,* Vol. I – Guidelines, Vol. II – Commentary. Prepared by ATC for the Building Seismic Safety Council, funded by the Federal Emergency Management Agency, Washington, DC.

FEMA 349 (2000) *Action Plan for Performance Based Seismic Design.* Federal Emergency Management Agency, Washington, DC.

FEMA 350 (2000) *Recommended Seismic Design Criteria for New Steel Moment-Frame Buildings.* Developed by the SAC Joint Venture for the Federal Emergency Management Agency, Washington, DC.

FEMA 356 (2000) *Prestandard and Commentary for the Seismic Rehabilitation of Buildings.* Federal Emergency Management Agency, Washington, DC.

Freeman, S.A. (1978) The prediction of response of concrete buildings to severe earthquake motion'. ACI Special Publication, SP-55, 589–605.

Gulkan, P. and Sozen, M., (1974), 'Inelastic response of reinforced concrete structures to earthquake motions', *ACI J.* **71**(12), 604–610.

Harrington, G. (2002) 'Performance based design option', *NFPA Journal,* Mar.–Apr. 2002, 28, 95.

Humar, J., Ghorbanie-ASL, M. and Pina, F. (2006) 'Displacement based seismic design of structures with significant higher mode contribution'. *Proceedings of the 1st European Conference on Earthquake Engineering and Seismology.* Paper no. 37. Geneva, Switzerland.

International Code Council (2001) *Performance Code for Buildings and Facilities 2001.* Washington, DC. ICC.

Kappos, A.J. and Manafpour, A. (2001) 'Seismic design of R/C buildings with the aid of advanced analytical techniques'. *Engg Struct.* **23**(4): 319–332.

Kowalsky, M.J. (2001) 'RC structural walls designed according to UBC and displacement based method'. *J. Struct. Engg.* **127**: 506–516.

Kowalsky, M.J., Priestley, M.J.N. and Macrae, G.A. (1995) 'Displacement-based design of RC bridge columns in seismic regions'. *Earthquake Engg Struct. Dyn.* **24**(12): 1623–1643.

Krawinkler, H. (1999) 'Challenges and progress in performance-based earthquake engineering', *Intl Seminar on Seismic Engineering for Tomorrow*, Tokyo, 1999.

Krawinkler, H. and Miranda, E. (2004) 'Performance-based earthquake engineering', In: Y. Bozorgnia and V.V. Bertero (eds), *Earthquake Engineering: From Engineering Seismology to Performance-Based Engineering.* CRC Press, Boca, Raton, FL.

Krawinkler, H. and Nasser, A.A. (1992) 'Seismic demand based on ductility and cumulative damage demands and capacities', in P. Fajfar and H. Krawinkler (eds), *Non-linear Seismic Analysis and Design of Reinforced Concrete Buildings.* Elsevier Applied Science, New York.

May, P.J. (2002) *Barriers to Adoption and Implementation of PBEE Innovations.* PEER Report 2002/20, Pacific Earthquake Engineering Research Center, University of California, Berkeley.

Moehle, J.P. (1992) 'Displacement-based design of RC structures subjected to earthquakes'. *Earthquake Spectra* **8**(3): 403–28.

Negro, P. and Mola, E. (2006) 'Performance Based Engineering Concepts: Past, Present and Future', *Proceedings of the 1st European Conference on Earthquake Engineering and Seismology.* Paper no. 77. Geneva, Switzerland.

National Fire Protection Association (2003). NFPA 5000: *Building Construction and Safety Code.* NFPA, Quincy, MA.

Panagiotakos, T.B. and Fardis, M.N. (1999) 'Deformation-controlled earthquake-resistant design of RC buildings'. *J. Earthquake Engg.* **3**(4): 498–518.

Paret, T.F., Sasaki, K.K., Eilbeck, D.H. and Freeman, S.A. (1996) 'Approximate elastic procedures to identify failure mechanisms from higher mode effects'. *Proceedings of the 11th World Congress on Earthquake Engg*, Acapulco, Mexico.

PEER Report: 2000/10 (2000) *Second US-Japan Workshop on Performance Based Earthquake Engineering Methodology for Reinforced Concrete Building Structures.* Japan Pacific Earthquake Engineering Research Centre, Report No. 2000/10.

Porter, K., Kennedy, R. and Backman, R. (2007) 'Creating fragility functions for performance-based earthquake engineering', *Earthquake Spectra* **23**(2): 471–89.

Priestley, M.J.N. and Kowalsky, M.J. (2000) 'Direct displacement-based design of concrete buildings'. *Bull. New Zealand Nat. Soc. Earthq. Engg.* **33**(4): 421–44.

SEAOC (1995) *Vision 2000: Performance based seismic engineering of buildings.* Structural Engineers Association of California (SEAOC) Sacramento, CA.

SEAOC (1999) *Recommended Lateral Force Requirements and Commentary.* Structural Engineers Association of California, Sacramento, CA.

Sullivan, T.J. Calvi, G.M., Priestly, M.J.N. and Kowalski, M.J. (2003) 'The limitations and performances of different displacement based methods'. *J. Earthquake Engg.* **7**, Special Issue 1, 201–41.

Tekie, P.B. and Ellingwood, B.R. (2003) 'Seismic fragility assessment of concrete gravity dams'. *Earthq. Eng. Struct. Dyn.* **32**: 2221–2240.

Index